D0008147

The Periglacial Environment

The Periglacial Environment

H M French

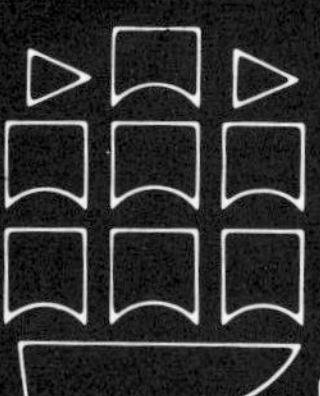

Longman London and New York

Longman Group Limited London

Associated companies, branches and representatives throughout the world

Published in the United States of America by Longman Inc., New York

First published 1976

Library of Congress Cataloging in Publication Data

French, Hugh M.
The periglacial environment.

Bibliography: p.
Includes index.
1. Frozen ground. 2. Glacial landforms. 3. Cold regions. I. Title.
GB461.F73 551.3'8 75-42200
ISBN 0 582 48078 7
ISBN 0 582 48079 5 pbk.

Set in IBM Univers Light 10 on 11pt
and printed in Great Britain by
Lowe & Brydone (Printers) Ltd, Thetford, Norfolk

Contents

Preface ix

Part 1 The periglacial domaine

1 Introduction 2
The periglacial concept 2
Periglacial processes 3

2 Periglacial climates 5
Boundary conditions 5
Periglacial climates 5
- High Arctic climates 8
- Continental climates 9
- Alpine climates 9
- Climates of low annual temperature range 10

Part 2 Present-day periglacial environments

3 Frost action 12
The significance of frost action 12
The freezing process 12
Freezing and thawing indices 13
The freezing and thawing regime of the soil 13
- The long-term temperature cycle 14
- Short-term temperature cycles 17

Frost-action processes 19
Thermal contraction cracking 21
- Frost cracks filled with ice 21
- Frost cracks filled with sand 23
- Soil wedges 25

Ice segregation processes 27
- Frost heaving 28
- Upfreezing of objects 30
- Stone tilting 32
- Needle ice 33
- Frost mounds 33
- Frost sorting 35

Volumetric increase in water as it freezes 37
- Frost wedging 37
- Mass displacements and cryostatic pressures 40

4 Permafrost 45
Historical perspective 45
Geothermal regime of permafrost 47
Distribution of permafrost 50
- Canada and Alaska 51
- The USSR 53
- Offshore permafrost 57

Relic permafrost 58

Permafrost and terrain conditions **59**
Surface features of permafrost and air photo interpretation **62**
Hydrologic aspects of permafrost **65**
  Groundwater supply **65**
  Groundwater, artesian pressures and open system pingos **67**
  Icings **68**
Engineering aspects of permafrost and frost action **69**

5 Ground ice 75
Ground ice description **75**
Types of ground ice **77**
Massive ice and massive icy bodies **80**
  Nature and extent **80**
  Origins of massive ice **81**
Ice-cored topography **82**
  Involuted hills **83**
  Ice-cored terrain **83**
Ice wedges **84**
  Origin **85**
  The development of the polygonal net **87**
  Climatic and palaeogeographic significance **91**
Pingos and other ice-cored mounds **93**
  Open system pingos **95**
  Closed system pingos **98**
  Pingo-like features **100**
  Hydrolaccoliths and frost mounds **101**

6 Thermokarst 105
Causes of thermokarst **106**
Thermokarst subsidence and thermal erosion **110**
Thermokarst subsidence **111**
  The alas thermokarst relief of central Yakutia **111**
  Alas relief in North America **115**
  Ice-wedge thermokarst relief **116**
Backwearing thermokarst **119**
  Ground ice slumps **119**
  Thaw lakes and depressions **122**
Fluvio-thermal erosion processes **125**
Man-induced thermokarst development **128**

7 Hillslope forms and processes 134
Mass wasting processes **134**
  Solifluction **135**
  Slopewash **141**
  Solution **143**
  Nivation **144**
  Rapid mass movement **145**
Periglacial slope forms **149**
  The free face and debris slope profile **150**

Smooth debris mantled slopes 152
Cryopediment forms 155
Cryoplanation terraces and stepped profiles 157
Slope evolution 161
Cryoplanation 162
Slope replacement 164
The rapidity of profile change 165
General conclusions on hillslope forms and processes 166

8 Fluvial processes and landforms 167
Fluvial processes 167
Hydrology and summer weather 168
Sediment movement 171
Thermal erosion 173
Channel forms 174
Periglacial valley sandar 176
Valley forms and slope asymmetry 178

9 Micro-relief features 184
Patterned ground 184
Description 184
Origins 189
Soils 196
The Northern Forest Zone 197
The Main Tundra Zone 197
The Polar Desert Zone 199

10 Wind action and coastal processes 202
The role of wind 202
Wind erosion 203
Wind deflation 204
Coastal processes 206
Sea ice and wave generation 207
The effects of ice on the beach 208
The influence of permafrost and ground ice 211

Part 3 Pleistocene periglacial environments

11 Pleistocene periglacial conditions 214
Introduction 214
The time scale and climatic fluctuations 214
Geomorphic considerations 218
Problems of reconstruction 221
Extent of Late Pleistocene periglacial conditions 222

12 Relic periglacial phenomena 227
Evidence for frost-action conditions 227
Frost-disturbed soils and structures 227

Blockfields and frost weathered bedrock 229
Tors 231
Stratified slope deposits and grèzes litées 234
'Head' and solifluction deposits 235
Evidence for permafrost conditions 236
Ice- and sand-wedge casts 236
Pingo remnants and related forms 245
Thermokarst forms 248
Evidence for Pleistocene wind action 249
Asymmetrical valleys 253

13 Periglacial landscape modification 257
The Chalk landscapes of southern England 257
Periglacial deposits 257
Landform modification 260
Asymmetrical valley development 260
The 'rock-streams' of Wiltshire and Dorset 262
Patterned ground features of the East Anglian Chalk 265
Periglacial valleys and dells 266
The periglacial legacy in southern England 268
The landscape of central Poland 269
The dry valleys of the Lódź plateau 271
Gora Sw Malgorzata 272
Walewice 275
Summary 277

References 278

Index 303

Preface

This book is intended for use by second- and third-year level geography students in universities or colleges of higher education in the United Kingdom. It is also suitable as a text for an undergraduate course on periglacial geomorphology at the honours level in Canada and the United States. On a more general level, the book may prove useful to high-school teachers and other individuals interested or specialising in the physical geography of cold regions. I have assumed, however, that the reader will already possess some understanding of the physical environment, such as might be provided by a first-year physical geography or elementary geomorphology course.

In writing this book I had two aims in mind. The first was to give a realistic appraisal of the nature of geomorphic processes and landforms in high latitude periglacial environments. The second was to provide some guide to the recognition and interpretation of periglacial features in the now temperate regions of North America and Europe. The regional emphasis is oriented towards areas of which I have personal field experience, notably the western Canadian Arctic, central Siberia, southern England and central Poland. Thus, the overall focus is more towards lowland, rather than alpine, periglacial conditions. Notwithstanding this comment, I have attempted to give a balanced world picture; important literature pertaining to other areas has been incorporated.

The reasons for writing such a book are also twofold. First, the majority of students will never have the opportunity to experience, at first hand, high latitude periglacial environments. However, since cold conditions prevailed over large areas of middle latitudes at several times during the last 1 million years, the appreciation of such conditions is essential for a balanced interpretation of these landscapes. Second, the vast northern regions of North America and Siberia are assuming an ever increasing importance in man's quest for natural resources. Their development will be possible only if we understand the terrain and climatic conditions of these regions. For both these reasons, I hope this book will serve a useful purpose.

I have divided the book into three parts. Part 1 is a general introduction to periglacial conditions in which the extent of the periglacial domaine and the variety of periglacial climates are briefly considered. Part 2 presents a systematic treatment of the various geomorphic processes operating in present-day periglacial environments. Wherever possible, I have attempted to show the relationship between process and form and to stress the multivariate nature of many landforms. The sequence of chapters is important since they are planned to be read successively. Part 3 serves only as an introduction to Pleistocene periglacial phenomena. Emphasis in this part is upon forms rather than processes and their interpretation in the light of our understanding of similar phenomena in present-day periglacial environments.

I have not attempted to be comprehensive in my treatment of the literature. By selecting information, I have attempted to give a viewpoint. Inevitably, this viewpoint is biased to reflect my own prejudices and field experience. For example, if I had worked extensively in alpine rather than high latitude lowland environments, probably I would not have given the same emphasis to permafrost, ground ice and thermokarst as I do. However, I believe a viewpoint is necessary since my experience with students is that they require some guidance in coping with the increasing volume of literature which appears each year.

I would like to acknowledge the help and encouragement given me by a number of individuals and organisations, without which this book would not have been written. The late Professor Jan Dylik of the University of Lódź, Poland, provided me with much inspiration and encouragement in the early stages, as well as friendship and hospitality. He was instrumental in planning the organisation of many of the chapters and it is to be regretted that his untimely death in 1973 did not permit him to see the final product. Professors Ron Waters and Stan Gregory of the University of Sheffield, England, were also extremely helpful in encouraging me to write this book and identifying its basic thrust. In Canada, the opportunity to work in the Arctic since 1968 has been made possible by the active support of the Geological Survey of Canada and the Polar Continental Shelf Project. Numerous individuals both in Canada and the United Kingdom have helped in many ways, by discussion, providing material, and reading some of the early draft chapters; they include R. J. E. Brown, M. J. Clark, J. G. Fyles, P. G. Johnson, D. Mottershead, A. Pissart, D. A. St-Onge, R. J. Small, and P. Worsley. To all, I extend my thanks.

Last, and most important of all, the unfailing encouragement and support of my wife, Sharon, is acknowledged with deep gratitude and affection.

PART
1
The periglacial domaine

Introduction 1

The periglacial concept

The term 'periglacial' was first proposed by a Polish geologist, Walery von Lozinski, to describe frost weathering conditions associated with the production of rock rubbles in the Carpathian Mountains. Subsequently, at the XI Geological Congress in Stockholm in 1910 he introduced the concept of a 'periglacial zone' to describe the climatic and geomorphic conditions of areas peripheral to the Pleistocene ice sheets and glaciers. Theoretically, this zone was a tundra region which extended as far south as the treeline.

For several reasons, Lozinski's initial definition is unnecessarily restricting. First, frost-action phenomena are known to occur at great distances from ice margins and to be unrelated to ice marginal conditions. For example, parts of eastern Siberia and the interior of Alaska are commonly regarded as being 'periglacial' in nature. They owe their distinctive nature, however, not to their proximity to ice sheets but rather to factors such as permafrost or low mean annual temperatures. Second, although Lozinski used the term to refer primarily to areas and not to processes, it has increasingly been understood to refer to a complex of geomorphic processes. These include not only the relatively unique frost-action processes but also the much wider range of processes which do not demand either a peripheral ice marginal location or a cold climate for their operation (e.g. fluvial processes, wind action).

Thus, modern usage of the term 'periglacial' refers to a wide range of cold climate conditions, regardless of their proximity to a glacier, either in time or space. For our purposes, periglacial environments may be defined simply as those in which frost-action processes dominate and the periglacial domaine refers to the global extent of these climatic conditions.

There are two diagnostic criteria, one or both of which are common to all periglacial environments. The first is the freezing and thawing of the ground, often in association with water. According to Tricart, '. . . the periglacial morphogenetic milieu is that where the influence of freeze–thaw oscillations is dominant' (Tricart, 1968, p. 830). The second is the presence of perennially frozen ground or permafrost. For example, Péwé considers all areas currently underlain by permafrost to be the modern periglacial zone; '. . . permafrost is the common denominator of the periglacial environment, and it is practically ubiquitous in the active periglacial zone' (Péwé, 1969, p. 4).

It should be emphasised that there is no perfect spatial correlation between areas of intense frost action and areas underlain by permafrost, although the coincidence of the two is high. There are certain areas in subarctic, maritime, and alpine locations which experience frequent freeze–thaw oscillations but lack permafrost. Furthermore, the fact that permafrost underlies extensive areas of the boreal forest in Siberia and North America and is unrelated to present climatic conditions (see pp. 58–9), complicates any simple delimitation of periglacial environments. In practice, the relic permafrost of Siberia and North America extends the periglacial domaine beyond its normal (i.e. frost action) limits.

Thus, the actual periglacial domaine extends over two major vegetation types, (1) the subarctic or northern forests and (2) the Arctic tundra and ice-free polar desert zones. In addition, the periglacial domaine includes the various high altitude or alpine zones which exist in many of the larger mountain ranges of the world. In a general sense therefore, one may distinguish high latitude from alpine, and forested from non-forested periglacial environments.

The extent of the periglacial domaine is considerable. A conservative estimate is that it currently extends over one-fifth of the earth's land surface. Furthermore, during the cold stages of the Pleistocene period, certain areas of the temperate middle latitudes experienced intense frost-action conditions and reduced temperatures because of their proximity to the ice sheets. In all probability, an additional one-fifth of the earth's land surface has experienced periglacial conditions at some time during the past.

Figure 1.1 summarises the various concepts of the periglacial domaine.

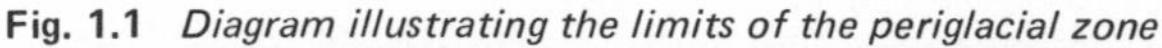

Fig. 1.1 *Diagram illustrating the limits of the periglacial zone*

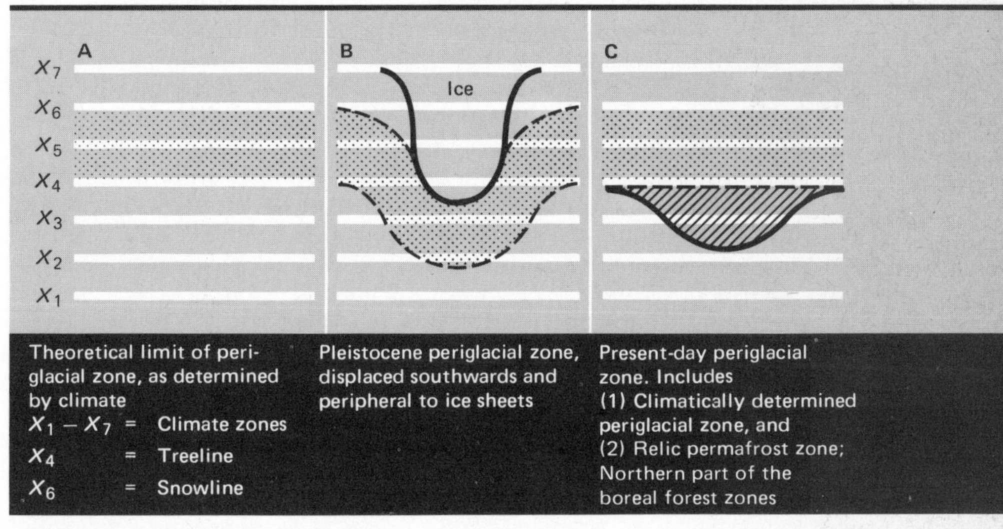

Periglacial processes

Ever since Lozinski first defined the periglacial zone, it has been widely accepted that frost action was the most important weathering process in periglacial environments. Since the effectiveness of chemical weathering generally decreases with decreasing temperatures, chemical weathering was assumed to operate at a slow pace in periglacial environments and to play a secondary role. While this is true in most instances, one must not infer that mechanical weathering is greater in periglacial environments than in non-periglacial environments. This has yet to be proven. Rather, it seems best to regard the prevalence of physical weathering as merely reflecting its strong dominance over chemical weathering. A second assumption has been that periglacial processes possess a unique and distinctive nature, giving rise to a readily definable periglacial morphogenetic region. In reality, however, many of these so-called periglacial processes are little different from those operating in other non-glacial regions of the world, and differ only in their frequency and/or intensity.

Within this context, it is relevant to summarise the various groups of processes which are thought typical of the periglacial environment. Bearing no order of magnitude or importance, they are as follows:

1. The development of permanently frozen ground with ice segregations, combined with the thermal contraction of the ground under intense cold, and the thawing of such ground.
2. Frost weathering, including the disintegration of surface rock by frost wedging.
3. The complex of frost activity operating within the seasonally thawed layer, such as frost heaving, soil churning, soil creep, stone tilting, upfreezing, mass displacement, and particle size sorting.
4. Rapid mass movement; either the slow flow of water-saturated soil (solifluction) aided by slopewash processes, or gravity-controlled rockfalls, slumps or failures, all of which lead to the active development of slopes.
5. Fluvial regimes characterised by marked seasonal discharge patterns and by a high level of suspended and bedload sediments.
6. Strong wind action processes including nivation and snowpatch effects, which are accentuated by lack of vegetation, and large amounts of glacially derived and/or frost comminuted debris.

All of these processes are discussed in the subsequent chapters of Part 2. Some may be regarded as distinctive to the periglacial domaine, or zonal in nature; others are clearly azonal and operate equally well in areas outside of the periglacial domainc. According to Dylik (1964, pp. 116–17), those processes which are unique to periglacial environments are the formation of permafrost, the development of thermal contraction cracks, the thawing of permafrost (thermokarst), and the formation of wedge and injection ice. In a second category are those processes which are unique by virtue of their high intensity and efficacy under periglacial conditions. These are the processes of ice segregation, seasonal frost action and frost weathering, and the various forms of mass movement. In a third category are all the other processes which operate in an azonal fashion throughout the world, such as running water, slopewash, and wind action. While geomorphologically significant within the periglacial domaine, this final group of processes are equally, if not more effective outside of it.

Periglacial climates 2

A number of distinct climatic environments exist in which frost-action processes dominate. Their combined spatial extent constitutes the periglacial domaine. The extent of this domaine is somewhat difficult to establish, however, since there are all gradations from climatic environments in which frost-action processes dominate and where the whole or a major part of the landscape depends upon such processes, to those in which frost-action processes occur but are subservient to others.

Boundary conditions

Boundary conditions between periglacial and non-periglacial conditions are arbitrary to a large extent, and will vary considerably depending upon the criteria used.

The presence or absence of perennially frozen ground is a reasonably easy boundary definition, although the outer boundaries of the discontinuous permafrost zone are often hard to delineate. The location of the treeline is a second relatively unambiguous boundary definition, since most investigators equate severe periglacial conditions with tundra or polar desert ecozones. However, the treeline is not static but either advancing or retreating depending upon environmental and regional climatic conditions. Moreover, the treeline is not a line but rather a zone, over 50–100 km wide in places, lying between the biological limit of continuous forest and the absolute limit of tree species.

For our purposes, and in keeping with a simple frost-action definition of a periglacial environment, we can adopt an empirical definition of the boundary condition for the periglacial domaine. It is defined to include all areas where the mean annual air temperature is less than +3°C. This definition closely follows the limits proposed by Williams (1961) for the development of solifluction and patterned ground of a frost action nature. Since it includes not only areas where frost-action conditions dominate but also areas which are marginally periglacial in character, it gives some idea of the maximum extent of the periglacial domaine. We may further subdivide the periglacial domaine by the –2°C mean annual air temperature into environments in which frost action dominates (mean annual air temperature less than –2°C) and those in which frost-action processes occur but do not necessarily dominate (mean annual air temperature between –2°C and +3°C). In this book we will be concerned primarily with those environments in which frost-action processes dominate.

Periglacial climates

The concept of a distinct periglacial climate was first proposed by Troll (1944) in his global survey of frost-action conditions. It was subsequently incorporated by Peltier (1950) into a scheme of morphogenetic regions. According to Peltier, the periglacial climate was characterised by mean annual air temperatures of between –15°C and –1°C, precipitation of between 120 and 1 400 mm per annum, and . . . 'intense frost action, strong mass movement, and the weak importance of running water'. Today, this definition is not regarded as being very useful. By masking a wide range of climatic conditions, it favours the assumption of a regular, uniform type of periglacial climate. This is misleading since it is more realistic to stress the variety of cold climates that exist and in which frost action dominates.

More recently, Tricart (1970, pp. 19–27; Tricart and Cailleux, 1967, pp. 45–67) has distinguished between three types of periglacial climatic environments. The first type (Type I – Dry climates with severe winters) experiences seasonal and deep freezing while the third type (Type III – Climates with small annual temperature range) experiences shallow and predominantly diurnal freezing, with the second type (Type II – Humid climates with severe winters) intermediate. A second differentiating factor is that permafrost is characteristic of the first, is irregular in occurrence and distribution in the second, and absent in the third type. One problem of this threefold division of periglacial climates is that the first two types are identified primarily in terms of humidity while the third is identified in terms of temperature. As a result, Type I includes the two rather different climatic environments of the Canadian high Arctic and the subarctic continental areas of central Siberia. Moreover, an Arctic subtype of category II, as typified by Spitsbergen, is more similar to the environments of other high latitudes than to a mountain variety, which is the other subtype of that category.

For our purposes, a modified classification based solely upon insolation and temperature is proposed in which four climatic environments are identified. These are summarised below. It must be emphasised that there are gradations between the different types and that continental and maritime influences result in significant variations within any one category.

1. ***High Arctic climates*** – in polar latitudes; extremely weak diurnal pattern, strong seasonal pattern. Small daily and large annual temperature range. Examples: Spitsbergen (Green Harbour, 78°N); Canadian Arctic (Sachs Harbour, 72°N).
2. ***Continental climates*** – in subarctic latitudes; weak diurnal pattern, strong seasonal pattern. Extreme annual temperature range. Examples: Central Siberia (Yakutsk, 62°N); Interior Alaska and Yukon (Fairbanks, 65°N; Dawson City, 64°N).
3. ***Alpine climates*** – in middle latitudes in mountain environments; well-developed diurnal and seasonal patterns. Examples: Colorado Front Range (Niwot Ridge, 40°N); Alps (Sonnblick, 47°N).
4. ***Climates of low annual temperature range*** – in azonal locations:
   (a) Island climates in subarctic latitudes. Examples: Jan Mayen (71°N); South Georgia (54°S).
   (b) Mountain climates in low latitudes. Example: Andean summits, Peru (16°S).

Climatic data for some of the locations mentioned above is given in Table 2.1. The nature of the typical frost-action conditions are graphically illustrated in Fig. 2.1, where the number of frost-free days, days of alternating freezing and thawing, and days in which temperatures remain below 0°C are plotted for four localities. The graphs clearly show that the least number of freezing and thawing days occur in the high Arctic and continental climates, dominated by seasonal patterns, while the greatest number occur in the climates of low annual

Table 2.1 *Climatic data for selected localities in different types of periglacial environments*

A – *Temperature and precipitation data*

		J	F	M	A	M	J	J	A	S	O	N	D
High Arctic climates													
Spitsbergen, Green Harbour	T (°C)	−16	−19	−19	−13	−5	+2	+6	+5	0	−6	−11	−15
	P (mm)	35	33	28	23	13	10	15	23	25	30	25	38
Canadian Arctic, Sachs Harbour	T (°C)	−29	−31	−27	−20	−8	+2	+5	+4	−1	−10	−24	−27
	P (mm)	2	2	3	2	5	4	24	16	16	12	4	3
Continental climates													
Central Siberia, Yakutsk	T (°C)	−43	−36	−22	−7	+6	+15	+19	+15	+6	−8	−28	−40
	P (mm)	11	9	6	11	18	33	43	42	26	20	16	12
Yukon, Canada, Dawson City	T (°C)	−31	−23	−16	−3	+8	+13	+14	+12	+6	−4	−17	−25
	P (mm)	20	20	12	18	23	33	40	40	43	33	33	28
Alpine climates													
Alps, Sonnblick, 3 060 m	T (°C)	−13	−14	−11	−9	−4	−1	+1	+1	−1	−5	−9	−11
	P (mm)	124	124	160	167	157	140	142	129	116	129	117	134
Rockies, Niwot Ridge, 3 750 m	T (°C)	−10	−10	−8	−4	+3	+9	+12	+11	+7	+2	−5	−9
	P (mm)	137	91	105	102	68	70	80	57	72	39	112	88
Climates of low annual temperature range													
Jan Mayen	T (°C)	−3	−3	−4	−2	−1	+3	+5	+5	+3	−1	−2	−4
	P (mm)	40	38	30	29	13	15	23	23	48	43	33	30
South Georgia, Gruytviken	T (°C)	+5	+5	+4	+2	0	−2	−2	−2	+1	+2	+3	+4
	(P mm)	84	104	129	134	139	127	139	129	86	66	86	86

B – *Summary data*

			Mean annual temperature (°C)	Annual range	Total precipitation (mm)
High Arctic	Spitsbergen	78°N	−8	25	298
	Sachs Harbour	72°N	−14	36	93
Continental	Yakutsk	62°N	−10	62	247
	Dawson City	64°N	−5	45	343
Alpine	Sonnblick	47°N	−7	15	1 638
	Niwot Ridge	39°N	−3	22	1 021
Low temperature range	Jan Mayen	71°N	0	8	365
	Gruytviken	54°S	+2	7	1 309

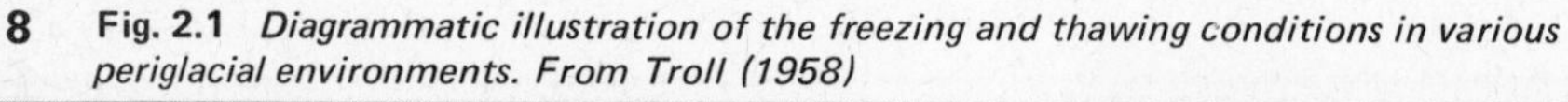

Fig. 2.1 *Diagrammatic illustration of the freezing and thawing conditions in various periglacial environments. From Troll (1958)*

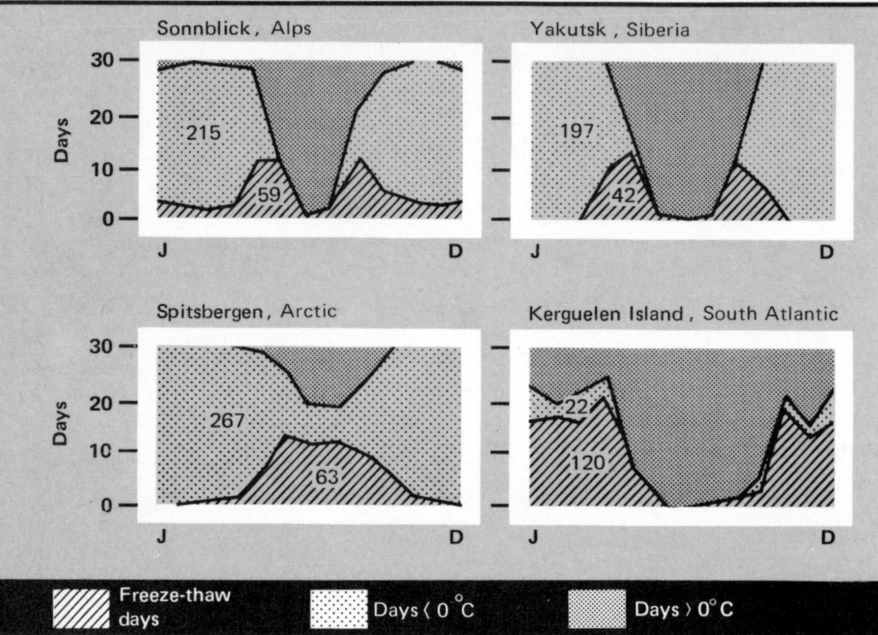

temperature range, dominated by diurnal or cyclonic (i.e. short term) fluctuations. In more detail, the characteristics of each climatic type are described below.

High Arctic climates Several characteristics make these the most distinctive of periglacial climates. First, extremely low winter temperatures occur for periods of several months, when there is perpetual darkness and temperatures may fall to −20 to −30°C. The ground freezes to form permafrost and the ground surface contracts under the intense cold. Second, temperatures rise above freezing for only 2–3 months of the year, when the surface thaws to depths varying between 0.3 and 1.5 m and average air temperatures rise 4–6°C above zero. Third, precipitation amounts are low. In parts of the Canadian Arctic, precipitation is less than 100 mm per annum, of which approximately one-half falls as rain during the summer period. Because of low evaporation rates, and the inability of the water to percolate through the soil on account of permafrost, the effectiveness of this precipitation is high. As a consequence, although arid in terms of total precipitation, these regions are surprisingly wet during the summer. In more maritime locations, such as Spitsbergen, precipitation amounts increase considerably to as much as 250–400 mm per annum. In these regions, snow assumes importance in protecting the ground surface from the extremes of cold and increases the magnitude of the spring run-off. The ice-free areas of Antarctica represent a unique type of high Arctic climatic environment. Temperatures are exceptionally low and precipitation is often negligible.

In all high Arctic climates, the snowcover during the winter is thin and often discontinuous, even in the moister regions. Upland surfaces and exposed areas are swept bare by wind allowing frost to penetrate deeply and snowbanks to accumulate only in hollows and lee-slope positions.

In the northern hemisphere, the approximate spatial limits of these climatic conditions may be arbitrarily defined by the glacial limit to the north and by either the treeline or the 8–10°C annual July isotherm to the south. For the most part, these areas are referred to as the tundra or polar regions of the world.

Continental climates South of the treeline in subarctic and continental locations are large areas which experience a wider range of temperatures than high Arctic areas. Although mean annual air temperatures are approximately the same or slightly higher than in areas north of the treeline, these mean values disguise excessively low temperatures in winter and remarkably high temperatures in summer. Yakutsk, in central Siberia, and Dawson City, in the Yukon Territory, illustrate this phenomenon (Table 2.1). At Yakutsk, for example, January temperatures of minus 40°C and July temperatures of plus 15–20°C are quite common. At Yakutsk, the annual average range of temperature is 62°C. This may be compared to that of Sachs Harbour, where the annual range is only 36°C.

Besides being hotter, the summers are longer than in high Arctic climates, and above-freezing temperatures occur for 5–6 months of the year. Seasonal thaw may penetrate over 2.0–3.0 m. Precipitation amounts are also greater than high Arctic regions since disturbances associated with the Arctic and polar fronts are more frequent at these latitudes. Between 250 and 600 mm per annum are typical, with the majority falling during the summer months. However, evaporation rates are also high during the summer and there is usually a soil moisture deficit.

The summer dryness of the northern boreal and taiga forests, together with the high air temperatures, constitutes a favourable environment for forest fires. The boreal forest is regarded by some as a fire climax in terms of vegetation successions. According to Veireck (1973), wildfires in Alaska alone burn an average of 400 000 ha/year. They constitute, therefore, a distinct feature of the continental periglacial environments and have no counterpart in the treeless and cooler high Arctic climatic environments. Winter snowfall is another differentiating characteristic of continental climates since amounts are considerably greater than high Arctic climates. The snowfall and forest cover interact to determine whether permafrost forms or not; in open areas where snow accumulates, it protects the ground from deep frost penetration and permafrost may not occur. Beneath trees, where depth of snow is less, there is deeper frost penetration and a greater probability of permafrost occurrence. Where continuous and thick permafrost is present, there is reason to believe that this is relic and unrelated to present climatic conditions.

Alpine climates Alpine periglacial climates are characteristic of tundra regions lying above the treeline in middle latitude locations. They are not as extensive as either the two previous climatic environments. In the European Alps and the Rockies, the treeline occurs at elevations varying between 2 000 and 4 000 m, but in other localities, such as northern Scandinavia, Iceland and northern Labrador–Ungava, the treeline approaches sea level. Irrespective of elevation, however, all these climatic environments experience a seasonal and diurnal rhythm of both temperature and precipitation.

None of these climates experiences the severe winter cold of either high Arctic or continental climates. On the other hand, the diurnal and seasonal rhythm imposed by their middle latitude locations results in a higher frequency of temperature oscillations around the freezing point. Precipitation is also heavy, the result of either orographic or maritime effects. Total amounts often exceed 750–1 000 mm per annum, much of which occurs as snow. Permafrost is usually lacking or discontinuous, the result of the higher mean annual temperature and the protection given to the ground surface by the winter snowcover.

Climates of low annual temperature range There are certain very restricted areas of the world which experience not only a mean annual air temperature below +2°C but also a remarkably small range of temperatures. These rather unique climatic conditions occur in two types of localities.

The first is a subarctic oceanic location, where the thermal influence of the surrounding water bodies exerts a moderating influence upon temperatures. In the northern hemisphere, Jan Mayen and Bear Islands experience this climate, while in the southern hemisphere, the islands surrounding Antarctica, such as Kerguelen, South Georgia, and the South Orkneys, are similar. In these climatic environments, the mean annual amplitude of temperature is of the order of only 10°C. Not surprisingly, these areas experience a high frequency of freeze–thaw cycles of short duration and slight penetrability into the ground. Because of their maritime locations, these islands experience considerable precipitation, often varying between 1 000 and 2 000 mm per annum, and unstable weather with much low cloud and fog.

The second is an alpine location in low latitudes where diurnal temperature variations dominate the weak seasonal influences. Numerous shallow freeze–thaw cycles occur throughout the year and precipitation may vary from near-arid to humid, depending upon location. These climatic conditions exist towards the summits of the various mountain ranges in the Andes of South America, and in East Africa, but clearly, are not widespread.

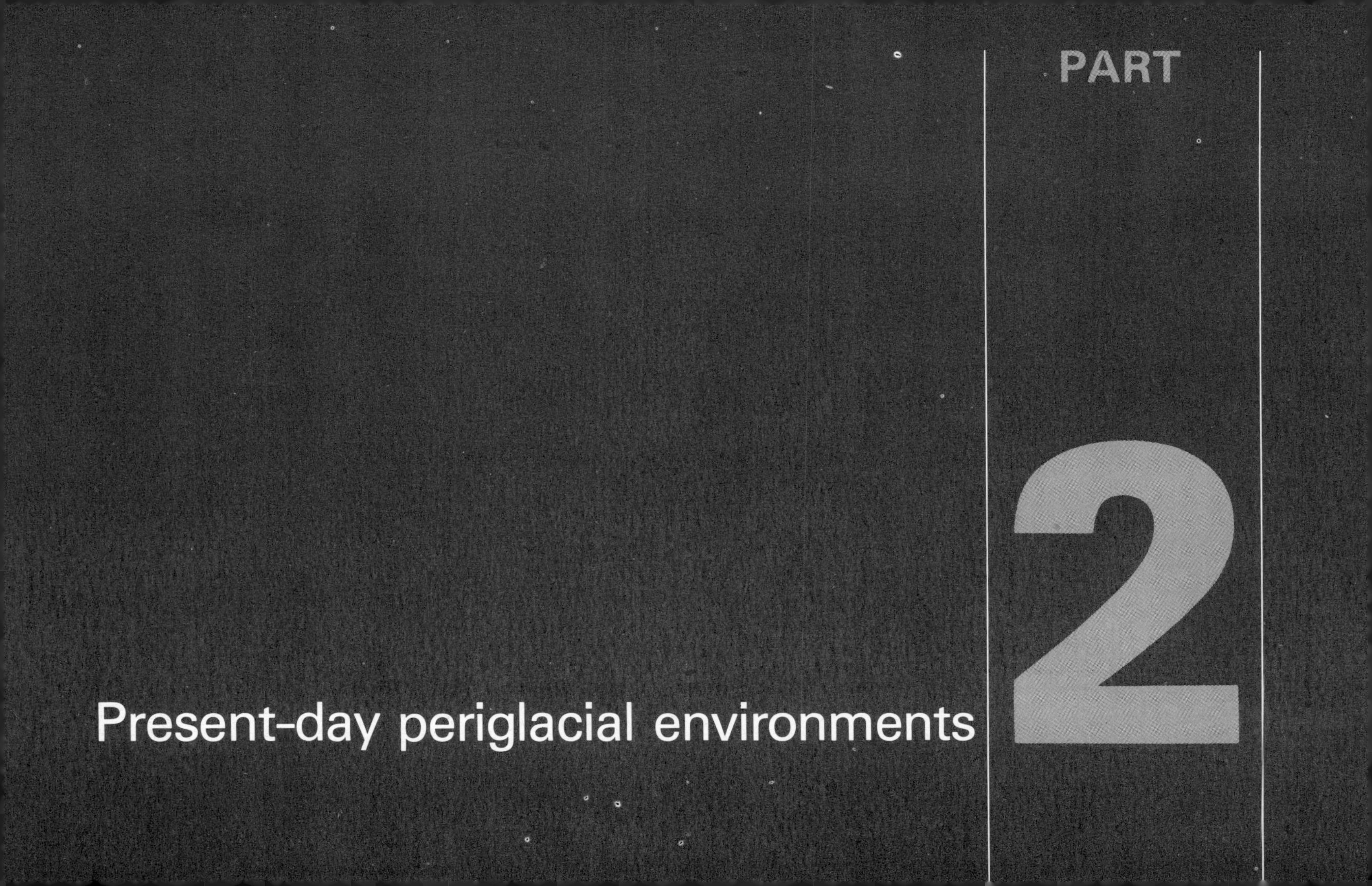

PART 2

Present-day periglacial environments

Frost action 3

The significance of frost action

Frost action is a collective term used to describe a number of distinct processes which result mainly from freezing and thawing. The wide-ranging nature of these processes, and the fact that they are presented first, necessitates some brief synthesis or overall perspective.

It is widely believed that frost action is the fundamental characteristic of present-day periglacial environments. Frost-action processes probably achieve their greatest intensity and relative importance in areas currently experiencing periglacial climatic conditions. It is this belief which has led to the organisation and arrangement of this and the following chapters. In this chapter the seasonal freezing and thawing of the ground surface is described and then the various frost-action processes associated with these temperature variations are outlined. In Chapter 4, perennially frozen ground, or permafrost, is examined since its formation reflects the duration and intensity of the annual temperature regime. The existence of ground ice and related thermokarst phenomena, which are described in Chapters 5 and 6, are directly dependent upon the presence of either seasonally or perennially frozen ground. In terms of the major morphological elements of the periglacial landscape, the presence of frost action and seasonally or perennially frozen ground largely determines the effectiveness of the various weathering and transporting processes operating on hillslopes and in river channels. These processes, and the landforms which result, are described in Chapters 7 and 8. Finally, many of the distinctive microrelief features of the ground surface owe their development to the freezing and thawing of the soil, and these are examined in Chapter 9. Wind action and coastal processes are not directly related to either frost action or frozen ground yet, because they assume certain distinctive characteristics in periglacial environments, merit separate attention in Chapter 10.

In terms of man's utilisation of periglacial environments, it is also widely believed that the freezing and thawing of the ground surface, and the presence of perennially frozen ground beneath, constitute the fundamental problem of economic development. Frost wedging and frost heaving within the seasonally frozen ground, and the thawing of perennially frozen ground constitute major engineering problems which affect nearly all aspects of life in these regions.

The freezing process

The nature of the actual freezing process has attracted much attention from individuals and research agencies such as the US Army Cold Regions Research and Engineering Laboratories (CRREL), the US Highway Research Board, and the National Research Council of Canada. The work of Williams (1967) provides an in-depth review of the nature of freezing soils.

The detailed physics of the freezing process need not concern us. On the other hand, a few comments are necessary to preface the following discussion of frost action. First, the freezing process is complex since different soils cool at different rates depending upon heat conductivity and moisture content, and thus freeze at different rates. Second, soils do not necessarily freeze when their temperatures fall to 0°C since they are known to exist in a supercooled state. A common condition is for saline groundwaters to lower the temperature at which the soil freezes. Third, the duration and intensity of a temperature drop below 0°C will affect the rate and amount of freezing of the soil. For example, the effect of a surface temperature remaining just below 0°C for a considerable length of time is probably the same as that of a more extreme drop of temperature below 0°C but of a much shorter time duration. It is clear therefore, that in discussing frost action one is dealing with a complex process, the details of which are still little understood.

Freezing and thawing indices

Two quantitative indices are frequently used in discussing frost action and related processes. For convenience, they are briefly outlined here.

First, the calculation of thawing and freezing degree days is based upon the cumulative total air temperatures above or below zero for any one year (Thompson, 1966). Mean daily air temperatures provide the basic data. Degree days provide an index of the severity of the climate and the magnitude of the thaw and freezing periods. Their calculations are useful, therefore, when comparing depths of seasonal frost penetration or rates of thermokarst development for example.

The frequency of freezing and thawing is of particular geomorphic significance with respect to frost wedging and rock shattering. Degree days give no indication of this and instead, the number of freeze–thaw cycles is a more relevant parameter. Unfortunately, a number of problems limit the usefulness of freeze–thaw cycles as a measure of frost-action effectiveness. First, there is the basic difficulty of defining the exact point of freezing across which the oscillations should be measured. Second, the use of air temperatures to define cycles (e.g. Fraser, 1959) is not completely satisfactory, since significant differences may exist between air and ground temperatures. Third, even when direct ground temperature measurements are available, just what constitutes a freeze–thaw cycle is debatable. Each separate occasion of freezing and thawing requires a different degree of heating and cooling, dependent upon such factors as the ambient temperature and water content of the soil, the duration of the fluctuation, the intensity of solar radiation, and the character and depth of any snowcover. Fourth, cycles may be of different intensities (i.e. different temperature ranges) and this makes any comparison of cycle frequencies difficult. Finally, the duration of any cycle may range from a few minutes to several days, and it is probably unwise to assign equal numerical weight to the various types of cycle.

The freezing and thawing regime of the soil

A description of the annual rhythm of freezing and thawing of the ground surface is an essential prerequisite to any discussion of the various frost-action processes. In recent years, direct field measurements of freezing and thawing of the soil have been made in a number of different periglacial climatic environments (e.g. Cook and Raiche, 1962a; Chambers, 1966; Washburn, 1967; Fahey,

1973). Studies such as these enable a proper appreciation of the role of frost action in present-day periglacial environments and give some insight into the nature of Pleistocene frost action.

The long-term temperature cycle The majority of periglacial environments experience some form of seasonal change through the year. It is possible to recognise a summer thaw period of limited duration and a long winter period during which temperatures remain below 0°C. The exception to this is the alpine periglacial environment of low latitudes which experiences a diurnal rather than a seasonal cycle.

The importance which should be attached to the seasonal or long-term temperature cycle is not fully understood since there are still relatively few studies with the sufficient time duration to give a long-term record over a number of years. One such study, however, was undertaken on Signy Island, Antarctica, in 1962 and 1963 by Chambers (1966).

In order to allow for slight variations in the actual temperature when the interstitial soil moisture changes from water to ice and then from ice to water, Chambers adopted a transitional zone of ±0.5°C as the freezing boundary indicator. This freezing boundary is sometimes known as the 'zero curtain' (Muller, 1945). Figure 3.1 shows soil temperatures at various depths during 1963 classified into four classes: above +0.5°C; below –0.5°C; the transitional freezing boundary; and a fluctuating class indicating temperatures to be alternating between the other categories. A number of important characteristics of the seasonal freezing and thawing regime are illustrated. The most obvious is that, at depths in excess of 120 cm, the ground remains frozen throughout the year. Moreover, the deeper sections of the seasonally thawed layer were in the transitional freezing boundary for a large part of the year. For example, at a depth of 120 cm the temperature never became consistently positive but remained in the transition zone for over seven months.

This 'zero curtain' phenomenon has been reported from other localities besides Signy Island, and is common in soils underlain by perennially frozen ground. It is the result of three factors. First, the onset of freezing during the autumn releases latent heat to the soil which temporarily compensates for the upward heat loss associated with the dropping of the air temperature. Soil temperatures are stabilised, therefore, during this period. Second, thawing of the ground in the spring absorbs latent heat and delays the rise in soil temperatures consequent on the rise in air temperatures. Third, the continued passage of water at temperatures just above 0°C across the top of the perennially frozen ground throughout the spring and summer reduces the temperature fluctuation.

Towards the ground surface the length of time during which the temperature remains within the freezing boundary progressively decreases. At Signy Island for example, only one week during the year showed a steady temperature in that category at the 1.0 cm depth. Instead, the temperature of the ground at depths of 1.0 to 5.0 cm underwent a large number of fluctuations. These reflect diurnal, cyclonic or other short-term influences. These are examined below. As a result, although maximum temperatures occur at the surface, the maximum continuous period of soil temperatures in excess of +0.5°C occurs at a depth of 2.5 cm.

The implications of these observations with respect to weathering processes

Fig. 3.1 *The ground temperature regime for Signy Island, 1963 illustrating the long-term annual freeze-thaw cycle and the presence of the transitional freezing boundary. From Chambers (1966)*

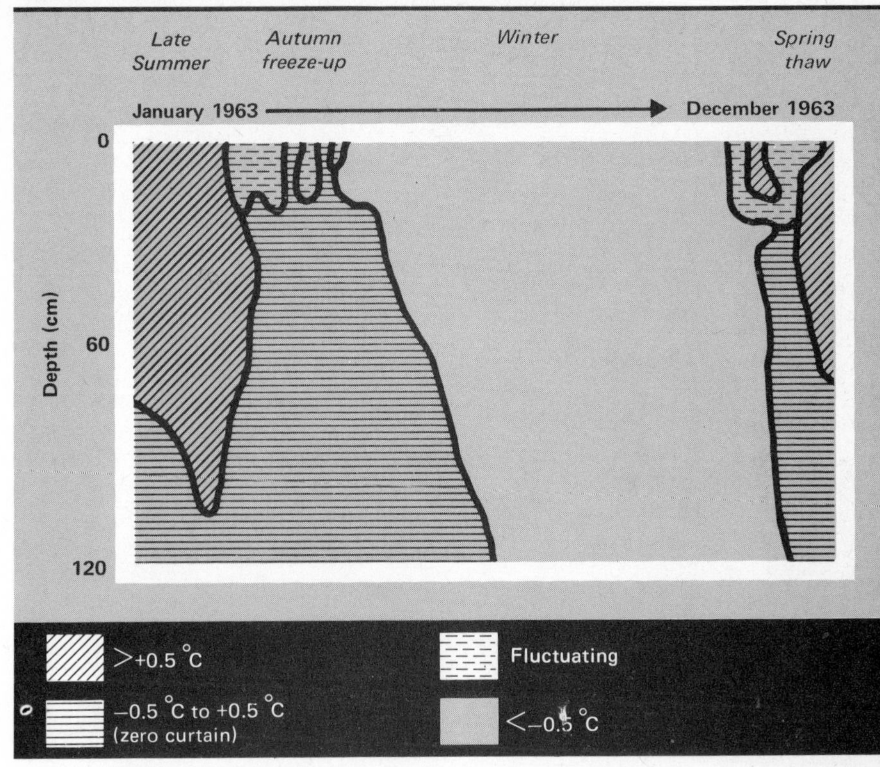

in seasonally frozen ground are still uncertain. It is clear that much of the soil profile remains within ±0.5°C for a long and continuous period. The restriction of short-term temperature fluctuations to the surface layers implies that they are incapable of inducing mechanical weathering at depth. It follows that the freezing boundary conditions, which are related to the annual cycle, are probably more relevant when considering weathering at depths in excess of 10–15 cm. However, it is an open question whether mechanical or chemical weathering is more important within a zone of near zero constant temperatures combined with high moisture content.

From a geomorphic viewpoint, the nature and rate of the spring thaw and the autumn freeze-up are of interest. The former influences the nature of spring run-off and the latter controls the nature of frost heaving and ice segregation in the soil. Usually the spring thaw occurs quickly and over 75 per cent of the soil thaws within the first month or five weeks of air temperatures rising above 0°C. For example, on Signy Island in 1963 the soil temperatures at all depths changed from being constantly below the freezing boundary on 7 November to above 0.5°C by 5 December. This rapidity of thaw is the result of the percolation of meltwater through the soil under gravity processes, transferring heat to the frozen material beneath. Coarse sediments are particularly suited to rapid thaw

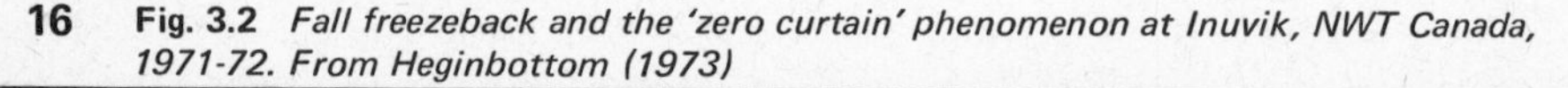

Fig. 3.2 *Fall freezeback and the 'zero curtain' phenomenon at Inuvik, NWT Canada, 1971-72. From Heginbottom (1973)*

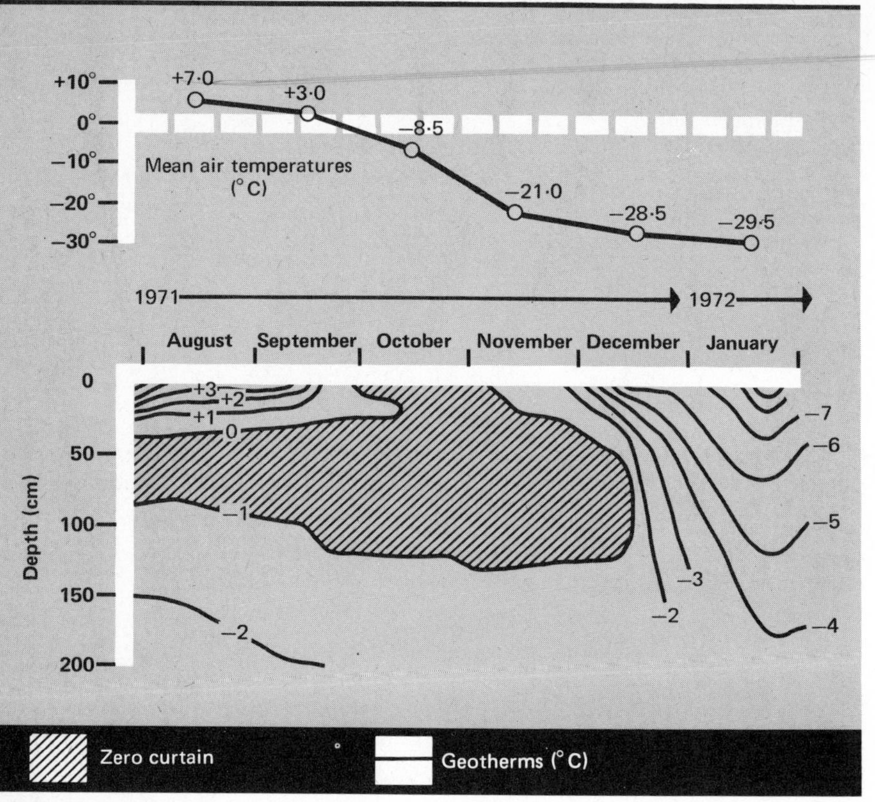

since percolation is easy and thermal conductivity of such sediments is generally high.

In contrast to the spring thaw situation, the fall freeze-up is rather more complex. There is both freezing from the surface downwards, and a certain amount of upfreezing from the perennially frozen ground beneath. The freezing period is also much longer and may extend over 8–10 weeks. At Inuvik, for example, in the Mackenzie Valley of northern Canada, freezing begins in late September and finishes in mid-December (Fig. 3.2). During the majority of this time, the 'zero curtain' persists.

A second feature of interest in the freeze-up period is the initially slow rate of freezing from the surface downwards and the dramatic speed up in freezing at depth. For example, at Inuvik the top 25.0 cm took over six weeks to drop below the freezing boundary but the remaining 100.0 cm took only four weeks. This increase in the rate of freezing at depth is the result of at least two factors. First, there is a decrease in soil moisture with depth since it has already been drawn upwards towards the freezing plane to form segregated ice lenses (see pp. 27–8). The amount of latent heat released as a result of the water–ice phase change at depth, therefore, is reduced considerably and no longer offsets the temperature fall. Second, upfreezing from the perennially frozen ground combines with downfreezing from the surface to quicken the overall movement of the freezing plane at depth.

There are certain areas for which the annual cycle, as described above, is not completely representative. Alpine environments of low latitudes are the obvious example since they experience diurnal rather than seasonal conditions. A second example is that of certain subarctic locations which experience seasonal freezing and thawing but are not underlain by permafrost. This is particularly common in the discontinuous permafrost zones (see Chapter 4, pp. 52–7). In such areas, the absence of an impermeable frozen layer beneath results in (1) the ability of melt-water to percolate to the ground water-table, and (2) the absence of upfreezing in the autumn freeze-up period. The net effect of both is to reduce the importance of the zero curtain.

Short-term temperature cycles Short-term temperature fluctuations are imposed upon the long-term temperature cycle previously described. In contrast to the annual cycle, the nature and importance of these short-term cycles is better understood. Although it is often difficult to distinguish between the two, these fluctuations may be classified as being either diurnal or cyclonic in nature.

Diurnal variations relate to changes in insolation and surface heating brought about by the variations in angle and azimuth of the sun. In high latitudes, the diurnal influence is relatively weak in comparison to lower latitudes since the sun is above the horizon for much of the Arctic day. Furthermore, although south- and west-facing slopes should receive greater insolation than slopes of other orientation, these differences are so weak that they are sometimes over-ridden by local factors. For example, on the Beaufort Plain of north-west Banks Island, the south- and west-facing slopes are not always the warmest (French, 1970). Instead, a marked surface inversion occurs on these slopes as the result of their exposure to dominant westerly winds causing evaporation and latent heat loss. At depths beneath 10 cm, the north- and east-facing slopes are consistently warmer by 1–3°C than others since they are in the lee of the dominant winds and suffer less heat loss. Thus, diurnal variations of insolation with orientation do not necessarily occur in extreme high latitudes.

In the alpine environments of temperate latitudes however, diurnal ranges are much better developed. For example, the range between mean daily maximum and minimum air temperatures at 3 050 m in the Colorado Front Range can reach 15°C in July (Fahey, 1973). It is also in alpine environments that the effects of orientation become well developed, with south- and west-facing slopes in the northern hemisphere experiencing significantly higher soil temperatures.

Numerous local events of a cyclonic or atmospheric nature may also produce very marked fluctuations in soil temperatures. This is especially the case in high Arctic regions where a period of direct solar radiation may heat the ground surface to give a 20–30°C temperature difference between the air and the ground. In Antarctica, for example, Souchez (1967) has observed thawing of snow when the air temperature was –20°C caused by the localised heating of dark rock surfaces. At Signy Island, Chambers (1966) has documented a case when daytime air temperatures were –0.6°C but ground temperatures were +5.6°C, as a result of direct solar insolation. The passing of a cloud or the falling of a small amount of precipitation will also affect soil temperatures. For example, on Banks Island, the author has documented the rapid changes in soil

Fig. 3.3 *Air and ground temperatures, Signy Island, 3-10 January 1963, and their relationship to cyclonic and diurnal conditions. From Chambers (1966)*

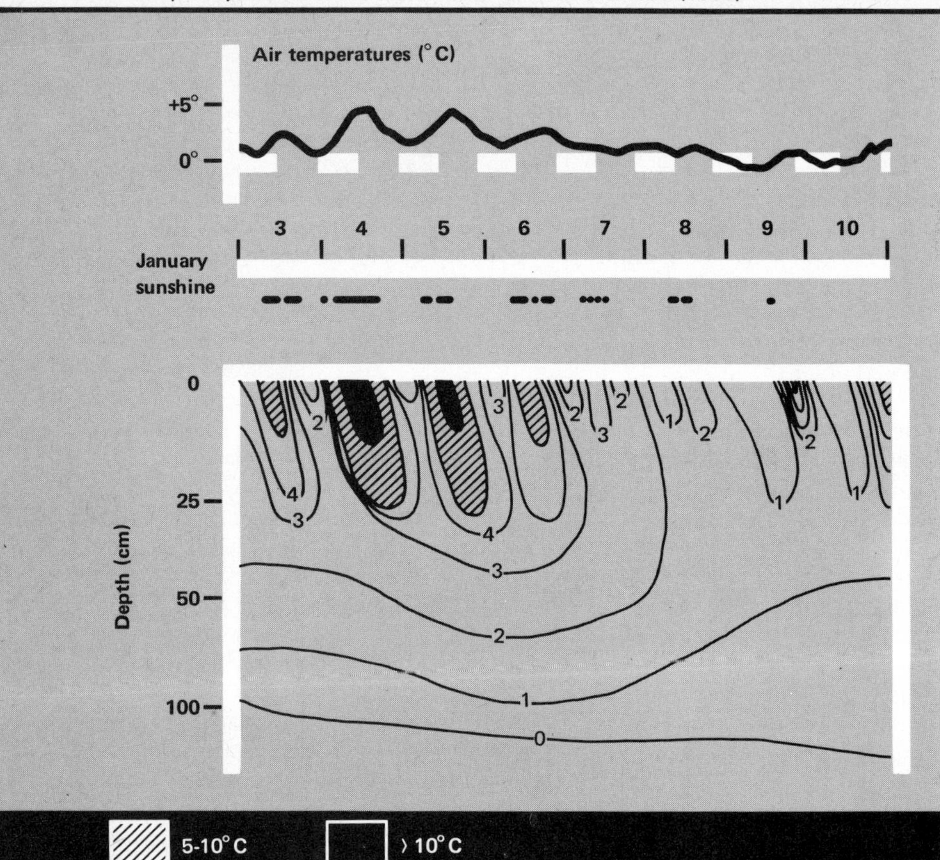

temperatures which follow a summer snowfall of 2.5 cm (French, 1970); within 3–4 hours, the whole of the active layer had attained a near isothermal state due to the rapid percolation downwards of the snowmelt. A final example is provided by the effects of a period of predominantly sunny days followed by several days of dull, overcast and cyclonic weather on Signy Island (Fig. 3.3). During the sunny days, there were distinct diurnal ground and air temperature cycles, with temperatures rising to +14.6°C at the 1.0 cm depth at one point. On the latter two days, when there was almost continuous low and medium cloud cover, the 1.0 cm soil temperature did not rise above +3.4°C, and the heating and cooling of the soil was less dramatic than those resulting from direct solar radiation.

Because of their supposed geomorphic significance with respect to frost wedging (see pp. 37–40), the frequency of short-term freeze–thaw cycles has been investigated by a number of workers. Fraser (1959) used air temperatures to estimate average frequencies of freeze–thaw cycles in northern Canada and concluded that freeze–thaw cycles increased southwards from approximately 15 per annum in the arctic islands to near 50 in southern regions. As such, he questioned the effectiveness of freezing and thawing as a geomorphic agent in high Arctic regions. More recently, direct ground surface measurements in a number of different climatic environments throws doubt upon the assumption that numerous freeze–thaw cycles exist in such environments. Observations indicate the number of freeze–thaw cycles is surprisingly few (Table 3.1). The greatest number occur at the ground surface and these cycles are twice as numerous as air cycles. With depth, there is a rapid drop in frequency such that beneath 5.0–10.0 cm only the annual cycle takes place.

Oceanic periglacial environments have traditionally been assumed most suited for freeze–thaw processes. Even in these conditions, however, it has been found that freeze–thaw cycles are not very numerous. For example, at Signy Island where air temperatures fluctuate constantly within a few degrees of zero, a total of only 42 freeze–thaw cycles were recorded at the 1.0 cm depth during the two years of 1962 and 1963 (Table 3.1). At the 2.5 cm depth, the number fell to 25, at the 5.0 cm depth the number was 13 and at depths in excess of 10.0 cm only the annual cycle occurred. According to Chambers, the freeze–thaw cycles at Signy Island are usually caused by cyclonic disturbances leading to a drop in surface temperatures. They are not due to diurnal influences and rarely last more than a few hours.

Mid-latitude alpine environments are also regarded as being particularly suited to numerous freeze–thaw cycles since they experience marked diurnal temperature fluctuations. Fahey (1973) provides data for the Front Range of the Colorado Rockies which does not confirm this assumption. The two subarctic–alpine sites (Saddle and Niwot Ridge) experience approximately twice the number of freeze–thaw cycles as Signy Island (67 and 50 over a one-year period – Fahey, 1973; 42 over a two-year period – Chambers, 1966) and approximately four times the number that occur at either Resolute or Mesters' Vig. Having stated this however, the maximum number of cycles involved in all areas is still very small. Moreover, all studies indicate that only the annual cycle occurs at depths in excess of 5.0–10.0 cm. For these reasons, one must seriously question the effectiveness of short-term freeze–thaw processes in promoting rock weathering both near the surface and at depth in all present-day periglacial environments. It may be that the annual cycle is a more effective geomorphic agent than has been thought previously.

Frost-action processes

Not withstanding the comments of the previous section, it is undoubtedly true that frost-action processes are of fundamental importance in most periglacial environments. The freeze–thaw cycle is but one of a number of frost-action processes. When considered as a group, these frost-action processes produce a wide range of distinctive phenomena. In many cases, several frost-action processes operate in conjunction with each other, while some operate in association with additional processes of a non-frost action nature. The majority, but not all, of frost-action processes operate in conjunction with the freezing of water.

The following section is organised around a description of each of the frost-action processes together with a brief discussion of the phenomena which result from them. To this end, we can identify three basic frost-action processes:

Table 3.1 *Recorded frequencies of freeze–thaw cycles in various periglacial environments*

Location	Climatic type	Year of record	Air cycles	Ground cycles	Cycles at 1.0 cm	Cycles at 2.5 cm	Cycles at 5.0 cm	Definition of cycle	Reference
Resolute Bay, Canada 74°N	High arctic	1960	15	23	n.d.	1	0	Range 28–32°F	Cook and Raiche (1962a)
Mesters' Vig, Greenland 72°N	High arctic	1961	21	23	n.d.	n.d.	18	Amplitude ⩾0°C	Washburn (1967)
Signy Island, South Orkney Islands 61°S	Low temp. range	1962	n.d.	n.d.	23	18	8	Range +0.5 to −0.5°C	Chambers (1966)
		1963	n.d.	n.d.	19	8	4		
Front Range Colorado, USA 39°N	Alpine Tundra (3 750 m)	1969–70	34	n.d.	9	n.d.	n.d.	Range >0°C and <0°C within 24 hours	Fahey (1973)
	Subalpine (3 500 m)	1969–70	44	n.d.	67	n.d.	n.d.		
	Subalpine (3 050 m)	1969–70	100	n.d.	50	n.d.	n.d.		

n.d. = not determined.

1. The thermal contraction of the ground leading to frost cracking.
2. The formation of segregated ice.
3. The increase in volume of water as it changes to ice.

It must be emphasised that, although treated separately, these processes are often intimately linked. For example, the formation of segregated ice is invariably associated with a volumetric expansion of the water as it freezes.

Thermal contraction cracking

The mechanics of thermal contraction cracking have been investigated by Dostovalov (1960) and Lachenbruch (1962), although the process was first proposed many years ago (Leffingwell, 1919). In essence, the lowering of the temperature of ice-rich frozen soils leads to a thermal contraction of the ground and the development of fissures. These are called frost fissures or frost cracks in the literature (e.g. Dylik, 1966; Dylik and Maarlveld, 1967). They develop because pure ice has a coefficient of linear expansion of 52.7×10^{-1} at 0°C and only 50.5×10^{-1} at −30°C, and because the rates of contraction and expansion of ice-rich sediments are probably little different to that of pure ice.

The conditions for cracking however, will vary for different soils with different ice contents. Both the rate of temperature change and the absolute temperature drop appear to be important. For example, when ground temperatures fall below −18 to −20°C, even soils with only average ice contents suffer contraction. On the other hand, Black (1963) has suggested that a very sharp drop of only 4°C in the below zero range is sufficient for frost cracking to take place in an extremely ice-rich soil.

Thermal contraction cracks divide the ground surface up into polygonal nets, often with junctions at right angles (Fig. 3.4). According to Dostovalov (1960), there is a certain regularity to this process. The size of the polygons reflects the severity (i.e. temperature range) of the climate; the smaller the dimensions of the polygons, the greater is the temperature range. It follows that frost fissures may be of different orders or generations, reflecting different temperature ranges, and that the ground may be progressively divided up by frost fissures of higher orders. A more detailed discussion of the mechanics of polygon development is to be found in Chapter 5 (pp. 84–91).

Since frost fissures are of great value as palaeoclimatic indicators of both perennially frozen ground and severe winter temperatures, it is useful at this point to discuss the different types of frost cracks that can occur and their physical appearance. In present-day periglacial environments one can distinguish between three types of frost cracks. First, there are those frost cracks which are filled with ice. These are called ice wedges and are discussed in detail in Chapter 5. Second, there are frost cracks which are filled with sand, known as sand wedges or 'tesselons'. Third, there are frost cracks filled with mineral soil known as soil wedges. While ice wedges and sand wedges develop not only in the seasonally thawed zone but also in the perennially frozen ground beneath, and may extend downwards for several metres, soil wedges are usually confined to the seasonally frozen layer and rarely extend downwards for more than 1.0–2.0 m.

Frost cracks filled with ice The infilling of the frost crack with water and snow which percolates down the crack results in the development of ice wedges. These

Fig. 3.4 *Oblique air view of fissure polygon terrain, eastern Banks Island, Canadian Arctic. The polygons are between 30 and 50 m in diameter.*

are wedge-shaped bodies of vertically foliated ice which build up over a number of years (Fig. 3.5). Their mode of formation is illustrated in Fig. 3.6A. Because the water freezes as it percolates down the wedge, the enclosing sediments are not able to resume their original position. As a result, the sediments on either side of the ice body may become deformed and bent upwards.

When the ground thaws and the ice melts, the wedge is infilled with material which slumps in from the sides and from the surface (Fig. 3.6B). Such ice wedge casts are widely recognised in those temperature regions which experienced periglacial conditions during the Pleistocene (e.g. Johnsson, 1959; Black, 1969b). In order that they be interpreted correctly Pissart (1970, p. 12) has stressed a number of points. First, the infilling material should contain not only material which entered the wedge after the thaw of the ice but also material which constituted the seasonally thawed zone during the period of its growth (as seen in the dark sediment bands of modern ice wedges (see Fig. 3.5). While it is true that some ice wedges are composed of remarkably pure ice, it is equally true that others possess considerable quantities of sediment. Thus, one should not necessarily expect uniformity in either texture or age of the infilling sediments. Second, the melting of an ice wedge is not always an indication of a climatic change since it can result from fluvio-thermal erosion operating preferentially along ice wedges. The nature of thermokarst processes and ice wedge relief is discussed more fully in Chapter 6. Third, much time can elapse between the development of the ice wedge and its melting, and the ice wedge can remain inactive in the ground for many years. This is the case for many ice wedges in central and southern Alaska (Péwé, 1966a). Fourth, the melting of the ice, the downward movement of material within the wedge, and the thawing of the enclosing sediments can induce considerable modifications to the original form of the wedge when it was occupied by ice.

Frost cracks filled with sand Sand wedges are a second type of frost fissure that develops from the thermal contraction of the ground. In this case, because of extreme aridity there is no moisture available to penetrate the crack and to subsequently freeze. Instead, the crack is filled with windblown sediments or other materials. Péwé (1959) first described sand wedges at McMurdo Sound in Antarctica which he termed 'tesselons', and subsequently similar features have been described from the polar regions (e.g. Nichols, 1966; Ugolini ***et al.***, 1973; Pissart, 1968). They occur in arid regions, with less than 100 mm of precipitation per year and the enclosing sediments are usually coarse grained sands and gravels. If the infilling material is derived essentially from the material surrounding the crack, then the surface expression of such a sand wedge is a very shallow depression, only some centimetres in depth. If, on the other hand, the infilling is by windblown sediments there is usually a certain amount of deformation of the surrounding sediments and shallow ramparts may be formed at the surface. Usually, the infilling material in sand wedges is in the loess size range. It presents a vertical stratification, with a tendency for a decrease in particle size with increased depth in the wedge, since only the smaller particles are able to penetrate the very narrow tip of the wedge. In no cases reported do sand wedges

Fig. 3.5 *A number of small and medium sized ice wedges exposed beneath peaty, organic sediments, Sachs Harbour, southern Banks Island, Canadian Arctic. The wedges are developed within ice-rich silts. Cryoturbation structures and small frost fissures are developed in the organic and sandy sediments overlying the silts.*

Fig. 3.6 *Schematic diagram illustrating method of formation of (a) ice-wedges and (b) ice-wedge casts.* **A1-A4** *— successive stages of frost cracking under cold climate;* **B1-B4** *— successive stages of ground thawing under warmer climate. From Péwé, Church and Andersen (196*

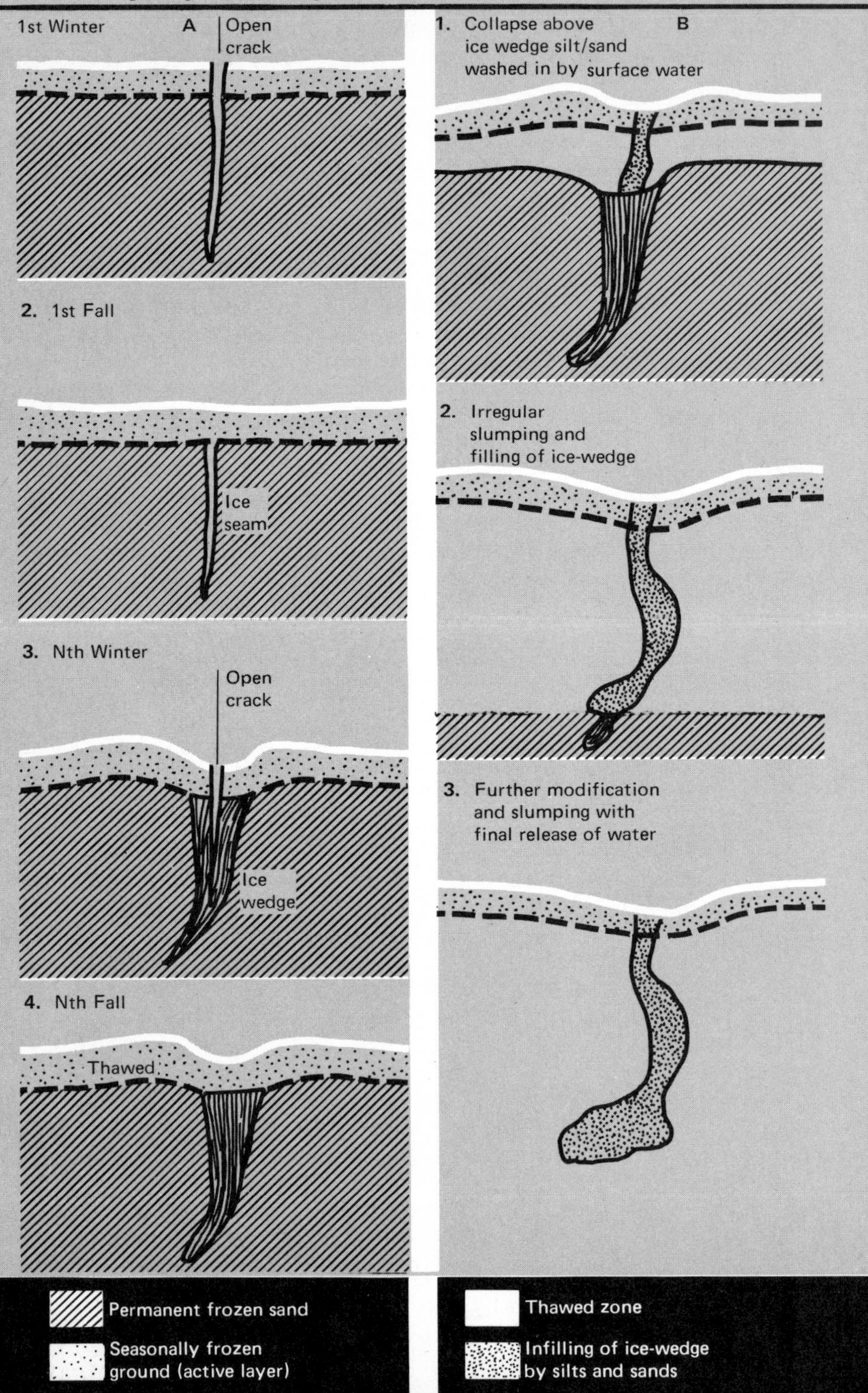

produce the same dramatic surface relief as ice wedges. Over a period of time, there is evidence that pedogenetic processes limit the availability of free sand to fill the cracks and the sand wedges may become less clearly defined (Ugolini ***et al.***, 1973; see p. 201).

The palaeoclimatic significance of sand wedges is similar to that of ice wedges, since genetically, they are the same. The recognition of fossil sand wedges is particularly rewarding since they indicate not only perennially frozen ground but also a very dry climate. Moreover, they indicate extremely low winter temperatures since the enclosing sediments contain little ice and possess a relatively small coefficient of linear contraction. Exceptionally low ground temperatures are required, therefore, for the thermal contraction cracks to develop in the first instance.

Soil wedges Frost cracks which develop primarily in the seasonally frozen layer have been described by a number of workers, notably Dylik (1966), Pissart (1968; 1970, pp. 15–19), Freidman ***et al.*** (1971), and several Russian investigators (e.g. Danilova, 1956; Katasonov, 1973; Katasonov and Ivanov, 1973). These cracks, which are infilled with mineral soil, are termed 'soil wedges', 'ground veins', or 'seasonal frost cracks', and are very different to the ice and sand wedges previously described.

Soil wedges have received little attention to date and the mechanics of their formation are little understood. They appear to be particularly well developed in parts of central Yakutia, USSR, possibly on account of the thickness of the seasonally thawed layer in that area. They have also recently been described from various parts of Iceland, where they form broad polygonal trough patterns at the ground surface (Freidman ***et al.***, 1971). One is not certain, however, whether the features occurring in Siberia are of the same general type as those described in Iceland since there are considerable differences in form and distribution. For this reason, the soil wedges of these two areas are described separately.

The majority of the Siberian ground veins are restricted to the seasonally thawed zone, rarely exceeding 1.5 m in depth. They are usually found in sands and gravels and possess a variety of distinct forms, varying from narrow, vein-like structures (Fig. 3.7) to broad, triangular forms. The infilling material shows some degree of stratification with coarser particles remaining near the surface. The surrounding sediments often show distinct deformation; in some areas there is a 'turning-down' of the enclosing sediments, in others, flame structures and other deformations are present. Finally, the wedges do not necessarily bear any relation to the adjacent frost fissures which penetrate the perennially frozen ground beneath; in fact, they often form polygonal nets which are smaller than the net produced by the fissures in the permafrost.

According to Katasonov (1973), the simple downturning or 'sag' veins are associated with tiny frost fissures and reflect layers that were dissected and pushed downwards along the fissure wall. The more complex structures involve varying degrees of infilling by mineral and organic matter from the surface together with deformation, and are less easily explained. The infiltration of water into the fissure and its subsequent freezing from both above and below may account for the contortions and upturnings.

In some respects, the frost cracks described in Iceland are similar to some of the Siberian structures since they are restricted to the seasonally frozen layer,

3

◀ **Fig. 3.7** *A soil wedge of the 'turning down' variety developed in sands and silts at Maimaga, Lena River, central Yakutia, USSR.*

and are all vein-like structures with a downturning of the enclosing sediments. In other respects they are very different. They occur under climatic conditions in which perennially frozen ground is unable to form, and they give rise to large polygonal nets more akin to sand and ice wedge polygons. Furthermore, no large triangular wedge forms or flame structures or contortions are reported.

It is clear that soil wedges pose a number of problems. First, it is not known what type of frost action environment is necessary for them to form, and whether perennially frozen ground is necessary for certain forms. Instances of seasonal frost cracking have even been reported from middle latitudes (e.g. Washburn, Smith and Goddard, 1963). Second, as stressed by Dylik (1966, p. 260), the reason for the downturning of some of the sediments adjacent to the soil wedges is not well understood. Third, the difference in form, as illustrated by the triangular wedge structures at the one extreme and the vein-like structures at the other, is so great that probably no single mechanism of development exists. It may be that some of the Siberian features are ice wedge pseudomorphs and result from the thawing of ice wedges formed in an earlier period. Finally, the different dimensions of the spacings between the fissures, as reported from Siberia and Iceland, are difficult to explain. Clearly the interpretation of soil wedges is not complete, and their use as indicators of perennially frozen ground not justified. A better understanding of their distribution would help towards solving many of these problems but at present, frost cracks developing in the seasonally frozen layer have not been widely recognised or studied.

Ice segregation processes

It is well known that as soil freezes any water present within the soil segregates into ice lenses. Much of the early work carried out upon the mechanism of segregated ice formation was by Taber (1929; 1930). He was concerned primarily with the phenomenon of frost heave within soils and his two important conclusions were (a) ice segregation is favoured in materials having a grain size composition of 0.01 mm diameter or less, and (b) the ice crystals grow within the material in the direction in which heat is being most rapidly conducted away, i.e. normal to the surface. These conclusions have been verified by a number of later studies.

Mackay (1971) has recently summarised much of the currently accepted view on the origin of segregated ice. In essence, segregated ice is favoured when the pore water pressure is high. For example, if one considers a layer of soil particles as representing a section of soil located approximately at the freezing level, when the ground freezes the freezing plane may either remain stationary above the soil particles, or tongues of ice may descend through the pores. If the freezing plane remains above the soil particles, water will move upwards through the soil pores from the unfrozen ground beneath and towards the freezing plane. Ice crystals will develop and, as long as the supply of water is maintained, an ice lens will quickly develop. This, in turn, will promote an upward heave of the overlying sediments. It would appear that the main control over whether the freezing level will remain stationary or descend is the pore water pressure which controls the supply of water to the freezing plane. If the pore water pressure is high and exceeds the overburden pressure, then it will maintain the freezing plane at its original level; if it is low, then a tongue of ice moves down through the soil by progressively freezing the pore water in place. This forms pore ice as opposed to segregated ice. When it is appreciated that fine grained soils possess small interstices, and that a high pore pressure can be more easily maintained in such a situation, it is clear that segregated ice formation and frost heaving is favoured in fine grained materials. Such sediments are often termed 'frost-susceptible' by engineers. Pore ice, by contrast, develops in coarser grained materials, often sands, where it is harder for strong pore water pressure to be maintained in the correspondingly larger soil interstices. According to Taber, segregated ice lenses may grow in the manner so described to upwards of 4 m in thickness.

Ice segregation in the seasonally thawed layer results in a number of distinct geomorphic processes such as frost heaving of soil, the upfreezing of stones and objects, frost sorting, the localised updoming or uplifting of surface sediments by ice or crystal growth, and the tilting of stones. These processes are outlined below.

Frost heaving Frost heaving is the predominantly upward movement of mineral soil during freezing caused by the migration of water to the freezing plane and its subsequent expansion upon freezing. Attempts to measure frost heave in the field usually involves the regular measurement of stakes, rods or cone-like targets resting on the surface or buried in the ground to varying depths.

Table 3.2 *Some typical frost-heave values recorded in a number of different localities*

Location	Climatic type	Year of record	Site characteristics		Total heave, per year (cm)
Mesters' Vig, Greenland[1] 72°N	High arctic	1958–64	Slopes (*a*) 'Wet'	10 cm depth	0.0–1.0
				20 cm depth	1.5–5.8
			(*b*) 'Dry'	10 cm depth	0.5–0.7
				20 cm depth	0.8–1.4
Signy Island, South Orkney Islands[2] 61°S	Low temperature range	1964	Sorted circle; highly frost susceptible		
			(*a*) Surface		4.0
			(*b*) Buried		0.4
			(*c*) On stones		3.6
			(*d*) At edge		2.0
Colorado Front Range, USA[3] 39°N	Alpine	1969–70	Frost boil; highly frost susceptible. Surface		25.0–29.5
Cape Thompson, Alaska[4] 70°N	High arctic		Frost boil; highly frost susceptible		32.5

(1) Washburn (1969); (2) Chambers (1967); (3) Fahey (1974); (4) Everett (1966).

Washburn (1969) has concluded that the critical variables influencing heave are (1) the moisture content of the mineral soil, and (2) the depth of insertion of the targets or stakes. The greatest heave is usually measured in those areas of abundant moisture, and takes place during the autumn period as the winter freeze-up occurs. Average amounts of soil heave vary considerably from locality to locality. Frost heave values of 1–5 cm p.a. are probably typical but much higher values have been recorded in frost-susceptible situations (Table 3.2). According to Washburn (1969), the amount of heave increases with depth, at least in the upper 30 cm of the ground, and is directly proportional to the 'effective height' of the object being heaved, i.e. the vertical dimensions of the buried portion frozen to and therefore heaved with the adjacent material. The greater heave of the more deeply inserted targets is primarily the result of the annual cycle.

In-depth studies of frost heave situations include those by Jahn (1961), Chambers (1967) and Fahey (1974). In all cases, a securely positioned metal frame or 'bedstead' was positioned over a highly frost-susceptible area such as a mud boil or a sorted circle. Rods were inserted at varying depths in the soil or rested on the surface. They were free to move up and down, and the subsequent movements were recorded against the fixed frame. The observations made by Chambers (1967) are reasonably typical and can be used to illustrate several important aspects of the annual frost heave cycle. The movement of some of the rods during the freeze-up period in 1964 is given in Table 3.3. For each rod, the data represents the entire soil heave for that winter. The pattern of heave can be summarised in a number of observations. First, the majority of total frost heave

Table 3.3 *Differential frost heave at surface and at depth within a large sorted circle during annual freeze-up, 1964, Signy Island, South Orkney Islands, Antarctica. Data from Chambers (1967)*

	March						**April**								**Total heave**
Date	**26**	**27**	**28**	**29**	**30**	**31**	**1**	**2**	**3**	**4**	**5**	**6**	**12**	**26**	**(mm)**
T (°C)	−4	−4	−5	−3	−1	+1	+2	−7	−8	−4	−4	−4	−7	−9	
Ground surface within circle															
Site A	+4	+6	+3	+3	0	+1	−1	+1	+2	+3	+2	+3	+6	+3	+36
Site C	+5	+5	+2	+2	+2	+1	0	0	+3	+3	+2	+3	+8	+2	+38
Site D	+4	+6	+3	+3	+1	+1	−2	+2	+3	+3	+2	+3	+8	+3	+40
Stone at surface within circle															
Site H	+2	+8	+3	+3	+2	+2	+1	+1	+2	+2	+3	+3	+9	+2	+43
Ground surface, edge of circle															
Site N	+4	+9	+2	−1	0	0	−3	+3	0	0	+1	0	+4	0	+19
Rods at depth (10–15 cm)															
Rod F	0	−1	−1	−1	−1	0	−1	+1	−1	−2	−1	−1	+3	+1	−5
Rod G	0	+1	−1	0	−2	0	−1	+1	−1	−2	−1	0	+6	+1	+1

 occurs during the initial two weeks of freeze-up when the freezing plane was descending from the surface to a depth of approximately 15 cm. Second, the beginning of heave of the buried rods did not begin until approximately two weeks after the initial movement of the surface rods, indicating the arrival of the freezing plane at that depth. Third, heave at depths below 28–30 cm was considerably less than at the surface and in some cases there was actually a negative movement. Fourth, no significant movement of any rod was recorded after that period when soil temperatures indicated that the top 40 cm of the seasonally thawed layer had been frozen. These observations suggest that frost heaving is primarily confined to the upper part of the seasonally thawed layer which possesses, therefore, a dual nature. There is an upper zone in which sorting, frost heaving and ice segregation occurs and there is a lower layer which is relatively passive as regards frost activity. Laboratory work confirms the dual nature of the thawed zone, and throws additional light upon the nature of the lower, passive zone. For example, Pissart (1964b; 1970) has conducted experiments in which a tray of water-saturated sediment some 10 cm thick was frozen from the surface downwards. Horizontal ice lenses only formed in the upper 6 cm of the sediments while the lower zone developed a number of vertical desiccation cracks.

A second important conclusion concerns the negative movement of some of the buried rods, and the generally small amount of heave recorded by the other buried rods. It would appear that, at the same time as the upward heave is taking place, a downward force is also exerted which is sufficient to cause compression of the unfrozen material below the freezing plane. Such freezing-induced pressures were termed 'cryostatic' by Washburn (1956) and are discussed more fully later in this chapter. The presence of such pressures combined with the absence of ice segregation at those depths probably explain the low frost heave values at depth.

Upfreezing of objects In addition to the general heaving of the mineral soil consequent upon ice segregation, the progressive upward movement of stones and objects is a further characteristic of frost heaving. This process is called upfreezing, and appears to be common in sediments which possess coarser particles in association with an appreciable content of fines.

The mechanics of upfreezing are still not completely understood. However, at least two different hypotheses have been suggested. The 'Frost-Pull' hypothesis involves the assumption that, as the ground freezes from the top downwards, the top of a pebble or coarser particle is gripped by the advancing freezing plane and raised in conjunction with the overall heave that is associated with the freezing of the overlying sediments. Upon thawing, the pebble is not able to return to its initial position because (a) lateral frost heaving (frost thrusting) during the freezing process would have compressed the hollow originally occupied by the pebble, and (b) during thawing, material would have slumped into the hollow.

Partial support from field experiments for this process is provided by Chambers (1967, pp. 18–19). In one location, a series of stakes and stones were inserted vertically to depths of 10–100 cm in a large sorted circle. At each depth, two stakes were inserted, one with a sharpened end and one with a blunt end. All pairs of stakes showed identical uplift, indicating that the force which

Table 3.4 *Values of upfreezing of wooden stakes and stones inserted to various depths in a large sorted polygon during 1962/63, Signy Island, Antarctica. Data from Chambers (1967)*

Depth (cm)	Upward movement, 1962 (cm)	Upward movement, 1963 (cm)	Mean movement (cm year^{-1})
Stones			
10–15	–	5.0	5.0
Stakes			
10	4.0	5.0	4.5
15	5.0	6.0	5.5
20	5.0	6.0	5.5
25	2.0	4.0	3.0
50	1.0	3.0	2.0
100	1.0	2.0	1.5

uplifted the stakes came not from beneath but from the sides. On the other hand, the frost-pull hypothesis is not completely adequate to explain the mean annual movement with depth as shown in Table 3.4. Theoretically, the stakes and stones should be gripped by the fines at the extreme surface when freezing first started so that the stakes were drawn upwards as the surface was lifted. The amount of uplift of any stake should be related to the depth of insertion in the ground; as the freezing plane reached the base of each stake, differential movement ceased and both fines and stakes would have been uplifted together. The deepest stakes should, therefore, be uplifted the greatest. The results indicate, however, that the maximum uplift occurred at depths of 15–20 cm and that movement at depths below 20 cm actually decreased. This is difficult to explain in terms of the frost-pull hypothesis.

A second explanation for the upfreezing of objects has been termed the 'Frost-Push' hypothesis.

According to this theory, upfreezing is the result of the greater thermal conductivity of stones resulting in the formation of ice around and beneath the stones. This ice would force the pebble upwards. Upon thawing, as in the frost-pull hypothesis, the infill of fines beneath the pebble would prevent its return to its original position. Experimental work has shown that this process is a viable one, especially in instances of numerous freeze–thaw cycles, when there is rapid upfreezing (Corte, 1966). Field support is provided by Washburn (1969) who has observed very rapid upfreezing of stones which sometimes break the overlying vegetation cover. This fact suggests considerable ice segregation at the base of the stone and favours the frost-push hypothesis rather than the frost-pull hypothesis.

It is clear, therefore, that the upfreezing of stones is a complex process. In all probability, both the frost-pull and frost-push hypotheses may be valid but under slightly different conditions. Moreover, the upfreezing of objects is intimately linked to the frost heave process in general, and cryostatic pressures may also be involved.

Fig. 3.8 *Diagrammatic illustration of the mechanism by which a stone becomes tilted through frost heaving of the ground. Positions 1–5 represent successive positions of the stone-soil interface during freezing and thawing.*

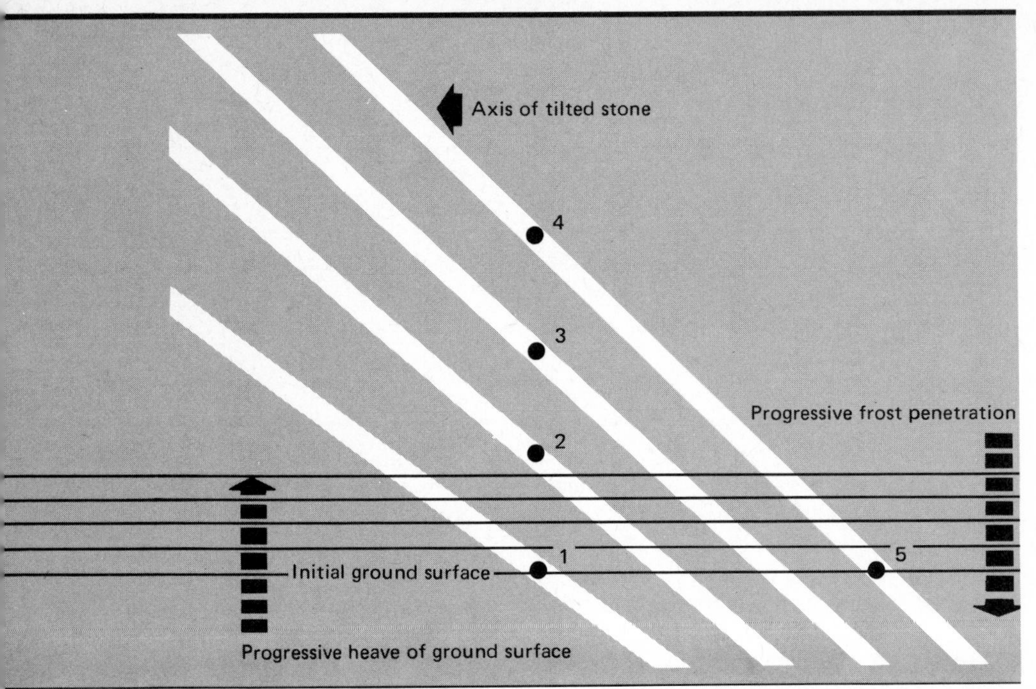

Stone tilting A common microfeature of the ground surface of many areas of frost action is the presence of tilted stones, often standing on end. This phenomenon is the result of differential frost heave at the top and bottom of the stone resulting in a rotation and tilting of the axis. The mechanism is illustrated in Fig. 3.8.

As the ground freezes from the surface downwards, the top of a stone at or near the ground surface is gripped by the freezing layer and subject to frost heave (i.e. 'frost-pull'). The lower part of the stone remains within the unfrozen zone and is not subject to heave. Thus the stone experiences differential movement with the greatest vertical movement near the surface. Depending upon the viscosity and other strength parameters of the unfrozen material, the axis of the stone undergoes rotation and there will be some readjustment of the object at depth. When all of the stone is totally within the frozen layer, there will no longer be any differential heave and any rotation and tilting will cease, although frost heave and upward movement of the soil in general may continue as the frost penetrates deeper. Upon thawing of the ground, again from the surface downwards, the stone is prevented from returning to its original inclination by the settling of thawed sediments. Thus, with repeated cycles of freezing and thawing, there is a progressive increase in the angle of the axis of many stones towards the vertical, as well as a progressive upward movement of the stones in general.

It follows that the upward tilting of pebbles is a characteristic of periglacial slope deposits. Angles of between 25 and 45° from the horizontal are most common since higher angles are prevented by the mass movement of material under gravity processes (mainly solifluction, see pp. 135–41) in the seasonally thawed layer. The fabric of periglacial slope deposits is distinctive therefore, in that (a) the long axes of pebbles is oriented downslope and (b) the pebbles are tilted upwards towards the surface at their downslope end.

Needle ice Needle ice or 'pipkrake' is another localised and small-scale heave phenomenon produced by ice segregation at or just beneath the surface. It reflects a delicate balance between temperature, moisture and soil conditions. Direct cooling of the ground surface leads to the formation of ice crystals which grow upwards in the direction of heat loss and lift small pebbles and soil particles. The needles can range in length from a few millimetres to several centimetres.

Under field conditions, the growth of needle ice is usually associated with diurnal freezing and thawing (Outcalt, 1969), and is particularly common in alpine locations where the frequency of freeze–thaw cycles is at its greatest. The thawing and collapse of needle ice has been suggested to be an important factor in periglacial creep and downslope movement in such environments (Troll, 1958; Soons and Rayner, 1960). Needle ice activity also assists the sorting process through small-scale differential heaving on the surface of miniature polygons, and may give rise to the development of microhummocks (Troll, 1958; Fahey, 1973). The importance of needle ice as a disruptive soil agent has probably been underestimated. In certain cases, it may be responsible for damage to plant materials when freezing causes vertical mechanical stress within the root zone (e.g. Brink ***et al.,*** 1967).

Needle ice occurs widely wherever temperatures fluctuate near the freezing point, and occurs in both Arctic and temperate regions in addition to alpine environments. Wherever wet, heave-susceptible soil surfaces are exposed to frequent diurnal freeze–thaw events, needle ice sorting and heaving must be considered a potent geomorphic agent. The nature and magnitude of periglacial creep, of which needle ice is the extreme example, is discussed more fully in Chapter 7 (pp. 135–7).

Frost mounds If frost heaving is concentrated into discrete areas, the resulting soil movement often produces small mounds or hummocks. Where they occur widely, these frost mounds represent a form of patterned ground (see Chapter 9). They vary from a few centimetres to several metres in height, with all gradations in between (e.g. Bird, 1967, pp. 201–3). The initial cause of the localised frost heaving is usually related to the presence of an insulating vegetation cover. During freeze-up, moisture from the adjacent area migrates, causing differential frost heave and the segregation of local ice bodies. Where no segregated ice bodies exist and the hummock is composed predominantly of mineral soil, it is probable that freezing-induced or 'cryostatic' pressures (see below, pp. 40–2) are important in the movement of the soil and the development of the hummocks.

The smallest feature within the frost mound category are tundra hummocks. In Canada, these are well developed in high Arctic regions north of the treeline.

They form initially from tussock forming grasses and sedges in wet, meadow tundra environments, or beneath Dryas on gently sloping terrain in upland

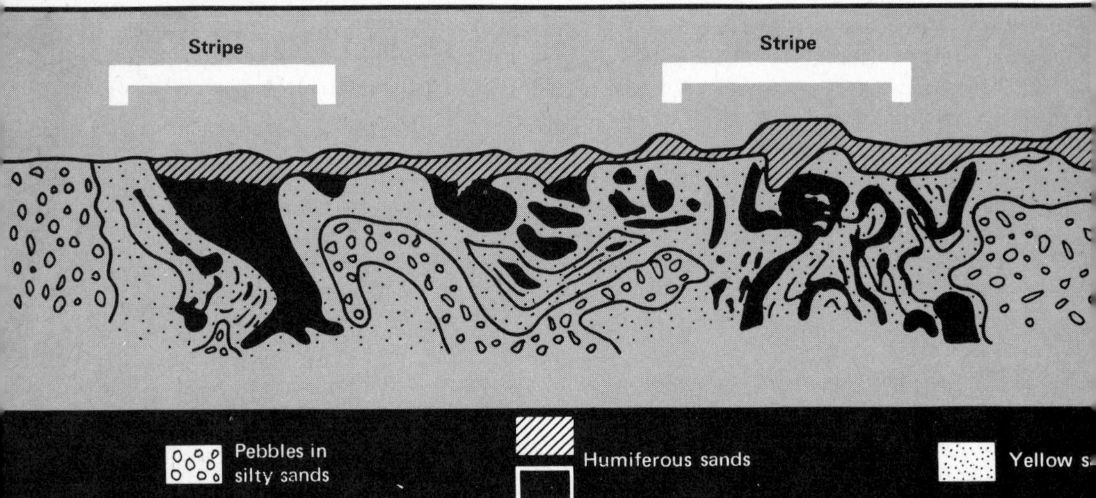

Fig. 3.9 *Section through non-sorted stripes, central Banks Island, showing cryoturbations. From Pissart (1975)*

tundra environments (Fig. 9.3). They are roughly spherical, 10–50 cm high and about 0.3–1.0 m in diameter, and are composed predominantly of mineral soil. Larger frost mounds usually possess lenses of pure ice (e.g. Fig. 5.15) and there is no clear division between these and other ice-cored features such as pingos (see Chapter 5). Earth hummocks are widespread south of the treeline, particularly in the northern Mackenzie Valley and northern Yukon (Zoltai and Pettapiece, 1974). They are 1.0–3.0 m in diameter and 40–50 cm in height and larger, therefore, than the tundra hummocks north of the treeline.

The soils of frost hummocks are characterised by broken and dislocated horizons and an organic rich subsurface layer at the base of the seasonally thawed zone. The term 'cryoturbation' is commonly applied to these irregular soil structures. Figure 3.9 illustrates a number of such structures observed in a series of non-sorted stripes on Banks Island.

Carbon-14 dating of the organic material within and below cryoturbated structures in the Mackenzie Delta and Alaska indicate that soil mixing and movement has occurred at all times over the last 8 000 years (e.g. Mackay, 1958; Tedrow, 1966a). In the Mackenzie Valley, periods of relatively high or low cryoturbation activity can be recognised over the last 200–300 years from tree ring studies of the spruce growing on the hummocky terrain (Zoltai and Tarnocai, 1974). Many of the trees are tilted as the result of this ground movement forming so-called 'drunken forests', and this tilting is recorded in the unequal growth of a tree ring. Differences of cryoturbation activity need not necessarily imply changes in the intensity of frost action or of regional climatic change, however. Spruce stands which have originated after a forest fire indicate that most trees begin to tilt 30–40 years after their establishment. This coincides with the progressive decrease in the thickness of the seasonally thawed zone, the result of the insulating qualities of the spruce–lichen cover, until the pre-fire thickness has been attained. Thus, with the growth of a boreal forest, the moisture content of the seasonally thawed zone increases and its thickness

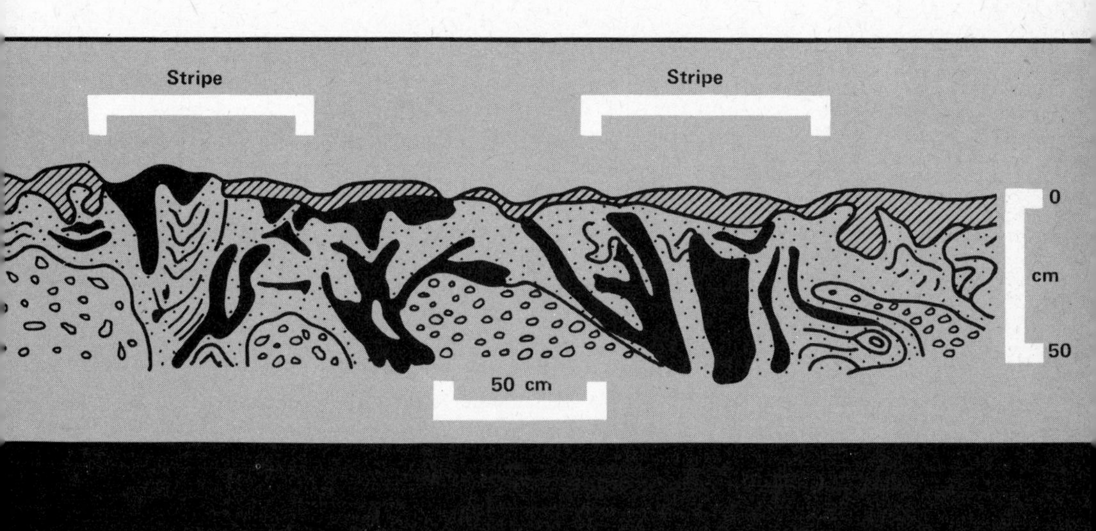

decreases. As a result, its susceptibility to both frost heave and cryostatic forces increases considerably. Therefore, given the prevalence of forest fires in the boreal forest zone, distinct changes in the intensity of cryoturbation may occur without any corresponding climatic change.

A number of experimental studies (e.g. Corte, 1971; Corte and Higashi, 1971) have demonstrated how repeated freezing and thawing results in deformation, mass displacement, and localised updoming of sediments. However, the role of frost heaving and its relation to freezing-induced pressures in the formation of mounds and other patterned ground features is still unclear. A fuller discussion of freezing-induced pressures and the structures which result are given later in this chapter (p. 40) as well as in the discussion of soils and patterned ground features in Chapter 9.

Frost sorting Frost sorting is a complex process by which migrating particles are sorted into uniform particle sizes. Often, a number of other frost related processes, such as needle ice and frost heaving, contribute to the sorting process. The complexity of the sorting process is probably best illustrated by the overwhelming varieties of patterned ground that exist. These are discussed in Chapter 9.

A number of controlled laboratory experiments have been undertaken to examine the frost sorting process. According to Corte (1966), the movement of particles depends on the amount of moisture present, the rate of freezing, the particle size distribution and the orientation of the freeze–thaw plane. Three types of sorting mechanisms were simulated under laboratory conditions:

(a) Sorting by uplift (i.e. frost heaving), when freezing and thawing occurs from the top. The larger particles move upwards, the smaller move downwards. This is called ***vertical sorting***, and gives horizontally bedded layers of different particle size.
(b) Sorting by migration in front of a moving freezing plane, when freezing and

thawing occurs from either the top or the sides. Corte's experiments show that under such conditions, the finer particles migrate in a parabolic path away from the advancing freezing plane, and that the coarser particles are left nearest the cooling side. This is called ***lateral sorting***, and if the freezing takes place from the sides produces vertical layers of different particle sizes.

(c) ***Mechanical sorting*** occurs when mounds and frost heaved structures are produced. The coarser particles migrate under gravitational influences to form borders of coarser material surrounding finer sediments.

Vertical sorting has already been discussed under frost heaving and upfreezing, and needs little further explanation. Likewise, mechanical sorting is an easily understood concept. However, neither of these processes is able to explain all of the sorting phenomena observed in nature in periglacial regions, and attention must be focused upon lateral sorting. This has been investigated in a number of detailed laboratory experiments by Corte (1966, pp. 196–209). These experiments have shown that fine particles migrate under a wider range of freezing rates than coarser particles. Below a certain freezing rate, finer particles travel more than coarser ones. This means that a heterogeneous material inevitably becomes sorted by freezing. It was also illustrated that, as the particles migrate in front of the freezing plane, air bubbles are formed as dissolved air in the water is expelled during the freezing process. Sometimes these air bubbles are trapped beneath the migrated particle position. Since these air bubbles are oriented normal to the freezing plane, their presence in massive icy bodies (see pp. 80–2) is a good indicator of the direction of movement of the freezing plane, and evidence of a segregational origin as opposed to either a vein ice or buried ice origin.

Corte has also demonstrated how side freezing can change vertical sorting of materials into lateral sorting. In one series of experiments vertical sorting was simulated in shallow trays by placing four layers of different particle sizes on top of each other. The samples were then saturated with water and subject to thawing and freezing from the sides and the top. After 33 cycles, mounds had developed at the surface, the coarser sands had moved from the surface of the mounds towards the sides, and silt particles had moved towards the centres of the trays. According to Corte, the change from vertical to lateral sorting is explained by freezing from the top, bottom and sides. Moisture was drawn from the centres towards the sides of the tray where the freezing plane was advancing. As ice lenses were formed, particles were forced towards the centre away from the freezing plane. This increase in compaction in the centre results in an updoming of the sediments. Subsequently during thawing, larger particles move from the mound under gravitational processes to the space left by the melting ice next to the wall of the tray. During the next freezing cycle, these larger particles are frozen while the finer particles migrate in front of the freezing plane towards the centre. Thus, a progressive sorting and updoming of the sediments occurs with each successive freeze–thaw cycle.

A large number of other experiments have been reported by Corte, and the interested reader is referred to them. In summary, Corte's conclusions on the sorting mechanism are as follows:

1. Sorting decreases as the moisture supply decreases.
2. Sorting develops best under conditions of slow freezing rates and saturated soils. Heterogeneous sediments subject to fast freezing rates will produce

little sorting in comparison to the same sediment subject to very slow freezing.

3. Fine particles can be sorted under a wider range of freezing rates than coarser particles.
4. By changing the orientation of the freeze–thaw plane, it is possible to change vertical sorting (i.e. horizontally bedded and sorted layers) into lateral sorting (i.e. vertically sorted layers).
5. Horizontal layers of fine particles produced by vertical sorting become dome-shaped and produce mounds when subject to a vertical freeze–thaw plane (i.e. freezing takes place from the sides).

These conclusions should be treated with caution, however. As stressed by Corte, there may be other processes of sorting by freezing which have not been examined, and further experimentation is needed. Moreover, the role of frost sorting in the formation of frost mounds and patterned ground structures needs detailed field investigation. This is difficult, however, since many frost-action processes are intimately linked to each other, and to the sorting process, and there is often no clear distinction between cause and effect.

Volumetric increase in water as it freezes

When water freezes at 0°C, it increases in volume by approximately 9 per cent. This characteristic of the phase change of H_2O results in two important frost-action processes. These are (1) frost wedging and (2) mass displacement and the development of cryostatic pressures.

Frost wedging The disintegration and mechanical breakdown of rock by the freezing of water present within the pore spaces, joints and bedding planes has long been thought of as a particularly potent geomorphic agent in the periglacial environment (Fig. 3.10). The presence of extensive upland surfaces of angular frost shattered rocks and boulders in both present-day and Pleistocene periglacial environments is the most dramatic morphological feature formed by intense frost wedging. They are sometimes referred to as 'block fields' or 'felsenmeer'. Other associated features include angular hillslope and summit tors, and slopes composed of vertical or near vertical free faces with extensive debris slopes developed below. On a smaller scale, frost shattered rocks and boulders protruding from the ground surface are a widespread illustration of frost wedging.

Theoretically, the maximum pressure set up by the freezing of water is 2 100 kg/cm^2 at −22°C. At temperatures below this, the pressure decreases because the ice actually begins to contract, as explained earlier. In reality, however, this maximum value is almost certainly never reached or even approached since a number of factors operate to reduce the pressures developed. First, the water or ice must be contained within a closed system for high pressures to develop; this usually means conditions of extremely rapid freezing from the surface downwards which seals the pores and cracks in the rock. Second, air bubbles in the ice, and pore spaces within the rock reduce pressures considerably. Third, and probably most important of all, the rock itself is not strong enough to withstand such extreme pressures, especially since it is a tensile force rather than a compressive force which is being considered. As a result, the actual pressures developed by the freezing of water in rocks is much less than the theoretical maximum. Tricart (1970, p. 73) for example, has estimated average

◀ Fig. 3.10 *An example of the result of frost wedging, Prince Patrick Island, Canadian Arctic. The rock is a fissile sandstone of Upper Jurassic age.*

frost shattering forces to be 14 kg/cm^2 and values of similar magnitude upwards to 100 kg/cm^2 are probably realistic.

The most important control over the effectiveness of frost wedging is, of course, the presence of moisture. Laboratory experiments by both Tricart (1956) and Potts (1970) have shown how the amount of disintegration in rocks supplied with abundant moisture was much greater than in similar rocks supplied with less water.

The second most important control over frost wedging is the nature of the rock involved. Based upon the field observation of frost shattered debris in Europe and the Canadian Arctic, several workers have ranked certain rock types in increasing order of susceptibility to frost wedging and mechanical disintegration. Their rankings are given in Table 3.5.

Table 3.5 *The susceptibility of different lithologies to frost wedging and mechanical weathering, based upon the field observation of frost-action phenomena*

	Ellef Ringnes Island, NWT, Canada, 79°N	**Ardennes, Belgium**	**Dartmoor, southwest England**
Most resistant	Gabbro	Quartzite	Coarse/medium grained granite
	Sandstone	Grits/sandstones	Diabase
		Limestones	Fine grained granite
Least resistant	Shale	Schists	Metamorphosed sediments
Source:	St-Onge (1969)	Alexandre (1958)	Waters (1964)

Other workers have subjected samples of different rocks to experimentally controlled frost shattering processes (Tricart, 1956; Wiman, 1963; Potts, 1970). For example, Potts (1970) concluded that igneous rocks were the most resistant to freeze–thaw action while shales disintegrated at a much greater rate. The varying rates of shatter were related to lithological and structural characteristics of the rocks. The most important lithological factors were the number of planes of weakness within a rock sample, and its grain size properties. The planes of weakness, called shatter or S-planes by Wiman (1963), depended on the proximity of the bedding or cleavage planes. A shale or mudstone with a high density of such planes was found to have a high rate of shatter. By contrast, sandstones which vary in compaction showed considerable variations in rates of shatter. Equally, the products of frost shattering reflect the planes of weakness and the grain size of the sediments. An important fact which emerged from Potts's experiments however, was the absence of material finer than 0.06 mm produced by frost action in any of his samples. If substantiated by other studies, this would indicate that processes other than frost action are responsible for the production of silt and clay particles in periglacial regions.

The relationship between the frequency and intensity of freeze–thaw cycles to frost shattering has also been investigated.

Two sorts of freeze–thaw cycles have been used; an 'Icelandic' cycle characterised by a duration of approximately 24 hours and a temperature range of

+7°C to −8°C, and a 'Siberian' cycle of 3–4 days duration and with temperatures ranging from +15°C to −30°C. The results are somewhat equivocal, since both Wiman and Potts concluded that greater amounts of weathering resulted from the 'Icelandic' cycles while Tricart concluded that the 'Siberian' cycle was more effective. This dichotomy may be partly explained by the fact that both Wiman and Potts regarded four Icelandic cycles as being the equivalent of one Siberian cycle, a completely empirical assumption. Probably more reliance should be placed upon direct cycle to cycle comparison. In this respect, Potts found that the percentage weathered from corresponding specimens after 100 cycles of both types was not statistically different. He concluded therefore, that the number of freeze–thaw cycles is more important than the intensity of such cycles. Similar conclusions have been reached by Lautridou (1971), reporting upon the laboratory work being conducted at Caen. There, controlled experiments have shown that the repeated freezing to −5, −10, −15 and −30°C of chalk and limestone samples have all given approximately the same amount of frost shattering for a given number of cycles.

If the laboratory studies are correct and the number of freeze–thaw cycles is more important than the intensity as regards rock shattering, the relatively small frequency of freeze–thaw cycles recorded in present-day periglacial environments suggests that frost wedging is not as important a process as is sometimes thought. This is not to deny the presence of block fields, talus slopes and shattered rock debris over large areas of the tundra and high Arctic regions. On the contrary, it seems best to interpret such phenomena as indicating either (1) a very long period of time during which the rock has experienced present-day freeze–thaw frequencies, or (2) the relative slowness or even absence of other major weathering processes, or (3) a period of more frequent freeze–thaw activity in the past, probably in association with regional Pleistocene climatic changes.

Mass displacements and cryostatic pressures Under certain circumstances, the freezing of soil may generate considerable but localised pressures which have been termed 'cryostatic' by Washburn (1956, p. 842). These pressures are essentially hydrostatic in nature since they develop in pockets of unfrozen material which are trapped between the downward migrating freezing layer and the perennially frozen ground below. Usually, differences in moisture content and/or grain size promote differential freezing rates such that localised 'closed systems' are created.

Pissart (1970, pp. 37–43) has simulated the development of freezing-induced pressures under laboratory conditions. In view of the importance commonly attached to cryostatic pressures in the interpretation of frost mounds (see above), patterned ground (see Chapter 9), and involutions (see below), it is useful to briefly consider Pissart's results. Trays of sediments 10 cm deep and 46 cm wide were frozen from the top downwards. The results of one experiment carried out over a 96-hour period are shown in Fig. 3.11. The temperature curve has characteristics similar to the annual temperature regime described earlier (see pp. 14–17) and the experiment may be regarded as a reasonable simulation of the annual freeze–thaw cycle. There is a rapid cooling to 0°C, followed by a period of near isothermal conditions (the transitional freezing boundary or 'zero curtain') reflecting the latent heat released by the freezing of water, and then a

Fig. 3.11 *Freezing-induced pressures recorded in a tray of sediment subject to freezing and thawing. From Pissart (1973)*

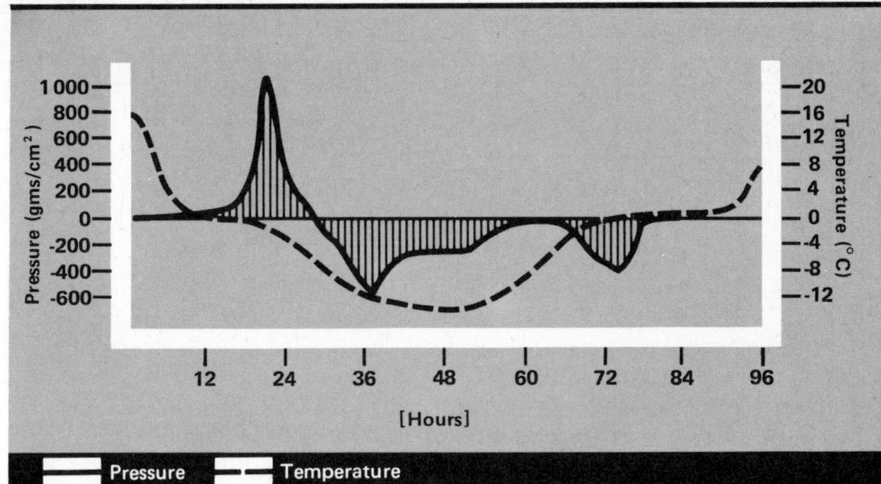

further cooling below 0°C. During thawing in the latter half of the experiment, the absorption of latent heat during the phase change from ice to water delays the rise in temperatures and also produces a period of near zero temperatures. The pressure changes recorded at a depth of 8.5 cm during these temperature fluctuations are also illustrated in Fig. 3.11. They are quite marked and are both positive and negative. Although small positive pressures are set up during the freezing transition boundary, it is only when temperatures drop below the 0°C level that pressures rise considerably and sharply. This is due to the increase in volume consequent upon the freezing of free water. Pressure then falls dramatically and is due to the plastic deformation of the frozen mass when the pressure exceeds the resistance of the enclosing material. This is followed by a slower drop in pressure to well below atmospheric values. This general pressure decrease is a combination of (a) the updoming of sediments and a lessening of pressure at the measuring gauge and (b) the contraction of ground as temperatures continue to fall and capillary water freezes. Ultimately, the drop in pressure within the soil surpasses the limit of resistance of the frozen sediments and a further plastic deformation takes place. After this movement, the residual pressure is equivalent to the tension which can support the frozen material at the temperature existing. With subsequent warming and a melting of capillary ice, conditions become reversed and there is an increase in pressure. The slight fall in pressure which then follows results from the decrease in volume accompanying the melting of the ice. The last variation of pressure is the return to atmospheric pressure when the sediment thaws.

This experiment illustrates that considerable pressures can be generated by the freezing of a soil. According to Pissart, pressures in excess of 4 kg/cm^2 have been recorded in certain experiments. It is also clear that pressures can be generated not only during the freezing process when a closed system exists (cryostatic pressure ***senso stricto***), but also when temperatures have dropped well below 0°C. These latter pressures are not restricted, as are the cryostatic pressures ***senso stricto***, to the seasonally thawed layer but may occur within any freezing body.

The development of freezing-induced pressures in the seasonally frozen zone can lead to the deformation and displacement of sediments. The structures produced are commonly termed 'involutions' (Sharp, 1942). In contrast to the cryoturbation structures previously described, involuted structures are generally more regular in their form and distribution. However, since 'involution' is a descriptive term, and identical structures can be produced by processes other than freezing (see below), it is preferable to refer to such freezing-induced structures as 'periglacial involutions' in contrast to other involuted structures produced by other processes.

The formation of periglacial involutions is not clearly understood, in spite of the common acceptance of cryostatic pressure as being the basic cause. A number of different hypotheses have been suggested. The most widely accepted mechanism is that suggested by Washburn (1956), where the downwards advance of the freezing plane in autumn would lead to the squeezing and contortion of the unfrozen zone in between it and the underlying perennially frozen ground. As a result, the involutions would develop completely within the seasonally thawed layer. Given that soils are not homogeneous, and that the quantity of water present (which may determine the speed of penetration of frost) varies from place to place, it is conceivable that pockets of unfrozen sediments will be created, and that deformations may occur both laterally or vertically. A second explanation involves the differential freezing of soil resulting from either differential vegetation cover or heterogeneous sediments. For example, Dylikowa (1961) has described structures developed in late Pleistocene valley bottom deposits of stratified clays, sands and gravels near Lódź. Since these sediments contain variable quantities of water, they will produce unequal deformations upon freezing. Moreover, depending upon the state of the water, be it free water or capillary water, will be the temperature at which each layer will freeze and deform. In a similar fashion, Sharp (1942) argued that ice segregation in silts and clays may result in the expansion of such sediments and their injection into adjacent unfrozen sediments.

Differential vegetation covers may also be the cause of the cryostatic pressure build-up. For example, Hopkins and Sigafoos (1954) have explained frost mounds or 'thurfurs' and the involutions which exist beneath them, by the more rapid freezing of ground adjacent to a grass tussock than beneath it. As a result, there is lateral pressure generated in a direction towards the unfrozen sediment beneath the grass tussock which forces it upwards. Raup (1966) and Chambers (1966) are two other investigators who have stressed the role of vegetation in promoting differential freezing and the development of cryostatic pressures in the case of earth mounds and patterned ground respectively.

There are problems associated with the cryostatic hypothesis which suggest that it is not the only cause of involuted structures. There is the basic problem that intrusion is usually from unfrozen materials into frozen material, and the mechanism for this is clearly difficult to envisage. Moreover, while considerable cryostatic pressures may develop under saturated soil conditions, if voids exist in the soil or it is not completely saturated, the pressures generated will be small and the expansion of the water upon freezing will only put pressure upon the air held within the capillary soil spaces.

In recent years, several investigators (e.g. Butrym ***et al.,*** 1964; Kostyaev, 1969) have stressed the importance of moisture controlled density differences, consequent upon the thawing of ice-rich sediments, as being the cause of many involuted structures. 'It is no coincidence that periglacial involutions are best developed in water laid sediments. [They are] . . . obviously soft sediment deformations inherent to all water saturated soils' (Butrym ***et al.,*** 1964). Under this hypothesis, involutions are interpreted as being similar to the 'load cast' structures well known in the geologic literature. It is argued that when the melting of ice-rich sediments occurs, significant pore water pressures may develop and the shearing resistance of particles caused by friction will be reduced. Thus, in a situation where coarse gravel sediments overly fine grained ice-rich sediments, distinct pore water pressure differences will be set up upon thawing of the sediments. The underlying sediments will possess a higher pore water pressure than the overlying sediments, and the shearing resistance of the underlying sediments will be reduced because of the excess water. If the overlying sediments are sufficiently heavy, a gravity controlled readjustment will result, with an upward movement of the finer, less dense, sediment and a downward movement of the coarser, denser, sediment, until an equilibrium condition is reached.

Such a process does not require cryostatic pressure and perennially frozen ground for its operation but merely water saturated sediments. A further advantage is that intrusion is into unfrozen materials. It is, therefore, a process particularly applicable to the seasonally thawed layer of many periglacial environments. Ice segregation in the autumn freeze-up and percolating snowmelt during the summer combine to keep the seasonally thawed layer at high moisture levels. On slopes, artesian and hydrostatic effects will also accentuate pore water pressure differences.

The delicate density controlled equilibrium which exists within water saturated sediments in the seasonally thawed layer may easily be upset by very minor disturbances, such as a man walking or a vehicle passing nearby. For example, at an oil exploration rig site on Banks Island in August 1973, the writer observed a number of small volcano-like mudpiles of fines which developed during one week following summer rains in a shallow depression adjacent to the living quarters of the camp (Fig. 3.12). According to an eye witness, one mudpile appeared over a 15–20 minute period following the passage of a light vehicle. Artesian effects as well as saturated soil conditions were probably responsible for this mass displacement. A similar occurrence has been documented by Washburn (1973, pp. 88–9) and it is probable that as the number of persons daily traversing saturated tundra terrain increases more examples of this phenomenon will be reported.

One must conclude that a critical revision of accepted views upon the interpretation of periglacial involutions is necessary. Although the existence of cryostatic pressures is undeniable, and periglacial involutions do exist, involuted structures may also be the result of moisture controlled density differences and completely unrelated to cryostatic pressures. As such, the latter type of structure may occur in environments other than the periglacial. Furthermore, Corte and Higashi (1971) have demonstrated experimentally how differential and repeated frost heaving can also produce similar deformations. A fundamental problem, therefore, is the identification of periglacial involutions from other

Fig. 3.12 *Example of localised mass displacement which occurred following heavy rain and man-induced disturbance at an oil company site, Big River, central Banks Island, in August 1973.*

involuted or cryoturbation structures. The use of involution structures as indicators of Pleistocene periglacial conditions (e.g. Poser, 1948) is extremely hazardous unless other indicators of either frost-action or permafrost conditions are also present.

Permafrost 4

In many periglacial regions, a consequence of the long period of winter cold and the relatively short period of summer thaw is the formation of a layer of frozen ground that does not completely thaw during the summer. This perennially frozen ground is termed permafrost. This word was first used by S. W. Muller (1945) to describe the thermal condition of earth materials when their temperature remains below 0°C continuously for a number of years. In theory, the occurrence of permafrost has little relation to the presence of moisture in the ground. On the other hand, ice is an important component of permafrost in many areas, giving rise to distinct relief features (see below, pp. 82–104). A further complication in defining permafrost on a strictly temperature basis is that unfrozen water may exist at temperatures below 0°C due to the presence of impurities or if the water is under pressure (see Chapter 3, p. 13). Despite these considerations, a thermal definition of permafrost is generally thought the most appropriate (Lachenbruch, 1968; Brown, 1970).

Several other terms need immediate definition. The permafrost table is the upper surface of the permafrost, and the ground above the permafrost table is called the supra-permafrost layer (Fig. 4.1). The active layer is that part of the supra-permafrost layer that freezes in the winter and thaws during the summer, that is, seasonally frozen ground. Although the seasonal frost usually penetrates to the permafrost table in most areas, in some areas it does not and an unfrozen zone exists between the bottom of the seasonal frost and the permafrost table. This is called a talik. Unfrozen zones within and below the permafrost are also termed taliks. If the permafrost thickness is increasing, permafrost is said to be aggrading, and if decreasing, permafrost is said to be degrading.

In practical terms, the importance of permafrost cannot be overemphasised since it has widespread geologic and biologic significance and influences virtually all aspects of economic activity in the regions which it underlies (e.g. Péwé, 1966a; Price, 1972). Moreover, as stated in Chapter 1, the presence of permafrost is considered to be the fundamental characteristic, along with intense frost action, of the true periglacial environment. Clearly therefore, an in-depth examination of permafrost is essential for a proper appreciation of the periglacial environment.

Historical perspective

Despite the numerous reports of frozen ground made by early eighteenth- and nineteenth-century travellers and explorers, our understanding of permafrost has only developed relatively recently. Russian and Soviet scientists have traditionally been more advanced than their North American counterparts in this field, primarily because of the earlier history of settlement of Siberia. The earliest known scientific investigations of permafrost were carried out at Yakutsk in eastern Siberia by the Russian scientist A. F. von Middendorf in 1844–46. He studied the temperature regime of Shergin's Well, a shaft 116 m deep, which was dug for a wealthy merchant in a vain quest for underground

 Fig. 4.1 *The relationship between permafrost, the permafrost table, the active layer and supra-, intra-, and sub-permafrost taliks. Modified from Ferrians et al. (1969)*

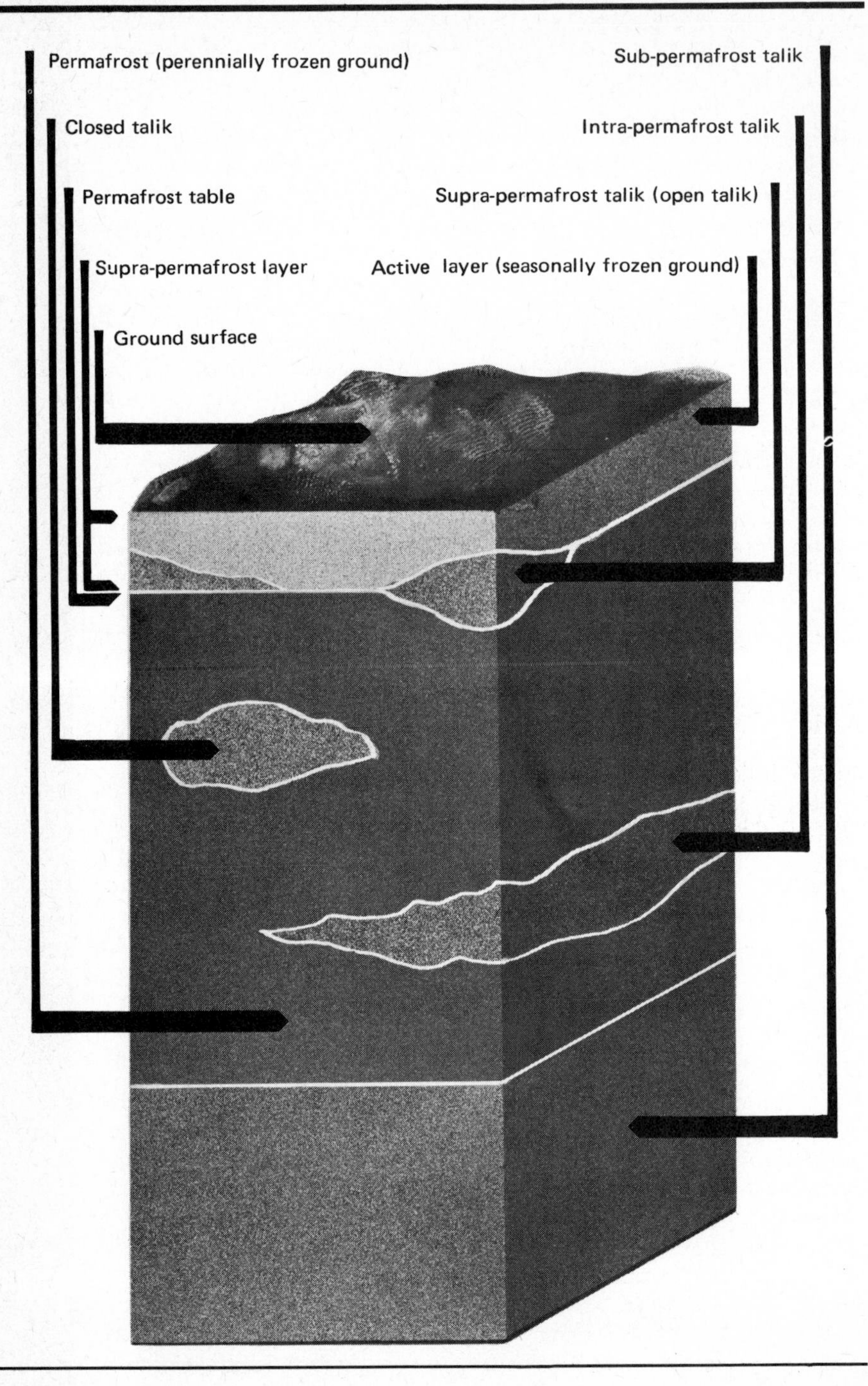

water. Von Middendorf found that not only did the temperature of the permafrost decrease with depth but that the seasonal fluctuations of temperature did not extend beneath 20 m. Subsequently, other Russian scientists developed the study of permafrost and by the late nineteenth century attempts were being made to plot permafrost boundaries in Russia. After 1920, the pace of investigation began to increase and today, for example, the Siberian Division of the Soviet Academy of Sciences maintains a large Permafrost Institute at Yakutsk employing several hundred people.

In North America, the study of permafrost only began in the early 1940s when a number of United States government agencies, including the Geological Survey, became increasingly interested in Alaska and other northern areas for military and strategic reasons. Up until then the problems posed by frozen ground upon construction and settlement had not been fully appreciated. In particular, the building of the Alaska Highway from Dawson Creek, B.C., to Fairbanks, Alaska, in 1942 demonstrated that traditional methods of construction and maintenance of roads, airstrips and buildings were inadequate in permafrost terrain. In addition, the difficulties of sewage and water provision, and the limitations placed upon agriculture and mining by permafrost meant that large-scale settlement of the north was impossible until the problems posed by frozen ground were understood.

The last thirty years has seen dramatic and significant advances in permafrost research by both North American and Soviet scientists. An extensive literature now exists in both languages. Probably the most comprehensive summaries of permafrost research are to be found in the publications associated with the First and Second International Permafrost Conferences, held in the United States in 1963 and in Yakutsk, Siberia, in 1973 (National Academy of Sciences – National Research Council Publication 1287, 1966; National Academy of Science Publication 2115, 1973).

Geothermal regime of permafrost

The growth of permafrost reflects a negative heat balance at the surface of the earth in which the complete thawing of ground frozen during the previous winter does not take place. The minimum limit therefore, for the duration of permafrost is one year. Thus, if the ground freezes one winter to a depth of 60 cm and thaws in the following summer to a depth of only 55 cm, 5 cm of permafrost comes into existence. If the climatic conditions are repeated the following year, the zone of permafrost will thicken and grow downwards from the base of the seasonal frost. Ultimately, permafrost several hundreds of metres in thickness can be formed.

The thickness to which permafrost develops is determined by a balance or equilibrium between the internal heat gain with depth and the heat loss from the surface. According to Lachenbruch (1968), heat flow from the earth's interior normally results in a temperature increase of approximately 1°C per 30–60 m increase in depth. This is known as the geothermal gradient. Thus, the lower limit of permafrost occurs at that depth at which the temperature increase due to the internal earth heat (i.e. the geothermal gradient) just offsets the amount by which the freezing point exceeds the mean surface temperature. This is illustrated in Fig. 4.2a. If there is a change in the climatic conditions at the ground surface, the thickness of the permafrost will change appropriately. For example,

Fig. 4.2 *Relationship between permafrost, ground temperatures and depth. From Lachenbruch (1968). A - Determination of lower (bottom) limits and upper (top) limits of permafrost. B - Effects of an increase in mean surface temperature from t_1 to t_2. Permafrost thickness decreases from d_1 to d_2. C - Effects of a decrease in mean surface temperature from t_1 to t_2. Permafrost progressively aggrades to depth d_2.*

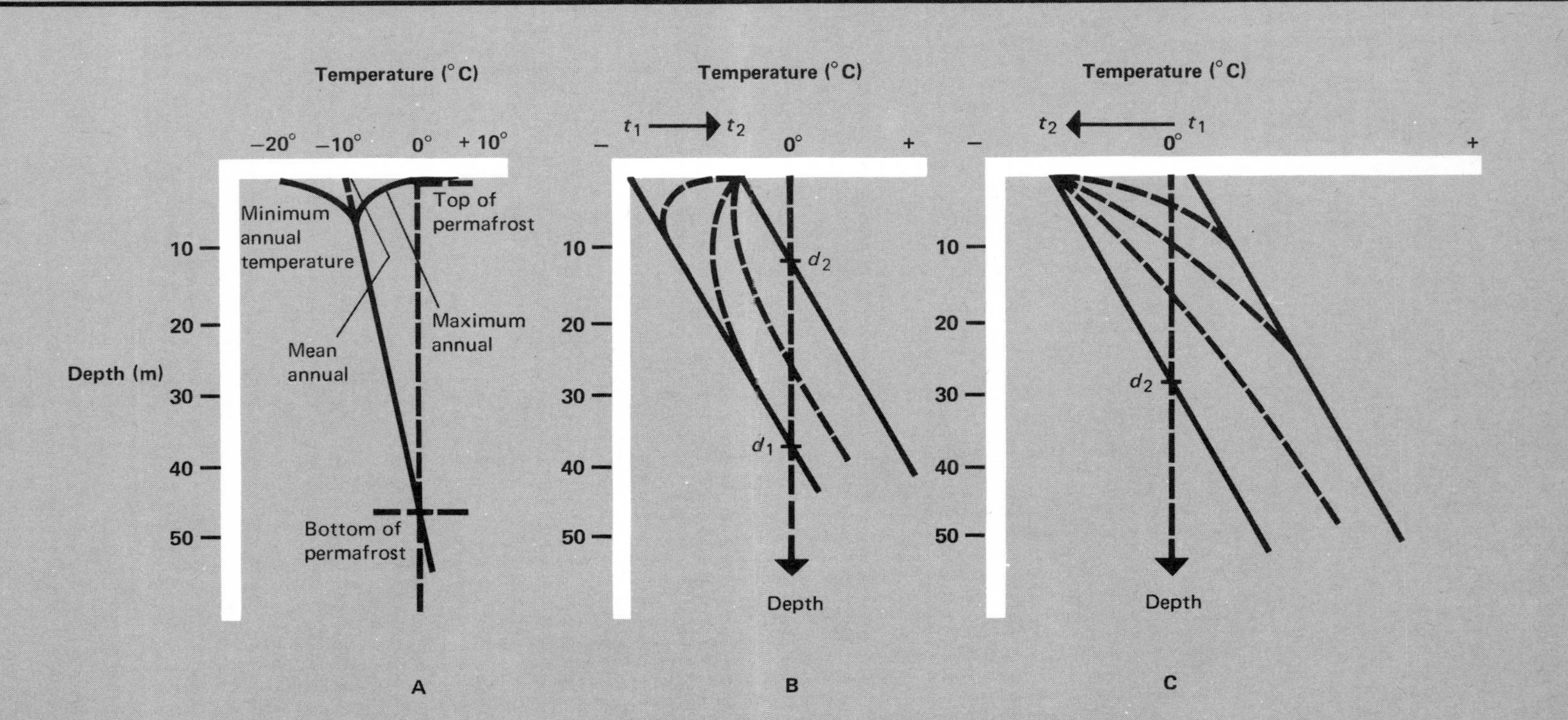

an increase in mean surface temperature will result in a decrease in permafrost thickness, as illustrated in Fig. 4.2b, while a decrease in surface temperature will give the reverse, as illustrated in Fig. 4.2c.

Above the permafrost is the active layer which is subject to seasonal freezing and thawing. Its thickness is controlled by a number of factors in addition to the seasonal temperature fluctuation. Vegetation, snowcover, thermal conductivity, aspect and albedo all influence the active layer thickness. Ignoring these complications for the present, the bottom of the active layer (i.e. the top of the permafrost in the continuous permafrost zone) is that depth at which the maximum annual temperature is 0°C.

Assuming a constant geothermal gradient of 1°C/50 m (Gold ***et al.,*** 1972) and stable surface temperatures, it is possible to predict the approximate thickness of permafrost at any locality, given the mean annual temperature. This is done by multiplying the negative of the mean annual air temperature by the geothermal gradient. For example, a locality with a mean annual temperature of −10°C might be expected to have a permafrost thickness of 500 m. However, when known permafrost thicknesses are tabulated against annual air temperatures (Table 4.1), it is clear that this rule of thumb is of only limited use, since

Table 4.1 *Permafrost depths and mean annual air temperatures at selected localities in the northern hemisphere*

Locality	Latitude	Permafrost zone	Mean air temperature (°C)	Permafrost thickness (m)
Canada[1]				
Resolute, NWT	74°N	Continuous	−12	390–400
Inuvik, NWT	69°N	Continuous	−9	100
Dawson City, YT	64°N	Discontinuous	−5	60
Yellowknife, NWT	62°N	Discontinuous	−6	60–100
Schefferville, PQ	54°N	Discontinuous	−4	80
Thompson, Man.	55°N	Discontinuous	−4	15
Alaska[2]				
Barrow	71°N	Continuous	−12	304–405
Umiat	69°N	Continuous	−10	322
Fairbanks	64°N	Discontinuous	−3	30–120
Bethel	60°N	Discontinuous	−1	13–184
Nome	64°N	Discontinuous	−4	37
USSR[3]				
Nord'vik	72°N	Continuous	−12	610
Ust'Port	69°N	Continuous	−10	455
Yakutsk	62°N	Continuous	−10	195–250

(1) Brown (1970), Table 1.
(2) Ferrians (1965) and Brown and Péwé (1973), Fig. 2.
(3) Quoted in Washburn (1973).

there are significant variations in permafrost thicknesses. At least two additional factors must be considered.

First, as Lachenbruch (1957) has demonstrated, large bodies of water exert a distinct warming effect upon adjacent landmasses. Permafrost is generally absent from beneath the arctic oceans and many of the larger water bodies. This probably explains why permafrost thicknesses at coastal locations are usually less than predicted. Second, the effects of past climatic changes must be considered. For example, if present-day mean annual surface temperatures exceed −15°C, permafrost thicknesses are so great (in excess of 800 m in theory) that they may still reflect little known climatic changes which occurred at the end of the Pleistocene glaciation, and the assumption of climatic stability is questionable. In parts of Siberia, permafrost extends to great depths (1 600 m). Varying geothermal gradients at varying depths indicate definite climatic fluctuations. It seems reasonable to assume that much permafrost is relic and unrelated to present climatic conditions (see pp. 58–9).

Distribution of permafrost

The importance of permafrost is best appreciated when it is realised that approximately 22–25 per cent of the earth's land surface is underlain by permafrost (Table 4.2). The majority occurs in the northern hemisphere (Fig. 4.3). Excluding those areas of frozen ground lying beneath glaciers and ice sheets, the USSR possesses the largest area of permafrost, followed by Canada and then Alaska. Other areas of permafrost occur in parts of central Asia, northern Scandinavia, Spitsbergen, Iceland, and at high elevations in many of the mountainous regions of the world. Substantial areas of offshore or sea bottom permafrost also exist, especially in the Beaufort Sea of the western Arctic and off the northern coast of Eurasia.

In the following discussion emphasis will be placed upon the permafrost conditions in North America and the USSR and the broad climatic controls over its distribution. The comparison of permafrost conditions between these two areas provides an introduction to offshore and relic permafrost.

Table 4.2 *Global distribution of permafrost according to I. Ya. Baranov (1959)*

Northern hemisphere (million km^2)		Southern hemisphere (million km^2)	
USSR	11.0	Antarctica	13.5
Mongolian People's Republic	0.8		
China (without Tibet)	0.4		
North American continent			
(*a*) Alaska	1.5		
(*b*) Canada	5.7		
Greenland	1.6		
Total	21.0		13.5
Total for both hemispheres	34.5 million km^2		
Total land area for both hemispheres	149.0 million km^2		
Area occupied by permafrost	23 per cent		

Fig. 4.3 *Distribution of permafrost in the northern hemisphere and location of the treeline. Modified from Péwé (1969) and Mackay (1972)*

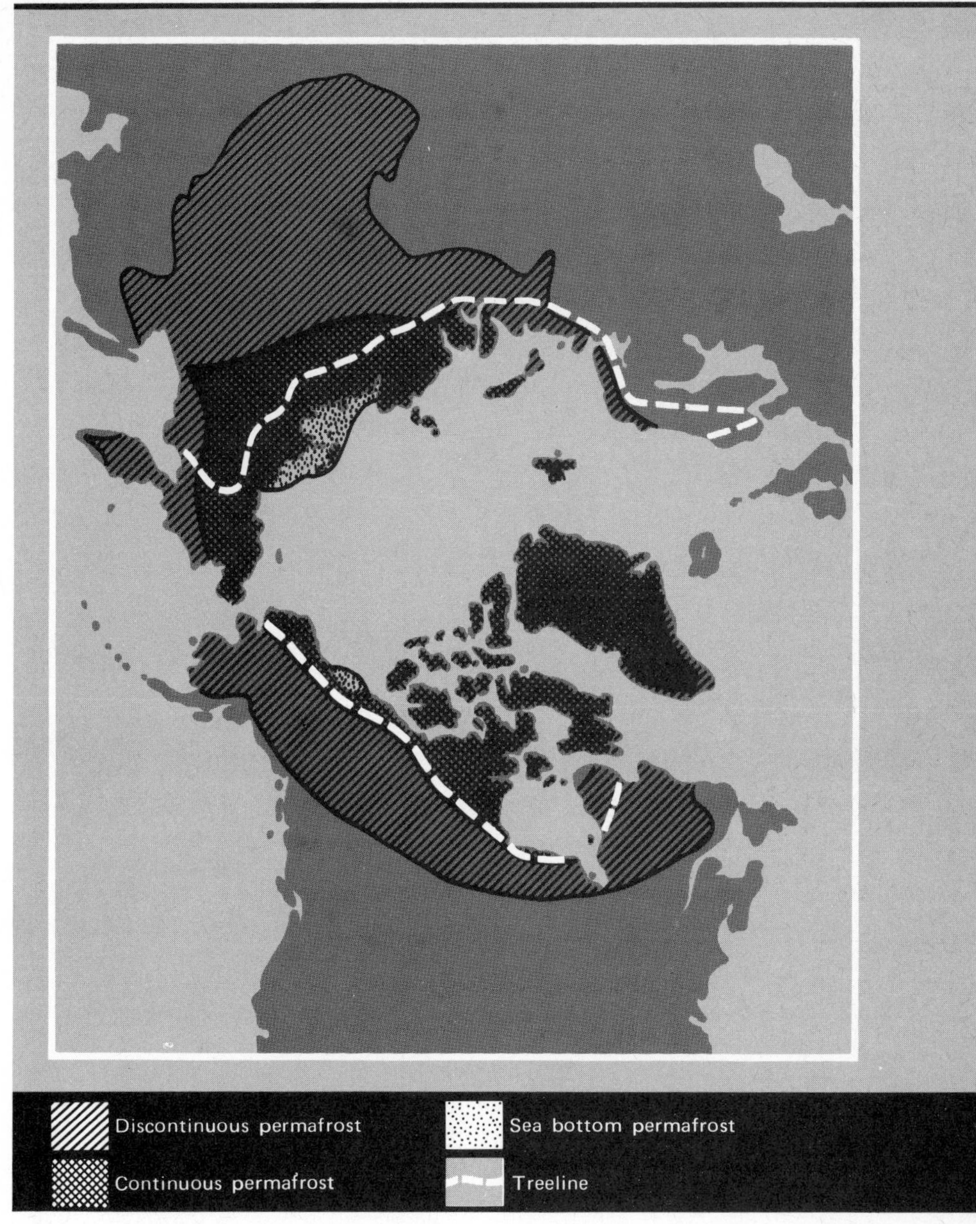

Canada and Alaska In North America, the broad outline of permafrost distribution is now well known, although detail is sparse in all but a few localities (Brown and Péwé, 1973). Permafrost maps are now available for both Alaska (Ferrians, 1965) and Canada (Brown, 1967). In Canada, nearly one-half of the

Fig. 4.4 *Permafrost distribution in Canada and its relation to air temperatures. Permafrost at high altitudes in the Cordillera is excluded. From Brown and Péwé (1973)*

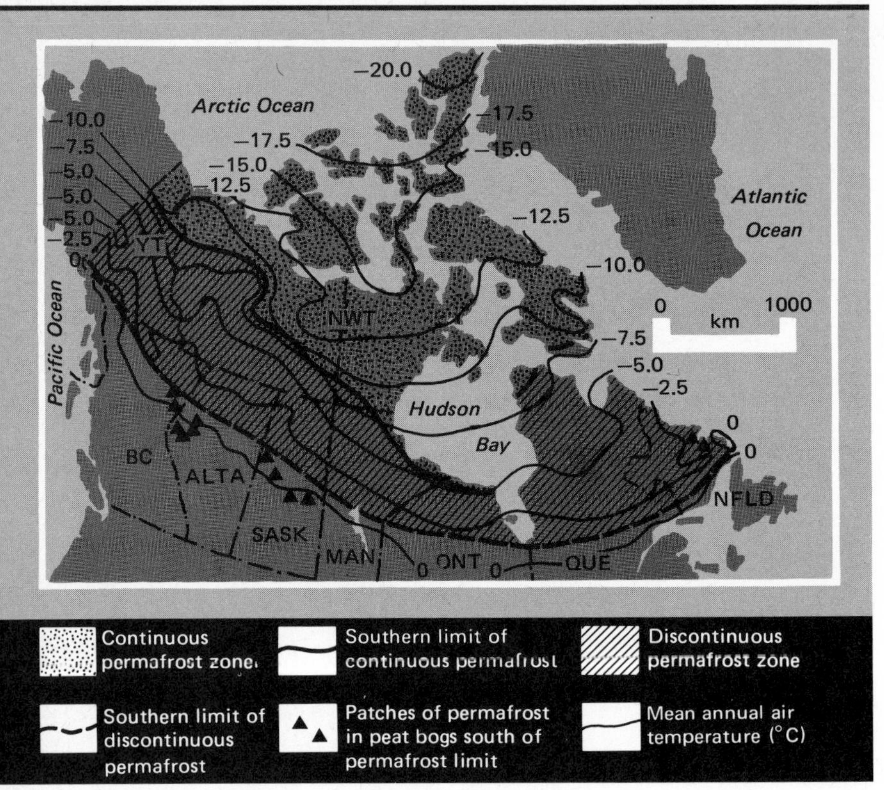

country is affected while in Alaska, 80 per cent of the land surface is underlain by various types of permafrost (Fig. 4.4).

Permafrost is classified as being either continuous or discontinuous. The distribution of both types of permafrost is broadly governed by climate since observations indicate a general relation between mean annual air and ground temperatures in permafrost regions. As a result of the complex energy exchange system at the ground surface, the mean annual ground temperature measured at the depth of zero annual amplitude is usually several degrees warmer than the mean annual air temperature. According to Brown (1966) a difference of 3–4°C is typical.

Field observations in both Alaska and Canada indicate that the southern limit of the continuous permafrost coincides with the general position of the –6 to –8°C mean annual air temperature isotherm (Brown, 1960, 1967a; Péwé, 1966). This relates to the –5°C isotherm of mean annual ground temperature measured just below the zone of annual variation (i.e. at the depth of zero annual amplitude). This –5°C ground temperature isotherm was first selected by Russian scientists and subsequently adopted by North Americans after numerous field observations indicated that discontinuities began to appear in the permafrost south of this isotherm.

In North America, the southern boundary of the continuous permafrost zone extends from the Seward Peninsula in Alaska, north through the Brooks Range foothills, and then southwards through Canada in a broad curve north of the Slave Lakes. Continuous permafrost reaches its greatest southerly extent at latitude 55°N where it fringes the southern shore of Hudson Bay. East of Hudson Bay, continuous permafrost reappears in the northern part of the Ungava Peninsula north of latitude 60°N. The reason for this latitudinal jump in the vicinity of Hudson Bay is problematical, but Brown (1967b) suggests that greater snow accumulation in the autumn to the east of Hudson Bay before the Bay freezes over in winter keeps ground temperatures relatively high.

In the continuous zone, permafrost occurs everywhere beneath the ground surface, except in newly deposited sediments and beneath large lakes and deep bodies of standing water. It is also possible that unfrozen ground exists beneath the channels of the large rivers such as the Mackenzie and Yukon, but this is not yet proven. The known thickness of the permafrost varies from 60–90 m at the southern limit of the continuous zone to depths of over 500 m in the Canadian Arctic Archipelago and northern Alaska. A maximum of 610 m has been reported from Prudhoe Bay in Alaska and 557 m from Winter Harbour, Melville Island, in Canada. Probably much greater thicknesses occur in the interior of the arctic islands and thicknesses in excess of 1 000 m may exist at high elevations on parts of Baffin and Ellesmere Islands (Brown, 1972).

The discontinuous zone lies to the south of the continuous zone. Here, unfrozen and frozen bodies exist together. According to Brown (1967a), the southern limit of discontinuous permafrost in Canada roughly coincides with the −1°C mean annual air temperature isotherm. Southwards of this isotherm, the climate is generally too warm for permafrost to form. Between the −1°C and −4°C mean annual air isotherms, permafrost is restricted to drier areas of peatlands, or to north-facing slopes or shady river banks. The delicate balance which exists between permafrost, terrain, drainage and vegetation in the southern fringe of the discontinuous zone has been described by Brown (1969) and is schematically illustrated in Fig. 4.5. From the −4°C air isotherm northwards, permafrost becomes increasingly widespread and thicker until the southern limit of the continuous permafrost, as defined above, is reached.

In the North American Cordillera, the distribution of permafrost varies both with altitude and with latitude. Field observations suggest that the lower altitudinal limit of permafrost rises progressively from 1 200 m.a.s.l. at 54°N (Brown, 1967a) to over 3 000 m.a.s.l. at 46°N (Ives and Fahey, 1971). Below this altitudinal limit and to the south of the southern limit of permafrost as depicted on Brown's map (Fig. 4.4), scattered permafrost islands only exist in specific and favourable types of terrain.

The USSR The extent and thickness of permafrost in the USSR is illustrated in Fig. 4.6. According to Baranov (1959), permafrost occupies over 11 000 000 km^2 or 49.7 per cent of the total area of the USSR. Most of it lies in the boreal forest zone east of the Yenesei River. As a generalisation, the permafrost is thicker than in North America. In the continuous zone the permafrost increases northwards from 300 m at the southern limit to over 600 m along the Arctic coastal plain in the Verkhoyansk–Kolyma region. Thicknesses in excess of 500 m also occur in central Yakutia while in other areas such as the Tamyr peninsula, permafrost reaches 400 m in thickness (Baranov, 1959, pp. 15–19). The thickest

Fig. 4.5 *Profile through typical peat bog in the southern fringe of the discontinuous permafrost zone in Canada, showing vegetation, drainage, micro-relief, and associated permafrost distribution. From Brown (1968)*

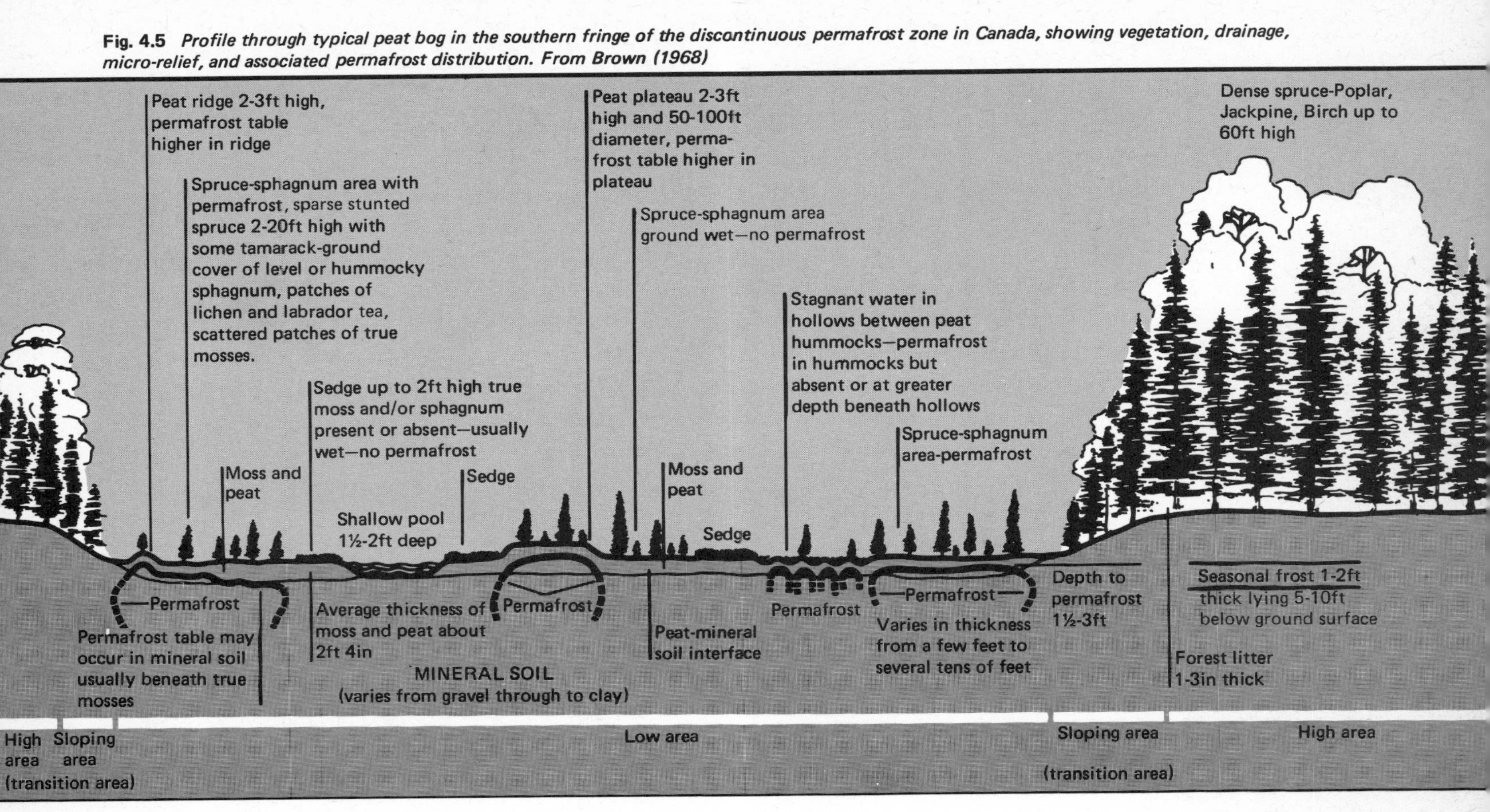

Fig. 4.6 *Permafrost distribution in the Soviet Union. Generalised after Baranov (1959)*

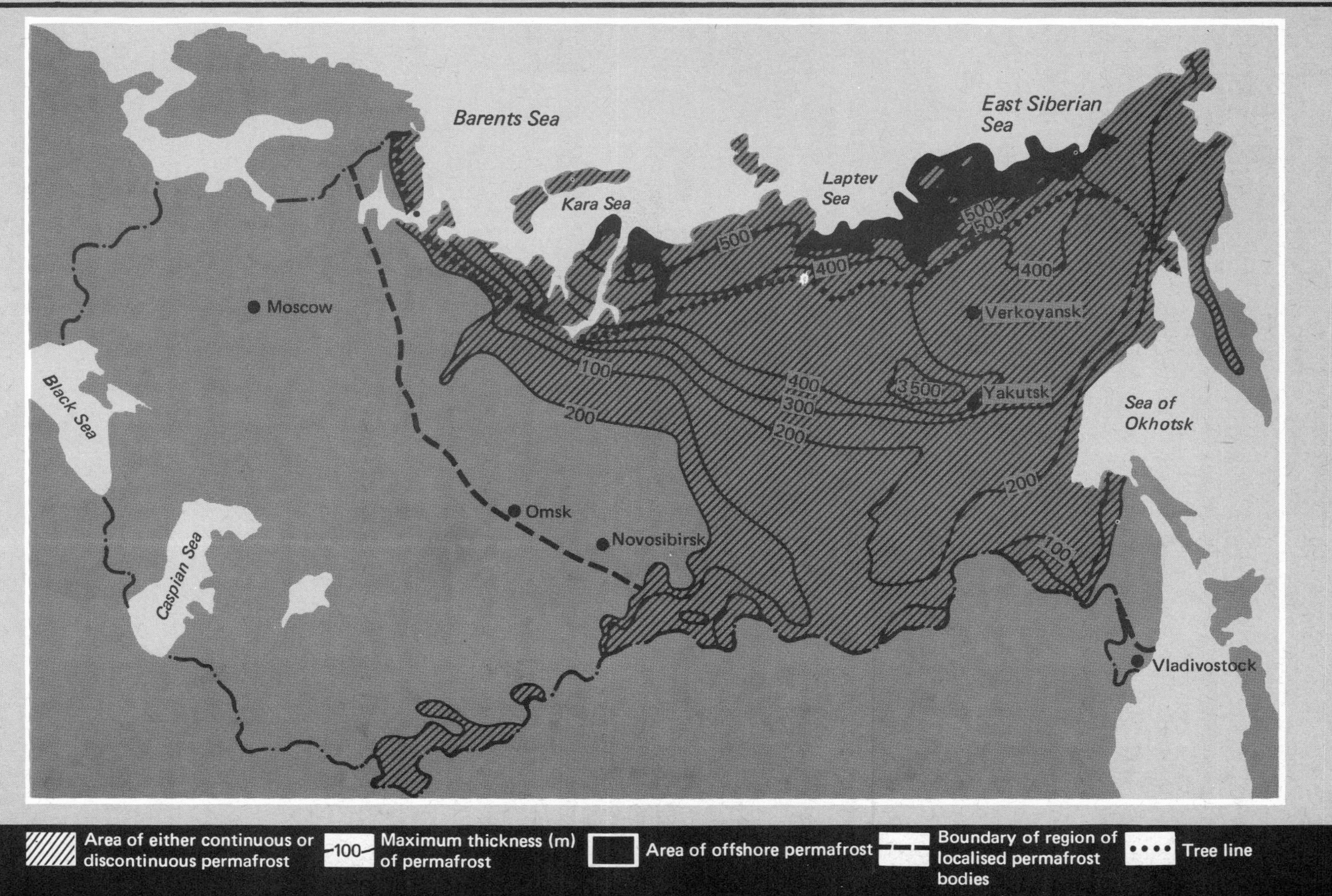

 known permafrost exists in northern Yakutia to a depth of 1 600 m, caused by the deep penetration of supercooled brine (Katasonov, in Brown, 1967b, p. 742).

Discontinuous permafrost is encountered in limited areas in the Kola Peninsula and in the tundra and coniferous forest zone between the White Sea and the Ural Mountains. East of the Urals, a broad zone of discontinuous permafrost exists in western Siberia. Here, the transition from discontinuous to continuous permafrost coincides with the northern boundary of the coniferous forest and is accompanied by a sharp increase in the thickness of permafrost. For example, within the coniferous forest discontinuous zone, average thicknesses of 25–30 m are typical, as compared with 300 m in the forest tundra and 400 m in the tundra. Thus, the continuous – discontinuous permafrost boundary is rather easier to identify in the USSR than in North America.

Given similar mean annual temperatures at the depth of zero annual amplitude, the Siberian permafrost is colder than the North American. For example, the geothermal gradient reflects the heat flow situation in the ground and therefore, influences temperature. In Yakutia, values of the geothermal gradient range from 40 to 178 m/°C in permafrost and from 30 to 135 m/°C in non-frozen rock (Melnikov, in Brown, 1967b, p. 742). The few observations available for North America indicate values ranging from 20 m/°C at stations adjacent to large rivers or oceans to 55 m/°C at interior locations, all of which indicate considerably steeper gradients in North America than in the USSR.

The greater extent, thickness and coldness of the permafrost in Siberia reflects a number of factors. Probably the most important is the difference in the history of continental glaciation of North America and Siberia. During the Pleistocene, ice sheets covered much of Arctic North America with the exception of parts of the Yukon, interior Alaska and the Western Arctic coastal lowlands (Prest ***et al.,*** 1968). In Siberia however, rather more localised ice sheets formed in the principal mountain units leaving the lowland plains largely free of ice.

Thus, although permafrost was widespread in North America, it was thin since it formed beneath ice sheets whose bottom temperatures were not far from 0°C. Furthermore, if the Wisconsinan ice sheet was typical, the retreat of the North American ice sheets were accompanied by either the development of extensive post-glacial lakes or marine inundations. The presence of these large water bodies may have led to the dissipation of the permafrost, such that it only began to reform several thousands of years later following the disappearance of these water bodies. The end result is that much of the North American permafrost has been subject to extremely low temperatures for only the post-Wisconsinan time. This is as short as 6 000 years in some areas.

The importance of these events has far reaching ramifications as regards the development of Pleistocene periglacial phenomena in southern Canada and the United States. When combined with the late glacial amelioration of climate and the rapid northward advance of the treeline, conditions were not conducive for the development of an extensive or long-lasting 'periglacial zone' (see p. 224). In contrast to North America, ice-free areas occurred widely in Siberia throughout the Pleistocene and deep and continuous permafrost developed in response to the very low air temperatures to which the ground was exposed (Gerasimov and Markov, 1968). The reason for the lack of extensive glaciation in Siberia seems best explained by the lower snowfall amounts which that area receives, since high mountain chains and the vastness of the landmass effectively prevent the penetration of moisture laden winds from either the Pacific or Atlantic Oceans.

The importance of the glacial limits as regards permafrost distribution and thickness can be demonstrated with reference to known permafrost thicknesses in glaciated and unglaciated terrain currently experiencing similar mean annual air temperatures. For example, 60 m of permafrost exist at Dawson City, in the unglaciated part of the Yukon Territory of Canada. The mean annual air temperature is −5°C. Approximately similar permafrost depths occur in southeast Siberia at Chita and Bomnak which also experience mean annual air temperatures of −5°C (Brown, 1967b, p. 746). However, at Fort Simpson, N.W.T., and Thompson, Manitoba, where air temperatures are similar to the other stations but which are located in glaciated terrain, the permafrost is only about 15 m in thickness (Brown, 1970, p. 10).

Offshore permafrost Since permafrost is defined exclusively upon the basis of temperature, permafrost may exist wherever mean annual sea bottom temperatures are below 0°C. A complicating factor is that, because of the lower freezing point which exists under saline conditions, the sediments are not necessarily frozen.

The major area of offshore permafrost occurs beneath the waters of the Laptev and east Siberian Seas (Baranov, 1959). However, the extent of sea bottom permafrost mapped by Baranov (Fig. 4.3) may be too large (Mackay, 1972a, footnote 5). Extensive areas of offshore permafrost also occur beneath the sea floor of the southern Beaufort Sea in the western Arctic (Shearer ***et al.***, 1971; Mackay, 1972b).

According to Mackay (1972b), offshore permafrost can be in either thermal equilibrium or disequilibrium with sea bottom temperatures and the geothermal heat flow. In areas where sea bottom temperature is negative, and has remained so over a considerable time period, equilibrium permafrost may exist (Fig. 4.7A). The thickness of the permafrost will depend upon such variables as the mean sea bottom temperature, soil properties, the geothermal heat flux, and sedimentation rates. Equilibrium permafrost is unlikely to exceed 150 m in thickness (Mackay, 1972b, p. 1552). Where permafrost exists beneath the sea floor and it is not in thermal equilibrium with a negative sea bottom temperature, the permafrost is in disequilibrium.

Most disequilibrium offshore permafrost is probably relic degrading permafrost. Most likely, it developed during the colder periods of the Quaternary when eustatic sea level fluctuations exposed areas of present sea floor as dry land for many thousands of years. This land was then subject to very low air temperatures and permafrost formed. The effects of the subsequent submergence of very cold permafrost land masses by waters of either a positive or negative temperature are illustrated in Figs 4.7B and 4.7C. In both cases, the permafrost suffers degradation either from both top and bottom (as with positive sea bottom temperatures) or from the bottom only (as with negative sea bottom temperatures). In the case of the latter, thawing is solely the result of the geothermal heat gradient. Rough calculations by Mackay (1972b, p. 1554) of the length of time for permafrost growth and thaw, derived from generalised sea level curves of the last 100 000 years, indicate that where present water depths are shallow, insufficient time has elapsed since submergence for thick permafrost to be completely degraded, and relic permafrost should exist.

Fig. 4.7 *Thermal conditions of equilibrium and disequilibrium of offshore permafrost. Modified from Mackay (1972b)*

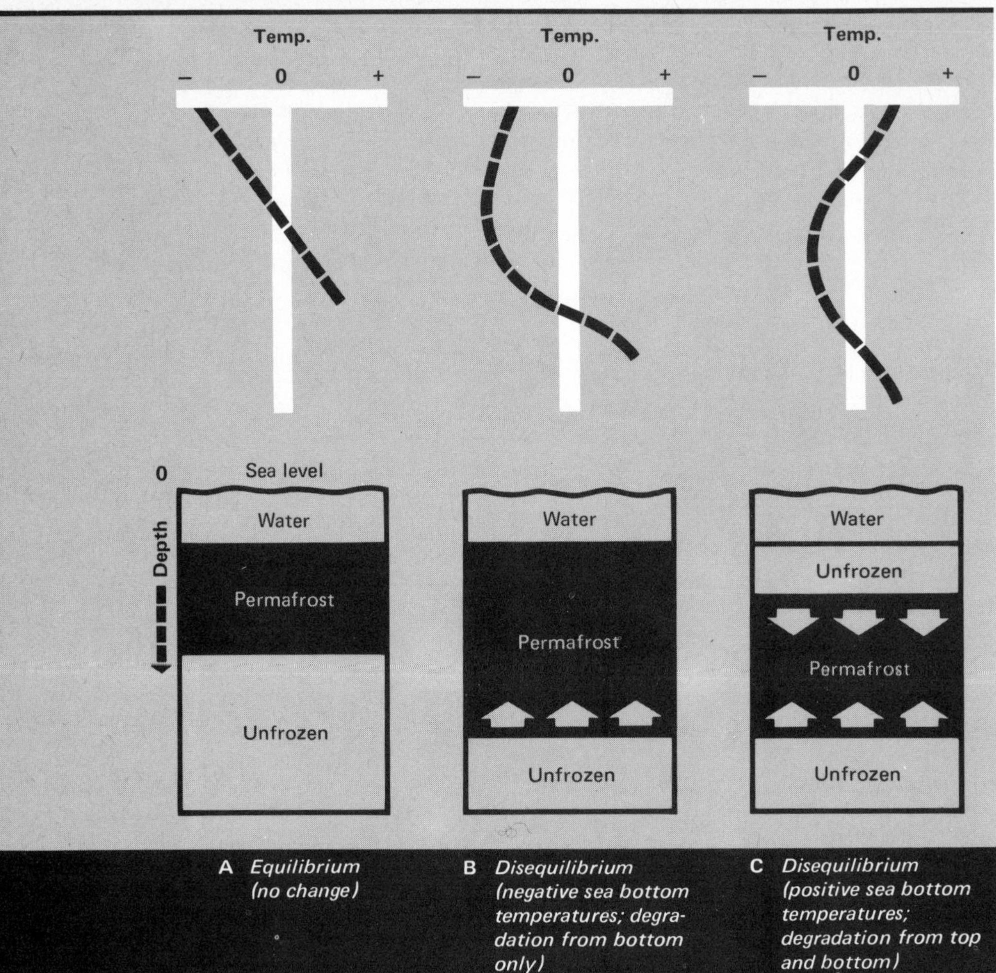

Offshore permafrost may originate in a number of other ways. Extensive and rapid coastal retreat of certain permafrost regions will result in permafrost existing in nearshore areas, irrespective of whether water temperatures are positive or negative. The progressive burial of permafrost by coastal sedimentation is a second method. Both of these mechanisms however, are of only local importance and eustatic sea level changes are more relevant for the majority of sea bottom permafrost.

Relic permafrost

It follows from what has been said in the previous section concerning permafrost conditions in Canada, Alaska and the USSR, and the existence of offshore permafrost, that much of the land permafrost is relic and unrelated to present climatic conditions. Various lines of evidence support this generalisation. At Cape Thompson, Alaska, for example, a permafrost thickness of over 360 m is found despite the fact that the mean annual air temperature is –5°C. Assuming a constant geothermal gradient of 1°C/50 m present air temperatures indicate that permafrost should only be approximately 250 m thick. By inference therefore, surface temperatures have been approximately 2–3°C colder at some time in the past for the permafrost at Cape Thompson to have formed (Lachenbruch, Greene and Marshall, 1966).

More striking examples of relic permafrost are to be found at the southern limits of permafrost in both Alaska and Siberia. Permafrost thicknesses of over 180 m have been recorded from the coastal plain of the Kuskokwim River valley in Alaska where mean annual air temperatures are between 0 and –2°C. In such a situation, it is probable that the permafrost is degrading and that the geothermal gradient is isothermal (see Fig. 4.2B). At the southern limit of permafrost in Canada however, the known thicknesses of permafrost seem to be in a reasonable equilibrium with the present climatic environment (Brown, 1970; Brown and Péwé, 1973). This may be explained by the insulating effects of the Wisconsinan ice which prevented permafrost formation until late and post-glacial times.

Some permafrost may be early Wisconsinan or Pleistocene in age. In the western Arctic of Canada, glacially deformed ground ice and icy sediments exist in an area beyond the maximum limits of late Wisconsinan ice (Mackay, Rampton and Fyles, 1972). The icy sediments have never undergone thawing and refreezing since thawing would have obliterated the primary sedimentary structures. ^{14}C dates and geomorphic evidence indicate the Arctic coastal plain was not glaciated during the last 40 000 years. It follows therefore, that the permafrost is early Wisconsinan or older in age and that it existed prior to the glacier ice thrusting. Evidence from Siberia also suggests that much permafrost originated during the past, probably the Pleistocene (Gerasimov and Markov, 1968). The most incontrovertible evidence is the presence of limbs of Pleistocene woolly mammoths preserved in permafrost. This can only indicate the presence of permafrost at the time of the animal's death, otherwise decomposition would have occurred. In other areas of Siberia towards the southern limit of permafrost, the upper surface of many permafrost bodies lies at a depth not reached by present winter freezing, another certain indicator that the permafrost is unrelated to the present climatic conditions.

Permafrost and terrain conditions

Although the broad controls over permafrost formation and distribution are climatic in nature, local variations in permafrost conditions are determined by a variety of terrain and other factors. Of widespread importance are the effects of relief and aspect, and the nature of the physical properties of soil and rock. More complex are the controls exerted by vegetation, snowcover, drainage and fire. Generally speaking, in the discontinuous zone variations in terrain conditions are primarily responsible for the patchy occurrence of permafrost, the size of the permafrost islands, and the thickness of the active layer. In the continuous zone however, the thermal properties of the ground as a whole, together with the climate, are more important (Brown, 1970, p. 22).

Relief influences the amount of solar radiation received by the ground surface and the accumulation of snow. Slope orientation also influences the amount of solar radiation received. The effects of differential insolation are particularly clear in mountainous regions such as northern British Columbia and the Yukon Territory where permafrost may occur on north-facing valley slopes and not upon adjacent south-facing slopes (Brown, 1969). Similarly, in the continuous zone, the active layer is usually thinner on north-facing slopes, but in certain instances, exposure to local weather conditions assumes greater importance. On the Beaufort Plain of northwest Banks Island for example, the active layer is thinnest on southwest-facing slopes (French, 1970). This is attributed to the influence of the dominant southwest winds in this part of the Arctic which promote evaporation and latent heat loss from that slope during the summer months.

Variations in the nature of rock and soil express themselves in differing albedo and thermal conductivity values. For example, the thermal conductivity of silt is only one-half that of coarse grained sediments and several times less than that of rock. This, and the albedo factor, attain their greatest significance in the continuous permafrost zone, where the climate is sufficiently cool to produce permafrost regardless of the type of terrain. Average albedo values for bare rock and soil can vary between 10 and 40 per cent. Thus, significant variations in active layer thicknesses and permafrost thermal regimes are to be expected in different rock and soil types (e.g. Brown, 1973a).

Probably the most complex terrain factor is that of vegetation. It affects permafrost in a variety of ways and is a significant factor in all areas of discontinuous and continuous permafrost with the exception of the vegetation-free polar deserts of extreme high latitudes. The most fundamental influence of vegetation is to shield the underlying permafrost from solar heat during the summer months. This insulating property is probably the single most important factor in determining the thickness of the active layer. Numerous observations from a wide variety of permafrost environments indicate not only a broad climatic relationship but also that the thickness of the active layer is thinnest beneath poorly drained and vegetated areas, and thickest beneath well drained bare soil or rock.

The presence of peaty organic materials at the ground surface is particularly effective in protecting permafrost from atmospheric heat. This is most clearly demonstrated in the southern part of the discontinuous zone where permafrost only occurs in peatlands and not in other adjacent terrain types (Fig. 4.5). The preservation of permafrost in peatlands is due to the rather unique thermal conductivity properties of peat (Brown and Williams, 1972; Brown and Péwé, 1973). Dry sphagnum has thermal conductivity values approximately one order of magnitude less than the lowest value for mineral soil (Brown and Williams, 1972, p. 20). During the summer therefore, when the surface layer of peat becomes dry by evaporation, warming of the underlying soil is limited. As a result, the depth of seasonal thawing in peatlands is considerably less than in other types of terrain. During the autumn and early winter however, the peat becomes saturated with moisture as evaporation rates fall off. When the peat freezes therefore, many of the interstices are filled with ice and the thermal conductivity of the peat increases considerably. Thus, the peat offers less resistance to the cooling of the underlying soil in winter than to the warming of it in summer. This leads to mean annual ground temperatures to be lower under peat than in adjacent areas without peat, and if the mean annual ground temperature beneath the peat remains below 0°C throughout the year permafrost results.

In addition to ground vegetation, trees are important controls over local permafrost conditions. Much permafrost terrain is forested, particularly in the USSR where the treeline extends in an east–west direction not far south of the Arctic coastline. In North America, the treeline extends to the south of Hudson Bay almost as far south as the discontinuous permafrost zone, and a greater percentage of the permafrost terrain is in the tundra and polar desert environments. The presence of trees shade the ground from solar radiation and intercept some of the snowfall in winter. Thus, the winter cold penetrates more deeply into the ground beneath trees than in areas of thick snowcover, and summer solar radiation is restricted. Actual vegetation differences further complicate local permafrost conditions. The Siberian taiga is composed predominantly of pine (***pinus silvestris***) and tamarack (***Larix dahurica***) whereas the spruce (***Picea glauca*** and ***Picea mariana***) is more common in North America. Since the spruce forest is more shady, a surface moss cover is widely developed and this tends to favour a thinner active layer in the boreal forest than in the taiga. In those areas where the treeline and the southern boundary of continuous permafrost are in close proximity, as is the case for much of North America, the presence or absence of trees assumes a critical importance as regards permafrost occurrence. In Alaska for example, Viereck (1965) has shown how even isolated white spruce may influence the energy exchange at the ground surface sufficiently for a small permafrost body to exist.

Snowcover is a further factor influencing local permafrost variations since it insulates the underlying ground from the extremes of winter cold. The snowfall regime and the length of time the snow lies on the ground are the critical factors. In general terms, a heavy snowfall in the autumn or early winter will inhibit frost penetration, while a winter of low snowfall will do the reverse. Also, if snow persists late into the spring, ground thawing will be delayed. In detail, variations in snowcover at any one locality are controlled by site characteristics such as micro relief and vegetation and their relation to the dominant snow bearing winds.

In the continuous permafrost regions north of the treeline snowfall amounts are limited and the effects of snowcover are less important than in areas further south. However, subtle differences in permafrost active layer conditions are to be found in the tundra and polar desert regions of high latitudes (Brown, 1972). It is not uncommon for the upland surfaces and interfluves to be swept clear of snow for much of the winter while extensive snowbanks accumulate in gullies and on lee slopes. Deeper frost penetration may occur on the uplands and interfluves therefore, than in the depressions and on lee slopes. On the other hand, ground thawing at snowbank localities may be delayed until late summer when the snowbank finally disappears. As a consequence, the active layer is often thinner at snowbank localities than on uplands. South of the treeline, in particular in the discontinuous permafrost zone, snowcover assumes greater importance. Studies at Schefferville in northern Quebec indicate that the pattern of accumulation of the seasonal snowcover is the controlling factor in the distribution of permafrost in that area (Granberg, 1973; Nicholson and Thom, 1973). Since mean annual air temperature in the Schefferville region is –4.5°C, extensive permafrost could be expected. However, permafrost is widespread only in the upland areas where the absence of trees allows snow to be blown clear and deep winter frost penetration to take place. In the adjacent lowlands, where thick snowcovers accumulate, permafrost is generally absent. It appears that a winter snow depth of 65–70 cm is sufficient to prevent the development of permafrost in the Schefferville region (Nicholson and Granberg, 1973).

Water is another factor which influences permafrost formation and distribution. This is because of the high specific heat possessed by a water body. In the discontinuous zone, drainage conditions are usually linked to vegetation, snowcover and other terrain factors, and is of secondary importance. In places however, as in the sedge and fenlands at the southern fringes of the discontinuous permafrost, drainage conditions may assume great importance in determining the presence or absence of permafrost (e.g. Zoltai, 1973; Brown, 1973b).

In the continuous permafrost zone, the effects of water upon permafrost are widespread. Numerous observations indicate an unfrozen layer often exists beneath water bodies that do not freeze to the bottom in winter. The extent and nature of these taliks varies with the area and depth of the water body, the water temperature, the thickness of the winter ice and snowcover, and the nature and compaction of the bottom sediments (e.g. Johnston and Brown, 1964).

A final terrain factor which may influence permafrost conditions is the effect of fire. Many forest fires start by lightning each year, particularly in the discontinuous zone. Much, if not all, of these regions have been burned over at least once and the boreal forest is, in many ways, a fire climax. Fires in the tundra zone are rare, although not unknown, on account of the relative absence of woody materials and the lower summer temperatures. The effects of a fire upon the permafrost will depend upon the nature and dampness of the vegetation, and the speed at which the fire passes through the area. If the fire passes rapidly, and if the ground vegetation is of peat, mosses or lichens, only the trees may burn and the ground vegetation beneath 2–3 cm may remain untouched. In this case, little change will occur in the permafrost conditions. If the ground vegetation is exceptionally dry however, and if the fire moves slowly, considerable changes in permafrost conditions may result. At Inuvik, N.W.T., for example, the effects of a 1968 forest fire have been documented by Heginbottom (1973). The destruction of much of the ground vegetation in addition to the trees resulted in the thawing of ice-rich sediments, rapid gullying and thermal erosion, and numerous earthflows. The more long term effect has been an increase in the thickness of the active layer in the burned over area (see Table 6.2, p. 109).

Surface features of permafrost and air photo interpretation

A number of important surface features are either the direct or indirect result of the presence of permafrost and their identification in the landscape, either from air photographs or in the field, are valuable indicators of the permafrost conditions at that site. Although the majority of these features are examined in detail in the following chapters, it is appropriate to identify them at this point.

In broad terms, permafrost related features can be divided into either (a) those associated with the aggradation of permafrost or (b) those associated with the degradation of permafrost (Brown, 1974a). In both cases, the landforms are often associated with the build up or degradation of ice in the ground. Features associated with the growth of permafrost include ice-wedge polygons, ice-cored mounds and pingos, and palsas and peat plateaus. Since ice wedges and pingos are treated in the following chapter under the heading of Ground Ice, they are not considered further at this point. Those features associated with the degradation of permafrost and the melting of ground ice bodies are generally classified as 'thermokarst'. Ground ice depressions and slumps, thaw lakes and sinks, beaded drainage systems, and thermokarst or 'cemetary mounds' are typical features. These are all examined in Chapter 6.

A number of other features are commonly associated with permafrost conditions. These include the many varieties of patterned ground that exist and the various forms of solifluction. It must be emphasised however, that although these features attain their best development in permafrost regions, particularly the continuous zone, these features are not restricted to permafrost regions, and may equally be the result of a number of other factors. Thus, nets, stripes and circles are not examined here but in Chapter 9 which deals with surface microrelief and soil characteristics, and solifluction sheets and lobes are considered within the context of slope evolution in Chapter 7.

Palsas and their associated peat plateaus find no easy place in the treatment given to either ground ice or thermokarst phenomena (Chapters 5 and 6). Since palsas are regarded as the only reliable surface indicators of permafrost in the discontinuous zone (Brown, 1974b, p. 6), they merit brief attention.

Palsas are mounds of peat involving predominantly mineral soil, 1–7 m high and 10–50 m in diameter (Fig. 4.8). Usually they occur in bogs and protrude above the level of the bog as low hills or knolls. Peat plateaus result from the growth and coalescence of adjacent palsas and may be several square kilometres in extent. It is generally agreed that palsas form by the combined action of peat accumulation and ice segregation in the underlying mineral soil. The latter results from the thermal properties of the peat previously described (p. 60).

Fig. 4.8 *Mature palsa in the southern fringe of discontinuous permafrost zone near Great Whale River, P.Q., Canada. The peat is 1 m thick overlying silty clay. Permafrost in the palsa is 3–6 m thick; no permafrost occurs in surrounding terrain. Photograph by courtesy of R. J. E. Brown, National Research Council of Canada.*

Exposure (reduced snowcover) is an additional factor in the maintenance of the permafrost. The ice bodies are usually thin, less than 2–3 cm in thickness, and interstratified with mineral soil. In this way palsas are thought different to the various ice-cored mounds and small pingo-like features discussed on pp. 101–4. On the other hand, there are reasons for classifying palsas as a type of ice mound. For example, Zoltai and Tarnocai (1971) suggest that palsa relief is more likely the result of the build up of ground ice than of an increase in the thickness of the surface peat layer. Part of the problem arises from the fact that in the past the distinction between palsas and ice-cored mounds rested upon the belief that palsas only occurred in the southern parts of the discontinuous zone. Recent work in Canada indicates that palsa-like peaty mounds also occur in continuous permafrost (e.g. Brown, 1974a, p. 29).

The most frequent occurrence of palsas is at the southern fringes of the discontinuous zone. They have been described from Canada, Iceland, northern Sweden and the northern parts of the European USSR (Lundquist, 1969, pp. 207–10). In Canada, they occur widely in all physiographic regions and are particularly prevalent in the Hudson Bay lowlands of northern Ontario and Manitoba (Brown, 1968). Towards the northern limit of the discontinuous permafrost, peat plateaus become more common. In Manitoba and Saskatchewan, Zoltai (1971) has mapped the southern limit of palsas and peat plateaus as coinciding with the 0°C mean annual air temperature isotherm. This is south of the limit of discontinuous permafrost as mapped by Brown (1967a). Zoltai proposes the term 'localised permafrost zone' be applied to these areas where permafrost is restricted to certain peatlands and is limited in areal extent and thickness. A characteristic of many palsas and peat plateaus in this zone is that they often show evidence of either past or present collapse. This is probably the result of ground ice melting following natural vegetation successional changes or forest fires. Since the permafrost in this zone is marginal, otherwise insignificant vegetational changes can alter the surface energy exchange such as to induce melting of the permafrost. In Sweden, the –2 to –3°C isotherm or the zone of 200–10 frost days per annum correspond to the southern limit of palsa occurrence (Lundquist, 1969, p. 209).

Permafrost interpretation using black and white aerial photographs is difficult yet, in view of the high cost and limited availability of colour and infra-red photography, is often the only available method barring extensive and costly ground investigations. It has been used widely over the last twenty years in Canada, the United States, the USSR and Sweden. Since permafrost is a temperature condition, it cannot be directly recognised from black and white photographs but must be inferred from secondary factors usually associated with terrain. Only pingos and palsas develop exclusively over perennially frozen ground. All other surface features cannot, in isolation, be used as indicators of present permafrost conditions. For example, frost fissure polygons are known to occur under conditions of seasonal frost as well as perennial frost. They may also exist as relics from a cooler period. Small frost mounds and hydrolaccoliths may also be the result of annual rather than perennial ice accumulations. Finally, the identification of thermokarst features merely indicates degraded or past permafrost rather than confirms its present existence.

Interpretation is particularly difficult in the discontinuous zone and in the forested regions of the continuous zone (Brown, 1974b). Pingos and palsas are often concealed by trees, and any one of a number of different vegetation: permafrost relationships may be applicable. For example, at the southern fringes of discontinuous permafrost, areas without trees often correspond to organic terrain and permafrost bodies. In the northern part of the discontinuous zone however, the situation is commonly reversed with permafrost bodies developed beneath tree stands and absent from adjacent areas of fen. In the continuous zone the recognition of deep rooted species such as jackpine, balsam, poplar and birch indicate deep active layers and relatively well drained subsoils, while the presence of stunted spruce indicates poorly drained subsoils and a high permafrost table. North of the treeline where permafrost is continuous, air photographs are used primarily to evaluate permafrost conditions rather than to establish its occurrence and distribution. The surface permafrost related features such as patterned ground and ice-wedge polygons are widely developed and clearly visible. Soil material and landform identification assist in predicting subsurface permafrost conditions. Adverse conditions (i.e. thin active layer, high ice content and frost susceptibility, poor drainage) are usually associated with fine grained and organic materials and terrain features such as low coastal plains, deltaic plains, alluvial flood plains, former lake-beds, and glacial till plains. More favourable conditions on the other hand, are usually associated with coarse grained sediments and elevated, well drained landforms. In general, the identification and prediction of permafrost conditions north of the treeline is relatively straightforward and merely needs the appropriate experience. South of the treeline interpretation is more subjective and field checking is required.

Hydrologic aspects of permafrost

It is generally assumed that groundwater movement in permafrost is restricted by the presence of both perennially and seasonally frozen ground. This is because permafrost, containing water in the solid state, is impermeable and acts as a confining layer. This assumption is only partly true however, since (a) permafrost may be 'dry' and contain no ice, (b) it may contain only liquid water if pressure and/or dissolved mineral contents are high and temperatures only slightly below 0°C, and (c) it is not uncommon for unfrozen zones or taliks to exist in permafrost in which normal groundwater movement may occur. A further effect of permafrost upon groundwater is that it limits the volume of unconsolidated deposits and bedrock in which water may be stored. Frozen ground eliminates many shallow depth aquifers and requires that wells be drilled deeper than in similar non-permafrost areas.

Evidence of groundwater flow in permafrost regions is to be found in the presence of springs and associated icings or aufeis, open water reaches in otherwise frozen streams and rivers, contributions to the base flow of streams, the highly mineralised nature of much groundwater, and the development of certain frost heaved mounds due to groundwater freezing as it moves towards the surface (Hopkins ***et al.,*** 1955; Williams and Everdingen, 1973). Here, attention will be focused upon the nature of groundwater supply in permafrost regions, the role of groundwater and artesian pressures in the development of frost heave mounds, and the nature and extent of icings.

Groundwater supply Of fundamental importance to groundwater flow in permafrost regions is the existence of taliks or unfrozen bodies which act as aquifers. The term 'talik' was first applied by Muller (1945) to the layer of unfrozen ground between the seasonally frozen ground and the permafrost (i.e. the supra-permafrost, Fig. 4.1). Today, the term is also used to describe both unfrozen layers within the permafrost (intra-permafrost taliks) and below the permafrost (sub-permafrost taliks).

Taliks may be of an open or closed nature depending upon whether the talik is completely enclosed by permafrost or whether it reaches to the seasonally thawed zone. Open taliks are common features even within the continuous permafrost zone. The majority result from local heat sources such as lakes, river channels and other standing water bodies. Closed taliks usually result from either a change in the thermal regime of permafrost, such as is brought about by the draining of a lake and the downward encroachment of permafrost, or, where they occur at depth in thick permafrost bodies, from long-past fluctuations in regional climate. In the discontinuous zone where the thickness of permafrost progressively decreases, unfrozen zones may perforate the permafrost layer and form 'through' taliks linking the unfrozen ground above and below the permafrost.

Despite the numerous lakes and ponds which are characteristic of large areas of permafrost, the continued supply of water for industrial and domestic purposes must ultimately be derived from groundwater sources, particularly of a sub-permafrost nature. Intra-permafrost water is rarely found except in the discontinuous zone and is not a stable supply. It may become exhausted and differs greatly in quality and quantity depending upon the lithological characteristics of the enclosing rock. Supra-permafrost water is also irregular in supply and may disappear before the end of the winter as the seasonal frost extends to the permafrost table. This water source is common in the discontinuous zone where the permafrost table may lie at some depth beneath the maximum limit of seasonal frost penetration. It is also subject to possible contamination from cesspools and sewage disposal fields on account of its near surface location. Surface water bodies are also inadequate for the most part. The majority of lakes are shallow and freeze completely during the winter. Even in deeper lakes which do not freeze to the bottom, the storage space is much reduced by surface ice and the freezing tends to concentrate the mineral and organic content of the lake in the unfrozen water beneath the ice. As a consequence, most engineers and town planners in Canada and Alaska work on the assumption that the majority of northern lakes do not provide adequate storage for a community of more than several hundred people.

Within this context, the importance attached to sub-permafrost water is considerable. Those localities where groundwater springs appear assume special significance. In central Yakutia for example, several major perennial freshwater springs occur in right bank tributary valleys of the Lena River (Anisimova, 1973). Water emerges from the lower layers of thick deposits of Quaternary sands near their contact with underlying Jurassic sandstone. The recharge of the springs occurs through water forming in both supra-permafrost and intra-permafrost taliks as well as by sub-permafrost waters. Conditions favouring the existence of water bearing taliks in this area are the relatively shallow thicknesses (25–60 m) and high temperatures (–0.2 to –1.5°C) of the permafrost, the permeability of the sediments, and the presence of considerable local surface relief caused by the proximity of the Lena River which ensures groundwater flow. The temperature of the spring waters remains practically constant, averaging between 0.1 and 0.3°C. A further characteristic is the high degree of mineralisation. This is because the restricted circulation imposed by the permafrost boundaries increases the concentration of dissolved solids. In Yakutia for example, dissolved solids varied from 0.9 to 34 g/l, being least in those springs supplied mainly from supra-permafrost water and greatest in springs resulting from sub- and intra-permafrost waters. In the Brooks Range of Alaska and the British, Richardson and Ogilvie Mountains of the northwest Yukon of Canada similar freshwater springs occur in association with fault zones and limestone bedrock (Williams and van Everdingen, 1973, p. 438). Discharge rates vary from 0.12 to 1.14 m^3/s but temperatures are between 4 and 6°C. Recharge of the springs appears to be associated with the percolation of meltwater and rainfall through unfrozen zones that perforate the permafrost.

Since the occurrence of groundwater springs is relatively rare, the location and exploitation of sub-permafrost groundwater is essential for the majority of permanent settlements. In the continuous permafrost zone, where permafrost thicknesses are greatest, the exploitation is extremely difficult since drilling becomes costly and impractical. The most economically developed sources of water in such regions are the unfrozen alluviums beneath large lakes and rivers. Even this however, is usually insufficient to support a large community. For these reasons, many ***ad hoc*** solutions have been adopted. At Inuvik, N.W.T., with an estimated population of over 4 500 (1974), water is pumped from the east channel of the Mackenzie River into a nearby lake for storage at periods of clean water (Brown, 1970, p. 92). The availability of this procedure was an important factor in the choice of the townsite. In many of the smaller communities water is hauled by truck from a nearby lake and distributed each day to personal water tanks. In the discontinuous permafrost regions of Alaska and Canada however, alluvial deposits provide an abundant source of groundwater for most large settlements (Williams and van Everdingen, 1973). Many of the major river valleys, such as those of the Tanana, Kuskokwim and Yukon rivers, possess unconsolidated sediments over 200 m thick, either as floodplain or terrace deposits, or as alluvial fans. Within these aquifers, permafrost thickness and distribution varies with the history of river migration, such that the older floodplains and terraces are underlain by thicker and more continuous permafrost than the younger terraces and floodplains. Thus, in the Fairbanks area of Alaska, water is supplied to residents by numerous small diameter wells (Péwé, 1966, pp. 28–9). Many are only 5–10 m deep and draw upon supra- and intra-permafrost waters. Others are 30–75 m deep and take water from beneath the permafrost. Large yields occur from both above and below the permafrost; records indicate yields as high as 3 000 gallons per minute for large diameter wells 30 m deep while shallower wells of 5 cm diameter commonly yield over 40 gallons per minute.

Groundwater, artesian pressures and open system pingos Groundwater in permafrost regions is subject to artesian pressure in the same way as groundwater in non-permafrost areas. The role played by permafrost is to act as an impermeable and confining layer for sub-permafrost waters. Where the permafrost is thin and/or discontinuous, artesian pressures may develop such that the overlying permafrost is ruptured. More commonly, the emergence of water under artesian head occurs at the downslope limit of permafrost bodies, usually in the central and lower parts of slopes. If the groundwater temperature is not far from 0°C and if air temperatures are below 0°C, the water may freeze as it rises through the permafrost zone or enters the seasonally frozen zone. The growth of an ice body with its associated frost heave pressures may result in the updoming of the overlying sediments and the formation of a conical ice-cored hill known as a pingo.

From a hydrologic point of view, two types of pingo are recognised; (a) the open system pingo produced by confined sub-permafrost water in the discontinuous zone, as discussed by Holmes, Hopkins and Foster (1968) and (b) the closed system pingo produced by confined intra-permafrost water in the continuous permafrost zone, as discussed by Muller (1959) and Mackay (1962). Since the nature of pingos is examined more fully in Chapter 5 (see pp. 93–104), only the presumed role of artesian pressure and permafrost distribution in the development of open system pingos will be discussed here.

A major requirement for open system pingos is the presence of thin and/or discontinuous permafrost and the confining of sub-permafrost waters. By far the largest concentration of open system pingos occur in the Yukon—Tanana uplands in central Alaska (Holmes, Hopkins and Foster, 1968), and in the Yukon of Canada (Hughes, 1969). They are found in narrow upland valleys particularly on the lower parts of south and southeast-facing slopes but are not found on the north-facing slopes or in the broader river valleys. This is probably because there are more opportunities for surface water to enter the ground in the permafrost-free areas of the upper sections of south-facing slopes than on north-facing slopes which are totally frozen (Brown and Péwé, 1973, p. 80). A second requirement for the formation of open system pingos is restricted groundwater flow since too large a flow rate would prevent freezing of the pingo. A third requirement is that the subsurface temperature be close to 0°C in order to provide the minimum tensile strength of frozen ground and prevent premature freezing of the groundwater supply. However, theoretical calculations by Holmes ***et al.*** (1968) suggest that the total force required both to maintain a pingo 30 m high and to overcome the tensile strength of ground and force it upwards is considerably greater than the highest artesian pressure yet measured in Alaska. It appears likely therefore, that the growth of ice lenses and the accompanying freezing pressures are mainly responsible for open system pingo growth. It follows that the essential hydrologic condition is a steady but slow supply of groundwater. While such a condition may be favoured by artesian pressure, this pressure is probably of only minor importance in the updoming and growth of open system pingos.

Icings Icings are sheetlike masses of ice which form at the surface in winter where water issues from the ground. They are also known as aufeis (e.g. Washburn, 1973, pp. 29–32) or 'nalyedi' (Brown, 1967b, pp. 747–9). Sometimes, the water source is of a sub- or intra-permafrost nature in which case the spring is usually a perennial one. More commonly, icings are the result of suprapermafrost waters. If the seasonal frost extends to the top of the permafrost table, flow may cease in mid-winter but if there is a sub-permafrost talik, flow can be perennial. The majority of groundwater icings which result from permafrost waters are small, usually less than 1.0 m in thickness and less than 0.5 km^2

in extent. Icings which are associated with perennial springs, however, may assume considerable dimensions.

River icings or naledi are a second category of icings which develop at localities where rivers freeze to the bottom and water is forced out of the bed. Depending upon the stream flow characteristics, river icings may be very large. In the Momskaya Depression of northeast Yakutia, complex icings as large as 62 km^2 are known to occur (Nekrasov and Gordeyev, 1973, pp. 37–40). In Alaska, an icing over 10 m thick and more than 1 km wide and 2 km long formed in the St John River valley in 1969 (Ferrians ***et al.,*** 1969, pp. 34–6).

Hydrologic studies of some of the large icings which are associated with the perennial springs of central Yakutia have been undertaken by the Permafrost Institute of the Siberian Division of the USSR Academy of Science (Anisimova ***et al.,*** 1973, pp. 37–47). Although the icings start to form as soon as the streams are frozen, the major period of icing growth is in the mid-winter to early spring period. The small monthly accretions of icing volumes during the early winter are explained by discharge continuing either in the channels or through the under river bed taliks which freeze only in late winter. Thus, maximum ice accretions involving sub- or intra-permafrost waters occur in February and March when the ground is fully frozen and before the spring rise in air temperatures which lead to the ablation of the icing. During the early summer months, icings melt away completely.

Similar icing accretion patterns occur in northern Alaska and Canada (Williams and van Everdingen, 1973) but where the groundwater is of a supra-permafrost nature, the icing may cease to grow in late winter. The amount of discharge stored in icings can be considerable and the release of this water in the ablation period may significantly alter the stream flow characteristics of the watershed concerned. For example, the water stored in the icings of the Caribou–Porcupine watersheds near Fairbanks, Alaska, is equivalent to 4 per cent of total runoff and 40 per cent of total winter streamflow. This is concentrated into the four weeks following ablation of the snowpack and provides a flow augmentation of 0.28 m^3/s. In other areas of Alaska, it has been calculated that the melt of icings over the ablation period may increase streamflow by as much as 20 m^3/s.

Icings are of great practical concern with respect to highway and railroad construction and maintenance. They were a problem in the construction of the Alaska Highway (Thomson, 1966; Brown, 1970, pp. 109–11) and are a hazard on most northern roads. They commonly develop where a roadcut intersects the supra-permafrost groundwater table and seepage occurs. Counter measures to reduce icing involve the limited use of roadcuts, the installation of culverts to divert water from the source of the icing, and the improvement of drainage adjacent to the road. Icings may also block culverts and, by diverting meltwaters, initiate washouts in the spring thaw period. River icings may also cause extensive damage to bridges through diversion of streamflow and burial of the bridge structure.

Engineering aspects of permafrost and frost action

Permafrost poses a number of important engineering problems which seriously hinder economic development of northern regions. As early as 1945,

S. W. Muller wrote . . . 'The destructive action of permafrost phenomena has materially impeded the colonisation and development of extensive and potentially rich areas in the north. Roads, railroads, bridges, houses and factories have suffered deformation, at times beyond repair, because the condition of permafrost ground was not examined beforehand, and because the behaviour of frozen ground was little if at all understood' (pp. 1–2). Since then, numerous detailed and comprehensive accounts of these problems have become available. Particularly useful are those by Brown (1970), Péwé (1966), and Ferrians, Kachadoorian and Greene (1969). The objective of this section however, is not to duplicate such reviews but to identify the reasons why permafrost poses such problems and to briefly illustrate some of the preventive measures that are taken.

The severity of the engineering problems will vary with the type of rock or sediment. They are worst in poorly drained, fine grained sediments which often possess large quantities of ice. Upon disturbance the ice may melt causing subsidence, excessive wetting and high plasticity. This results in a loss of strength which can lead to failure and flowage. In winter, the seasonal freezing of such sediments and the formation of ice results in frost heaving. Coarse grained and well drained sediments on the other hand, will present few, if any, construction or maintenance problems. A fundamental guideline therefore, is that structures be located, whenever possible, upon coarse grained sediments of low frost susceptibility.

It is generally agreed that the thawing of permafrost and the heaving and subsidence caused by frost action are primarily responsible for the majority of major problems. In many engineering situations, it is inevitable that the surface vegetation mat is destroyed, its insulating effect is lost, the permafrost begins to thaw and the active layer thickens. In fine grained sediments with high ice contents, this leads to the development of thermokarst terrain (see pp. 105–33). Greater frost heaving of the ground surface in winter is also associated with the thickening of the active layer in the summer since more moisture is able to accumulate in the active layer.

As a result, the majority of structures require some sort of pad or fill to be placed upon the ground in order to compensate for the increase in the thaw depth that will result. In this way the thermal regime of the underlying permafrost is not altered. It is possible to calculate the thickness of fill required, given the thermal conductivity of the materials involved and the mean air and ground temperatures at the site. Too little fill plus the increased conductivity of the compacted active layer beneath the fill will result in the thawing of the permafrost (Fig. 4.9). If the fill is too thick, the insulating effects of the fill and compacted layer is greater than the original active layer and the permafrost table will rise on account of the smaller amplitude of the seasonal temperature fluctuation. Where the calculated thickness of fill becomes too great as to be impractical, alternate procedures are necessary such as using insulating materials under the fill. Gravel is the most common aggregate used in northern Canada and Alaska since it is readily available and not as susceptible to frost heave as fine grained sediments. Where suitable aggregate materials are not readily available, the provision of such becomes a major and expensive problem. An example of what happens when a building is placed directly upon ice-rich permafrost is given in Fig. 4.10.

Fig. 4.9 *Effects of a gravel fill upon the thermal regime and thickness of the active layer. From Ferrians et al., (1969).* A — *Too little fill.* B — *Too much fill.* C — *Effects of cases* A *and* B *upon thermal regime; too little fill increases amplitude of seasonal temperature fluctuation at permafrost table.*

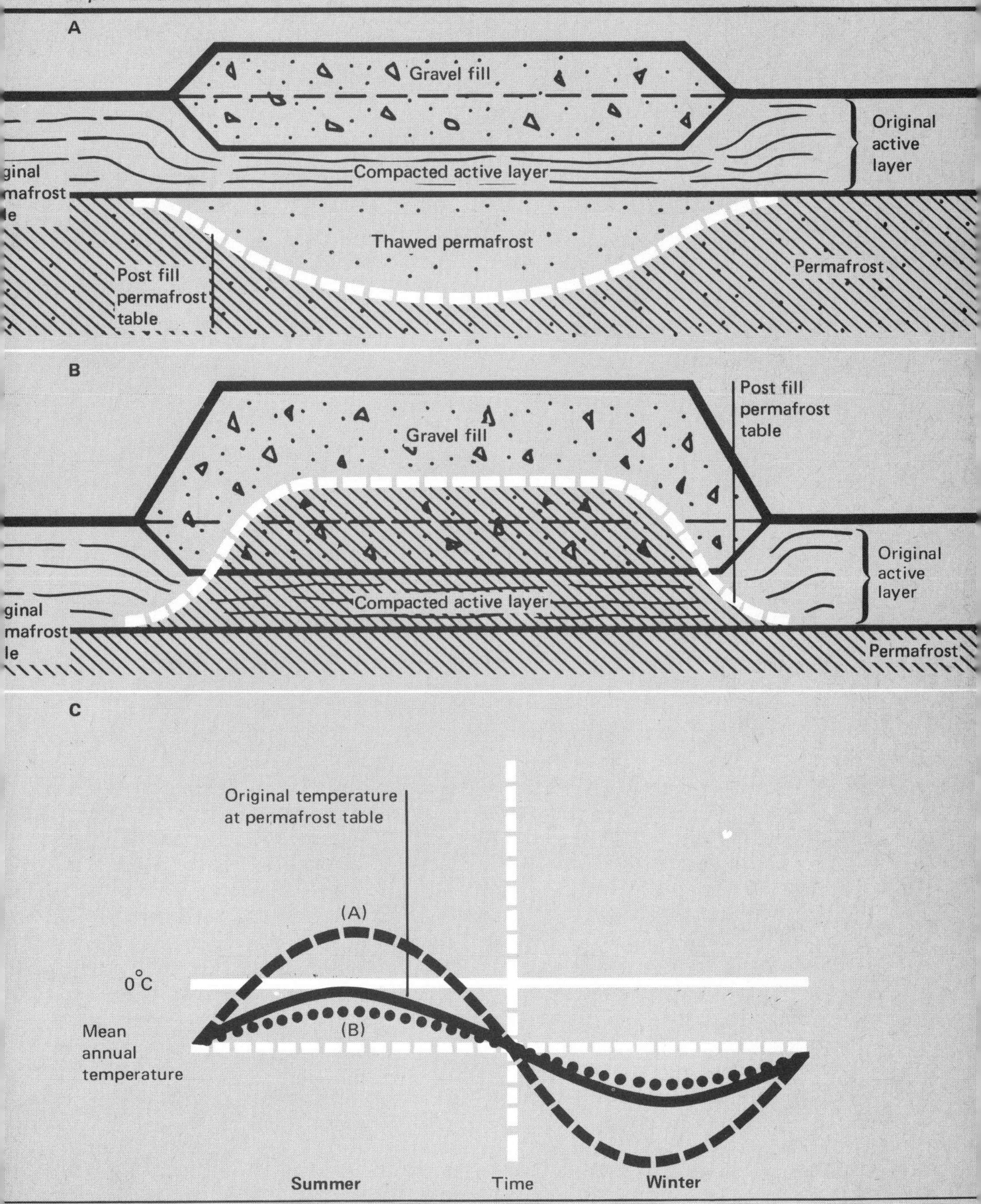

Fig. 4.10 *Example of subsidence of a building situated directly upon ice-rich permafrost, Yakutsk, USSR. No piles or gravel pads were used.*

A major reason governing the location of the Inuvik townsite in the Mackenzie Valley was the proximity of large quantities of fluvio-glacial gravels at the exit of a meltwater channel just to the south of the chosen site. Rear-end dumping of these gravels over the undisturbed surface terrain was the essential prerequisite to any construction activity and heavy equipment movement. A special design was also used for the construction of the airstrip (Brown, 1970, p. 134). The trees were cut by hand and laid on the ground after which a minimum of 2.4 m of crushed dolomite limestone fill from adjacent exposures was placed upon the undisturbed surface. Ten years after construction, the permafrost has aggraded only to the lower layers of the fill and many more years of operation of the airstrip are anticipated before the permafrost reaches the surface layers and maintenance problems develop. The experience at Inuvik may be contrasted with that at Sachs Harbour where a gravel airstrip was built at approximately the same time. In a search for suitable aggregate to level the proposed airstrip site, over 5 hectares of land were disturbed (see pp. 130–1 and Figs 6.13–6.14). Today, the borrow pits are still undergoing thermokarst subsidence since the underlying sediments possessed 15–25 per cent ground ice by volume and the active layer increased by an average of 10 cm (French, 1975b). The airstrip requires constant maintenance and regrading. Heavy summer rain often makes the airstrip slippery and unusable, on account of the relatively high percentage of silt size particles used in the aggregate, while gullying occurs periodically at springtime as the result of water accumulating in and overflowing the borrow pits. It is clear that borrow pits should not be sited upslope or downslope of structures since thawing may cause drainage problems or slumps.

In the case of heated buildings, additional precautions are taken to prevent the permafrost from thawing. Often, the building is supported upon wooden blocks or piles resting on, or inserted in, the gravel pad. An airspace is left between the floor of the building and the ground surface to enable the free circulation of air and minimise the transfer of heat to the ground. Where concrete wall foundations are used, large gaps in the wall act as air vents in a similar fashion. The provision of services such as water supply and sewage disposal are also complicated. Pipes to carry these services cannot be laid in the ground below the seasonal frost depth as is done in permafrost free areas since the heat from the pipes will promote thawing of the surrounding permafrost and subsequent subsidence and fracture of the pipes. The answer has been to carry these services from one building to the next in continuously insulated enclosures or boxes constituted of concrete, wood or metal. They are termed 'utilidors' and, like other buildings, are placed above the ground surface to enable air circulation beneath. The cost of utilidor systems are high, involving centralised plant, a high degree of town planning and constant maintenance. As a result, provision of these services can only be justified in a few of the largest centres. In Siberia where modern cities of over 100 000 population have developed in recent years, utilidor type systems are widely employed. In Yakutsk for example, all of the recently constructed high rise buildings are connected to a central large utilidor which runs beneath the main street and empties into the adjacent Lena River. In northern Canada, population densities are much less and only Inuvik, N.W.T., has a modern utilidor system. The majority of smaller settlements in the continuous permafrost zone of northern Canada possess no centralised water or sewage disposal systems. The dumping of garbage and waste is also a problem since landfill and burial procedures are unavailable in regions underlain by permafrost.

Frost heaving in the seasonally thawed layer is the second major problem to be overcome. The nature of this process has been outlined in the previous chapter. With respect to construction, frost heaving affects the use of piles for the support of structures. While in warmer climates the chief problem of piles is to obtain sufficient bearing strength, in permafrost regions the problem is to keep the piles in the ground since frost action heaves them upwards in the same way that stones are heaved to the surface. Variations in moisture content result in differential heave and buildings supported by the piles may tilt and sometimes collapse. Examples of the displacement of piles through frost heaving have been described from many parts of Alaska and northern Canada (e.g. Péwé and Paige, 1963). In theory, the upward heaving force is generated in the frozen active layer and is a function of the adfreezing strength between the pile and the active layer, the depth to which the pile has been inserted into the active layer, and the perimeter of the pile. This upward heave becomes progressively greater as the active layer freezes. It follows that the thicker is the active layer, the greater is the upward heaving force.

In the discontinuous zone, where the active layer is at its thickest, the frost heaving of piles assumes its greatest importance. In parts of Alaska, bridge structures illustrate most dramatically the effects of frost heave (Fig. 4.11). In those instances where the bridge support is in the form of piles, it is not uncommon for the piles within the stream bed to be inserted in unfrozen sediments. As such, they experience little or no frost heaving. However, piles adjacent to the river bank and on both sides of the river undergo frost heaving since they are located in the seasonal frost zone. As a result, uparching of both ends of the

Fig. 4.11 *Schematic illustration of how frost heaving of piles inserted in the layer of seasonal frost can result in bridge deformation. From Péwé and Paige (1963), after an Alaskan example*

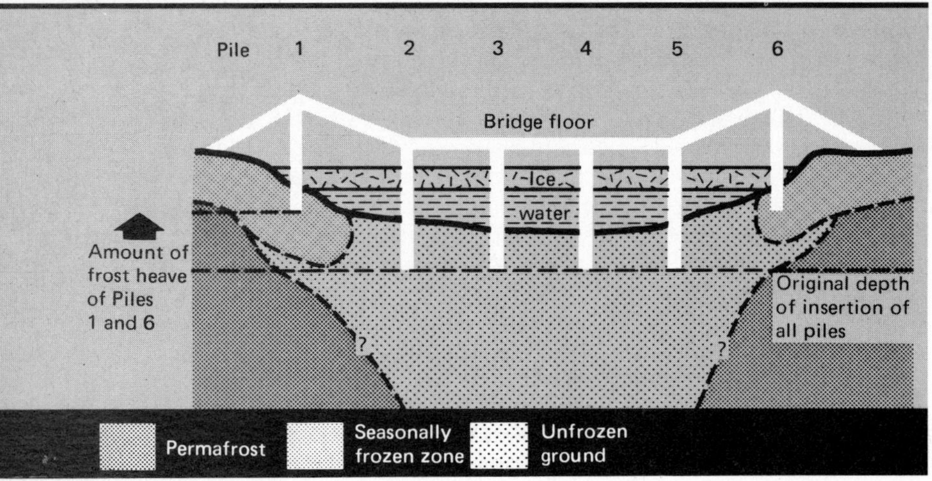

bridge may occur. The considerable magnitude of the upward forces involved has been illustrated by Péwé and Paige (1963, p. 364). Calculations applicable to silt of constant texture and water content indicate the upward frost heaving forces become greater as the freezing layer becomes thicker at a rate of approximately 21 600 lb/ft.

There are several preventive measures for frost heaving of piles. In the case of bridges, alternate bridge structures involving minimal pile support are often considered. In other cases, the frost heaving can be minimised by improving the drainage conditions at the heaving piles. With buildings, the most widely used method in northern Canada and Alaska is to anchor the piles securely and deeply in steamed or drilled holes in the permafrost. Where pile insertion is difficult, as in stoney soils, recourse must be made to gravel pads or fill as described above. However, most large buildings today warrant the extra costs involved in pile insertion. Telephone poles and other transmission poles are particularly prone to frost heaving because they are not usually deeply imbedded in permafrost. Constant maintenance appears to be the best answer in this case since pile insertion would be too costly in all but a few localities.

5 Ground ice

There has been a tendency to underestimate the importance of ground ice within the periglacial environment. However, in recent years, the stimulus given to northern development by the activities of oil companies in the western Canadian Arctic and on the north slope of Alaska has made it clear that ground ice is perhaps the biggest obstacle to northern development. As such, it poses relatively more distinct and difficult problems than the traditional and commonly recognised limitations imposed by the climatic, permafrost and terrain conditions of these regions. In coming to this conclusion, North American workers are increasingly looking towards the Russian experience for analogies and comparisons, since ground ice and all of its associated forms has long attracted attention in many parts of Siberia and the USSR (e.g. Shumskiy, 1964; Shumskiy and Vtyurin, 1966).

This chapter is specifically devoted to the nature and extent of ground ice, and it attempts to outline the more important surface features associated with the occurrence of ground ice. It is a natural starting point, therefore, for any consideration of the melting of ground ice, either by thermokarst processes ***senso stricto*** or by the various forms of thermal erosion. Both of these topics will be dealt with in the next chapter since they are significant and important geomorphic processes operating within the periglacial environment.

Ground ice description

Ground ice is defined as a 'body of more or less clear ice within frozen ground' (Pihlainen and Johnston, 1963).

The field description of ground ice requires standardised nomenclature. One system, proposed by Pihlainen and Johnston, has proved useful and is widely applied in northern Canada. It is a purely descriptive terminology based upon the form of the ice and is not intended to be a geotechnical assessment of frozen materials according to their performance properties. The system is presented in Table 5.1. Three major groups are identified in which the ice is (a) not visible to the eye, (b) visible by eye with individual ice layers less than 2.5 cm in thickness, and (c) visible by eye with individual ice layers greater than 2.5 cm in thickness. Further subdivision is based upon the distribution, shape, size and other descriptive features of the ice.

Two quantitative parameters are also used in describing ground ice conditions. First, the 'ice content' of a soil is defined as the weight of ice to dry soil, and is expressed as a percentage. For example, if a soil sample weighed 100 g when frozen and 40 g when oven fried, then the weight of ice (i.e. moisture) in that sample was 60 g. Therefore, the ice content would be 150 per cent. This simple calculation can often be performed in the field on a 200 g sample, using a small balance and air drying the sample. Since ice contents may vary considerably within a few centimetres, it is probably unrealistic to adopt more sophisticated weighing and drying techniques. To be really meaningful, a number of ice content determinations should be made at different depths and the moisture content of the seasonable thawed zone should also be determined. Low ice content soils are generally regarded as those having ice

Table 5.1 *A descriptive field system for ground ice, as proposed by Pihlainen and Johnston (1963)*

Main group	Symbol	Subgroup description	Symbol	Field identification
Ice not visible	*N*	Poorly bonded or friable	*Nf*	Hand examination
		Well bonded	*Nb*	Thaw sample to determine excess ice (supernatant water)
		No excess ice	*Nbn*	
		Excess ice	*Nbe*	
Visible ice – less than 2.5 cm thick	*V*	Individual ice crystals or inclusions	*Vx*	Visual examination. Observations upon: location orientation thickness length spacing size shape pattern of arrangement hardness structure colour
		Ice coatings on particles	*Vc*	
		Random or irregularly oriented ice formations	*Vr*	
		Stratified or distinctly oriented ice formations	*Vs*	
Visible ice – greater than 2.5 cm thick	ICE	Ice with soil inclusions	ICE+soil type	Visual examination. Observations upon: Hardness – hard; soft structure – clear, cloudy, porous, candled, granular, stratified colour admixtures
		Ice without soil inclusions	ICE	

contents less than 40–50 per cent. Soils with high ice contents are usually fine grained and have ice content values which commonly range between 50 and 150 per cent.

In certain instances, it is of considerable importance to determine the Atterberg Limits for the mineral sediment content of the permafrost. It is not uncommon for the natural water (ice) content of frozen soils to exceed their liquid limits. If this is the case, the ground will pass upon thawing from a solid state possessing considerable shear strength to a considerably lower liquid state shear strength. Buildings may subside and the thawed sediments will become relatively mobile, capable of flowing on very slight inclines. This is an important consideration in the development and maintenance of the ground ice slumping process (see pp. 119–22).

The amount of 'excess ice' is a second parameter commonly used in the description of ground ice conditions. Excess ice refers to the volume of supernatant water present if a vertical column of frozen sediment were thawed. In this case, the sample is allowed to thaw and the relative volumes of supersaturated sediments and standing water (i.e. excess ice) are noted. The volume of supernatant water is then expressed as a percentage of the total volume of sediment and water. For example, if upon thawing the relative volumes of supersaturated sediments and supernatant water were 300 and 500 cm respectively, then the excess ice value would be 62.5 per cent. The advantage of this index is that it provides some indication of the potential morphological change, or volumetric ground loss, consequent upon the thawing of the sediments. It is useful, therefore, in discussions concerning thermokarst development. Visual estimates of excess ice are deceptive since a body of sediment may not necessarily contain excess ice even though it may contain visible ice lense, since the sediment between the lenses may not be saturated with ice and upon melting all water would be retained in the voids. Sediments that contain excess ice are often referred to as 'icy sediments'. In unconsolidated sediments, excess ice values of between 15 and 50 per cent are reasonably common although exceptionally icy sediments may have values as great as 70–80 per cent.

The term 'massive ice' or 'massive icy bodies' is usually reserved to describe relatively pure ground ice whose ice content averages at least 250 per cent for thicknesses of several metres. One such exposure on Garry Island in the Mackenzie Delta, N.W.T., has been described by Mackay (1966). The average ice content of a 6.4 m high face was over 300 per cent and excess ice values exceeded 80 per cent. It was estimated that the annual thawing of the icy face releases about 1 m of saturated soil and 5.4 m of water.

Types of ground ice

Several attempts have been made to classify the various types of ground ice that exist. In general terms, ground ice may be either 'epigenetic' (i.e. develops inside the enclosing rock and after the latter is formed) or 'syngenetic' (i.e. forms at, or almost at, the same time as the enclosing sediments are deposited and usually associated, therefore, with surface aggradation).

In detail, however, there are a variety of ground ice forms. J. R. Mackay (1972a) for example, has proposed a classification in which the main criteria are (a) the source of water immediately prior to freezing, and (b) the principal transfer process which moves water to the freezing plane. The various types of buried ice (e.g. glacier ice; sea, lake or river ice; snowbanks; river and groundwater icings) are excluded. This classification results in ten mutually exclusive ground ice forms (Fig. 5.1).

In addition to its clarity, an advantage of this classification over previous attempts is that it emphasises the variety and complexity of the transfer process, recognising at least six basic mechanisms. Moreover, although the theoretical principles behind these transfer processes are beyond the scope of this book, the classification has the added advantage but it focuses attention upon three broad types of ground ice: (1) Wedge ice, categories 2, 3 and 4; (2) Segregated ice, categories 6 and 7; (3) Intrusive ice, categories 8 and 9. These three types of ground ice are particularly important from a geomorphological viewpoint since their localised occurrence gives rise to distinctive periglacial landforms and terrain, such as ice-wedge polygons, pingos, and various types of ice-cored topography.

Before these surface manifestations of ground ice are considered, the various types of ice and their characteristics require brief description.

Pore ice is the bonding cement which holds soil grains together. The distinction between pore ice and segregated ice is related to the water content of the

Fig. 5.1 *Classification of ground ice according to J. R. Mackay (1972a)*

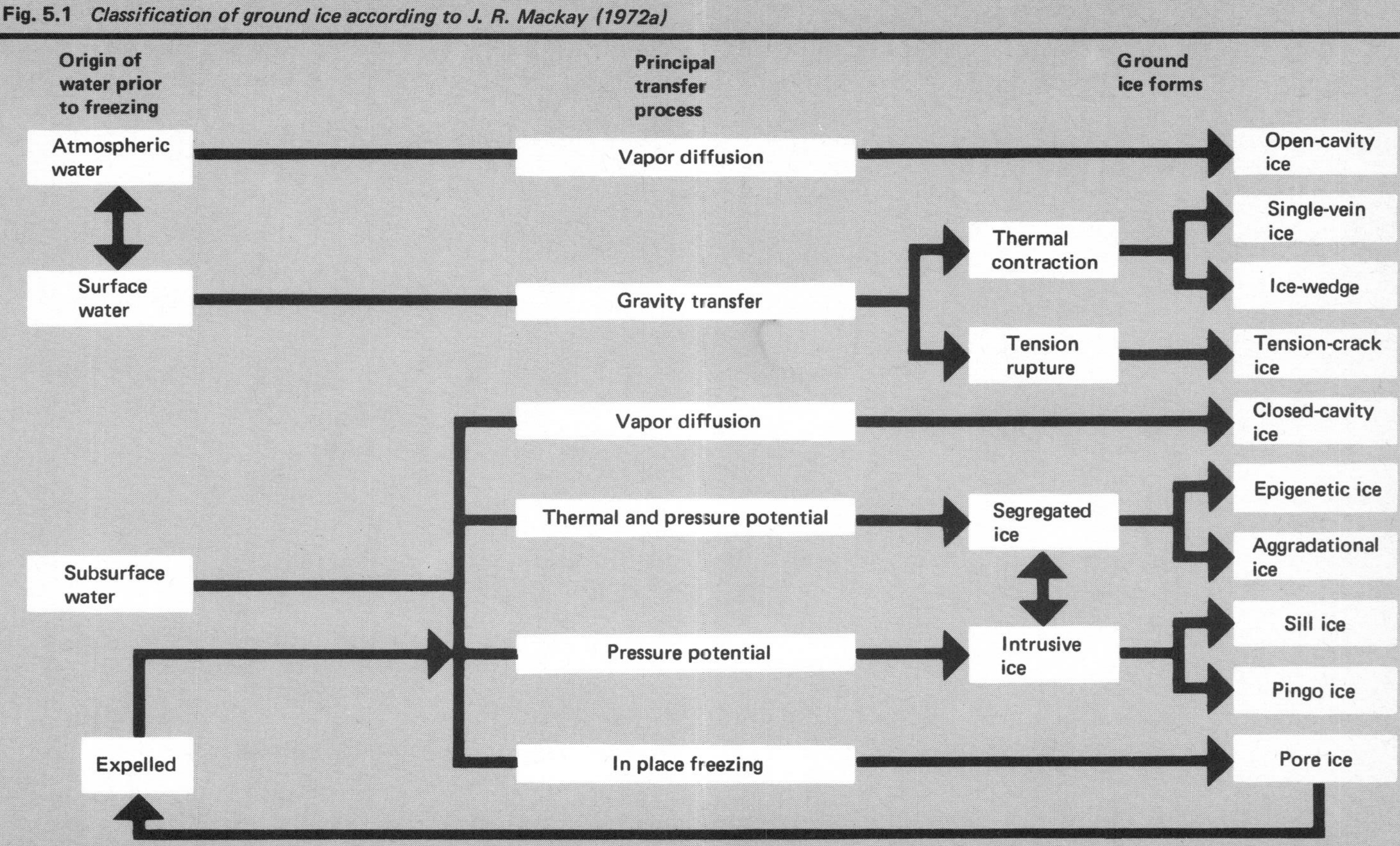

soil. It is best determined by thawing the soil and noting the presence or absence of excess ice or supernatant water. If supernatant water is present, this indicates that the frozen soil was supersaturated and that segregated ice was present. Open cavity ice is formed by the sublimation of ice crystals directly from water vapour in an open cavity or crack in the ground, and is similar to hoarfrost (Mackay, 1972a, p. 5). Closed cavity ice is restricted to underground cavities and is rarely exposed and of little importance. Mackay (1965) has described ice crystals growing in an underground cavity through the doming action of methane under pressure, but other reports are limited.

Segregated ice is a broad term for soil with a high ice content. The mechanism of formation of segregated ice has already been discussed in some detail in Chapter 3 (see pp. 27–37). Usually, the ice lenses are visible to the naked eye but in certain soils, particularly those which are fine grained, lensing is minimal. Segregated ice lenses vary in thickness from layers a few centimetres thick to massive ice bodies, sometimes tens of metres thick (Fig. 5.2). In theory, segregated ice may be distinguished from intrusive ice on account of the relative purity of the latter and the stratification and presence of soil particles and air bubbles oriented normal to the freezing plane in the former. However, when dealing with massive icy bodies, such as occur in pingos and other ice-cored topography, this distinction is often difficult to make.

Vein ice is formed by the penetration of water into open fissures developed

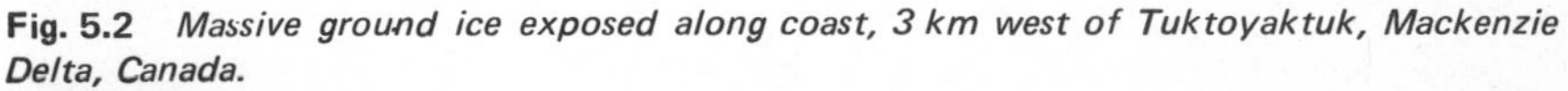

Fig. 5.2 *Massive ground ice exposed along coast, 3 km west of Tuktoyaktuk, Mackenzie Delta, Canada.*

at the ground surface. In contrast to segregated ice therefore, the origin of the water prior to its freezing is of a surface nature, usually meltwater from snow and summer rain. Vein ice can be distinguished from segregation ice on account of its vertical foliation and structures. Single vein ice develops in fissures which are formed in the permafrost by thermal contraction cracking. Usually, the uppermost 60 cm of permafrost is characterised by an intricate pattern of small intersecting ice bands. The majority are very thin, less than 0.2 cm in thickness. Repeated vein ice is a specific variety of single vein ice and is the result of successive ice formation in fissures periodically forming in about the same place for many years. The result is the formation of vertical or near vertical sheets of ice which, because of their form, are termed ice wedges (see Fig. 3.5). Since these are important features of periglacial conditions in general, they are dealt with separately in a following section.

Intrusive ice is formed by the intrusion of water, usually under pressure, into the seasonally or perennially frozen zone. Sill or sheet ice and pingo ice are the two types of intrusive ice which are identified. The former grows when water is intruded into a confining material and freezes in a tabular mass along the base of the active layer and parallel to the permafrost surface. It is extremely difficult however, to tell whether a particular tabular ice body grew as a sill, through slow injection possibly associated with artesian pressure, or by ice segregation. There appear to be relatively few observations made upon the occurrence of sill ice, although Mackay (1972a) has described sill ice at least 1.2 m thick in the Brock River Delta, N.W.T. It is probable that sill ice is most favoured in areas of strongly jointed and relatively impermeable rock within the discontinuous permafrost zones and with relatively strong relief. Under such conditions, water may move easily along bedding places to freeze near the surface.

Massive ice and massive icy bodies

Massive ice and massive icy bodies, often tens of metres thick, are the most spectacular of ground ice forms (Fig. 5.2). In recent years, these sediments have attracted considerable attention, not only because of their origin and the light this may throw upon Quaternary history in general but also because of the thawing and settling properties of terrain underlain by such sediments.

Nature and extent The thickest and most extensive massive ice bodies are to be found in the coastal and aggradational lowlands of northern Siberia and central Yakutia (Popov, 1962, 1970; Bobov, 1969; Shumskiy and Vtyurin, 1966). Ice exposures in excess of 70 m in height are reported from the cliffs bounding the lowland areas of the Novowibirskiye Ostrova islands (Bird, 1967, p. 200). In North America, segregated ice, in the form of icy sediments and massive ice, has been widely reported from the coastal lowlands bordering the Beaufort Sea. Thick exposures may be seen along the north slope of Alaska, the Arctic coastal plain of the Yukon Territory, and in the Mackenzie District of the N.W.T., particularly the Mackenzie Delta. Shot hole data provided by oil companies provides additional information for certain areas. Presumably, as more information becomes available, the occurrence of massive ice in the lowland areas of Banks Island and adjacent western Arctic localities will be reported.

J. R. Mackay has undertaken the most comprehensive investigation of these sediments in the western Arctic (e.g. Mackay, 1966, 1971; Rampton and

Mackay, 1971; Mackay, Rampton and Fyles, 1972). In general, the thicknesses reported are less than those in Siberia. The most detailed information available is from the Tuktoyaktuk Peninsula and Richards Island areas of the Mackenzie Delta where recent coastal exposures of massive ice and icy sediments have been supplemented by shot hole logs obtained during seismic operations by oil companies. For example, in examining over 5 000 shot hole logs, Rampton and Mackay (1971) found that 20–30 per cent of the holes penetrated icy sediments and 0–5 per cent penetrated massive ice. Of 264 drillholes which penetrated massive ice, over 70 per cent encountered massive ice at a depth of 12.1 m (40 ft), and 80 per cent of the shot holes reported icy sediments anywhere from the surface to a depth of 42.4 m (140 ft). In some instances, some holes were drilled over 40 m into solid ice without bottoming through it. It must be concluded therefore, that large bodies of massive ice and icy sediments underlie extensive areas of the Mackenzie–Beaufort coastal plain.

The nature of massive ice bodies in the western Arctic can be observed in numerous coastal exposures, especially east of Herschel Island and in the Mackenzie Delta. Usually the massive ice occurs beneath Pleistocene sediments, commonly till, reworked till, or glaciofluvial or lacustrine sediments, since much of the area has been glaciated, not necessarily in late Wisconsinan times. In appearance, the ice may vary from relatively pure, massive ice to thinner, clearly segregated lenses with mineral bands. While the majority of the icy sediments and massive ice bodies appear undisturbed and horizontally bedded, other sections possess fractures and small fissures and give the appearance of deformation and contortion by glacial pushing and overriding (e.g. Mackay, 1963a; Mackay, Rampton and Fyles, 1972).

Origins of massive ice It is debatable whether the relatively simple explanation proposed for thin segregated ice lenses, following the ideas of Taber, is satisfactory to explain massive ice over 20 m in thickness. Mackay has pointed out that such ice thicknesses cannot grow solely from the water beneath, because free water in the required volume is not readily available. Moreover, it is not clear as to the mechanism by which this water was forced to the surface.

There are at least three hypotheses concerning the origin of massive ice. The first, which has been called the snow-glacial hypothesis, involves the burial of ice, such as stagnant glacial ice, compacted snow drifts, and lake, sea and river ice. A second hypothesis involves the interpretation of ground ice as being exceptionally well developed vein ice, i.e. syngenetic ice wedges, which have grown under conditions of sub-aerial sedimentation, with the vertical growth keeping pace with the rate of sedimentation. The third explanation interprets massive ice as being of segregated origin, from the ***in situ*** freezing of drawn up pore water. Certain differences of opinion are apparent between Russian and North American workers. In general, many Russian authorities propose a vein ice hypothesis to explain the massive ground ice of northern Siberia (e.g. Popov, 1962; Shumskiy and Vtyurin, 1966). According to the vein ice theory, the horizontal layering observed in massive ice bodies is an illusion being only the vertical foliation of vein ice in long section. On the other hand, Mackay (1971) has argued strongly that the massive icy bodies found in the western Canadian Arctic are of a segregational origin. He has concluded that there is little or no evidence favouring a vein ice or buried ice origin for the massive bodies.

 Concerning the vein ice hypothesis, it is clear from field observations by many workers in North America, that there is not the same gigantic growth of ice wedges as there is in Russia, where wedges have grown in association with long continued and widespread river sedimentation. Moreover, the sizes of some icy bodies in the western Arctic are at least 100 times that of the largest known ice wedges, and the horizontal banding of certain exposures of massive ice is difficult to interpret as vertical ice foliations seen in an oblique section. Finally, many of the icy bodies of the western Canadian Arctic occur within Pleistocene interglacial sediments which were laid down within a marine environment. Syngenetic vein ice, however, can only form upon an aggrading alluvial floodplain, and this condition has demonstrably not occurred in much of the Mackenzie Delta area.

The possibility that some of the massive ice bodies of the western Canadian Arctic may be buried glacial ice masses is not generally accepted but on the other hand, difficult to completely disregard. The situation is complicated by the fact that first, many of the icy bodies are apparently glacially deformed, second, that Pleistocene sediments are involved in the deformations, and third, that the glacial limits of the Wisconsin and possibly earlier glaciations are not completely known in that area. Some of the permafrost and enclosing sediments are certainly older than 40 000 years B.P. (Mackay, Rampton and Fyles, 1972) and therefore predate the last ice advance.

Mackay (1971; 1972a) has illustrated how most massive ice bodies in the western Arctic can best be explained by existing ice segregation theory and has also suggested the probable source of the excess water required for such ice segregation. Bore hole log data for the Richards Island and Tuktoyaktuk Peninsula areas indicate over 85 per cent of all massive icy bodies were underlain by either sands and gravels. Mackay has argued that the freezing of such materials, within a closed system, would result in the expulsion of water under a high pore water pressure, and has linked the freezing of such materials to the downward aggradation of permafrost in an emerging coastal plain. He concludes (Mackay, 1972a, pp. 21–2) . . . 'that water expulsion . . . has supplied the excess water now found in many massive ice sheets of the western Arctic of North America. The downward growth of permafrost in . . . coastal areas is believed to have been associated with the sea level lowering which followed the onset of the Wisconsin glaciation about 100 000 years ago. In inland area, the growth of massive ice may have accompanied changes from warm interglacial or interstadial to cold glacial conditions.'

Ice-cored topography

Recent geomorphic and geophysical investigations in the Mackenzie Beaufort lowlands of the western Arctic (Rampton, 1974; Rampton and Walcott, 1974) suggest that much of the topography is the result of the aggradation of icy sediments and massive ice bodies. In particular, gravity profiling and the identification of Bouguer anomalies has resulted in the definite association of certain terrain features with the presence of excess ice and massive ice beneath.

The use of gravity profiling techniques relies on the difference in density between normal frozen saturated sediments (assumed density of approximately 2.0 Mg m^{-3}) and excess ice (density of approximately 0.9 Mg m^{-3}). Given baseline geologic data, the regional gravity anomalies that are observed in an area can usually be removed as linear trends on profiles over 1 km in length. Thus, the average amount of excess ice in the topography along the profile is obtained by removing the trend, obtaining the Bouguer density of the topography, and calculating the proportions of the frozen saturated sediments and ice required to produce this density (Rampton and Walcott, 1974). In general, the elevation of the topographic feature examined is inversely proportional to the Bouguer anomaly value along the profile.

Since it is highly probable that future investigations in other areas will reveal similar conditions, the nature of ice-cored topography needs brief description. Positive features due to massive ice include pingos and involuted hills; negative features include the whole range of thermokarst phenomena which are discussed in the next chapter.

Involuted hills Aside from pingos which occur widely throughout the Mackenzie–Beaufort region and which are discussed separately, the most striking geomorphic features which owe their existence to the presence of massive icy beds are 'involuted hills' (Mackay, 1963a). Generally, these hills have relatively flat tops and steep sides and rise 15–50 m above the surrounding terrain. They can vary in size from 0.5 to 2.0 km^2 and are common to the east of Tuktoyaktuk. A unique characteristic is their surface appearance which '. . . resembles, on air photographs, the wrinkled skin of a well dried prune' (Mackay, 1963a, p. 138). The involuted nature is the result of 'curving to branching ridges, ranging up to several hundreds of yards in length, several score of yards in width, and 20 feet in height' (Mackay, 1963a, p. 138). According to Rampton (1974), 4.0–5.0 m of reworked clayey till or diamicton commonly overlie the massive ice that forms the core of the feature, and sands and gravels generally lie beneath.

The results of gravity profiling across an involuted hill near Tuktoyaktuk, N.W.T., suggest that excess ice accounts for 70 per cent of the surface relief and drill holes reveal the ice to be as much as 22 m thick at the crest of the hill (75 per cent of its height) (Rampton and Walcott, 1974, p. 118).

Other evidence indicating an ice core to the feature is the common occurrence of fresh and stabilised ground ice slumps (see p. 119) along the flanks of the hill. The exact cause of the involuted surface appearance is still rather puzzling but it may reflect the development and melting of ice wedges in the clay-diamicton overburden combined with slumping and mass movement.

Ice-cored terrain Gravity profiles undertaken in other areas of the Mackenzie Delta also indicate that icy sediments and massive ice bodies underlie much of the surface topography. For example, two profiles across shallow circular depressions developed in outwash plains indicated the depressions to be free of ice but that thick massive ice underlay the higher flanking terrain (Rampton and Walcott, 1974, pp. 119–20). This implies that the depressions are simple thermokarst features, the result of the melting of ice-cored terrain. On adjacent Richards Island, where hummocky and rolling topography resembling morainic topography occurs widely, gravity profiles indicated that greater amounts of excess ice were present under the hills than the depressions. In one profile thought typical of the topography of Richards Island, the gravity traverse indicated that the relief (30 m) was due solely to excess ice. It follows therefore, that if the ice were to melt, most of the terrain presently known as Richards Island would be below sea level, and that much of the present topography is in fact, complex thermokarst terrain brought about by the partial thawing of the underlying massive ice bodies.

The formation of ice-cored terrain of the extent described is not easy to explain. In particular, the conditions necessary for the availability and movement of vast quantities of water to the freezing plane are difficult to identify.

One possible explanation involves permafrost aggradation following the retreat of an ice sheet whose bottom temperature was at the melting point. Basal melting of the glacier would have supplied large quantities of water through the sands overlying the glacier terminus to the proglacial zone where the permafrost was beginning to aggrade. The necessary pressure for the continued water flow towards the freezing plane would have been provided by the glacier ice overburden pressure (Rampton, 1974, pp. 52–7).

The extent of ice-cored topography in other periglacial areas is not known. It is probably restricted to lowland areas underlain by thick unconsolidated sediments, such as occur in the western Arctic and in parts of northern and central Siberia, and where the proximity of Pleistocene ice sheets provided the vast quantities of water necessary for the excess ice.

Ice wedges

Their widespread occurrence and distinctive surface manifestation makes ice wedges one of the most characteristic features of the periglacial landscape (Fig. 5.3). By definition, they require permafrost for their existence. There are numerous references to ice wedges in the literature and the polygonal surface patterns which they form have been called tundra polygons, fissure polygons, ice-wedge polygons and Taimyr polygons. However, unless an ice wedge origin for the polygonal pattern can be proven, it is probably best to use the term 'fissure polygon' since some polygonal nets have been reported in which vein ice is insignificant (e.g. Péwé, 1969; Freidman ***et al.,*** 1971; see Chapter 3, pp. 23–7).

The size of ice wedges varies considerably from locality to locality, depending largely upon the availability of water and the age of the ice wedge. The majority are probably no more than 1.0–1.5 m wide near the surface (see Fig. 3.5), and extend into the ground for 3.0–4.0 m as somewhat irregular wedge-like ice masses. In parts of northern Siberia and in the lowlands of the western Arctic, ice wedges may attain considerable dimensions, often 3.0–4.0 m wide near the surface and extending downwards for 5.0–10.0 m (see Fig. 5.4). Some wedges extending downwards to depths in excess of 50 m and being over 10 m wide near the surface have been reported from Siberia (Dostovalov and Popov, 1966), but these are probably not typical of many areas. On the other hand, the ubiquitous presence of average sized wedges suggests that, in many northern areas, as much as 50 per cent of the total volume of the upper 3 m is occupied by ice wedges (Brown, 1966). Many Russian authorities regard wedge ice as the dominant type of underground massive ice (e.g. Popov, 1962; Shumskiy and Vtyurin, 1966; Dostovalov and Popov, 1966) in contrast to North American workers (see pp. 81–2).

The most favourable conditions for the current development of ice wedges are to be found in the relatively poorly drained tundra lowlands within the

Fig. 5.3 *Ice-wedge polygons developed on alluvial terrace, Thomsen River valley, central Banks Island, Canada. They are of the low-centred type with upthrust rims bordering the polygons.*

continuous permafrost zone. In the polar desert regions, in the high Arctic islands and in Antarctica in particular, ice wedges are not so well developed, a reflection of the increased aridity of these regions and/or an absence of fine grained silts and alluviums. In these areas, sand wedges assume dominance. South of the treeline in the northern hemisphere, discontinuous permafrost becomes more common and the climate generally less severe. Many of the large ice wedges known to exist in the boreal and taiga forest of southern Alaska, northern Canada and Siberia are probably largely relic features formed under colder conditions in the past (Péwé, 1966b).

In northern Siberia, ice wedges reach their greatest thickness and vertical dimensions on ancient alluvial surfaces or river terraces. In many cases, their growth appears to have paralleled the slow accumulation of sediments. Such ice wedges are termed 'syngenetic' (Dostovalov and Popov, 1966). In the North American Arctic, where similar geomorphic histories and terrain conditions are largely lacking and where ice wedge dimensions are generally less, the majority of ice wedges are 'epigenetic', i.e. they have formed after the deposition of the enclosing sediments.

Origin The contraction theory of ice wedge development is now widely accepted (see Chapter 3, pp. 21–3). The basic mechanism is illustrated in Fig. 3.6A. In summary, when soils are cooled to below −15 to −20°C there is a shrinkage

of the volume of ice held within the soil and this leads to the development of cracks and fissures in the frozen ground. The following spring, moisture collects in the fissure and freezes, preventing the crack from closing as the ground expands as the temperature rises. The process is self perpetuating since the ice vein provides a plane of weakness which is reopened under stress the following year, allowing the deposition of a further layer of ice.

As a result, ice wedges are characterised by vertical foliation and faint banding caused by the repeated ice accumulation and the washing in of fine dirt particles in the spring with the thawed water. Several types of foliation can be identified, depending upon the nature of the ice wedge. Epigenetic ice wedges have foliations which extend the length of the ice wedge, while syngenetic wedges have foliations which often end within the wedge. Several gradations between these two types can be recognised. Often, the stratification of the enclosing materials is deformed and tilted upwards at the junction with the ice wedge. This is on account of the summer expansion of the ground and the resistance of the ice vein. In certain instances, this leads to the development of two distinct ridges or walls on either side of the fissure at the ground surface with a central depression running along the line of the crack (Fig. 5.3).

According to Lachenbruch (1962), a rapid drop in sub-zero temperatures is probably the best climatic environment for the development of thermal contraction cracks. The time of year when this condition is at its optimum is in the early winter and there are various reports that the cracking of the ground at this time of the year is an audible process. Black (1952; 1960) has examined ice wedge cracking in Alaska and has concluded that the present rate of growth of the ice wedges there is of the order of 0.5–1 mm per year. Using these figures, and bearing in mind the average size of ice wedges in that area, Black concluded that the majority of ice wedges (and therefore surfaces) are between 1 000 and 4 000 years old in the vicinity of Point Barrow. Similar studies in Antarctica for both sand and ice wedges over a 6–8 year period (Black, 1973) indicate net growth rates of between 0 and 1.5 mm/year and an age for the ground surface of less than 5 000 years.

J. R. Mackay has also carried out repeated observations on ice wedge cracking at Garry Island, in the outermost part of the Mackenzie Delta (Mackay, 1974). Although maximum annual growth over a six-year period was approximately 2 mm, most wedges grew far less than 1 mm. This was because not all wedges cracked each year; Mackay found that nearly 40 per cent of the ice wedges under observation cracked each year. Statistically, ice wedges that crack in any given year are more likely to crack the following year, and most cracks reoccur within 5–10 cm of the previous year's crack. He also observed that very few cracks are continuous for more than 3–4 m, and that the frequency of cracking is inversely related to snowcover.

The development of the polygonal net Ice wedges join predominantly at right angles and form polygonal, chiefly tetragonal, nets of patterned ground that cover extensive areas of the Arctic and subarctic. Commonly, the average dimensions of the polygons range from 15 to 40 m. They may also possess a complex subsurface structure or texture. Syngenetic wedges may be joined by horizontal sheets of ice from which further ice wedges extend downwards, thus producing a honeycomb pattern of vein ice distribution (Dostovalov and Popov, 1966).

The development of the polygonal ice wedge pattern has been theoretically examined by Lachenbruch (1962; 1966). He concluded that the angular intersection of a polygonal network of frost cracks will exhibit a preferred tendency towards an orthogonal (i.e. right angle) pattern (Fig. 5.5). However, this conclu-

Fig. 5.5 *Types of polygonal nets in permafrost terrain*

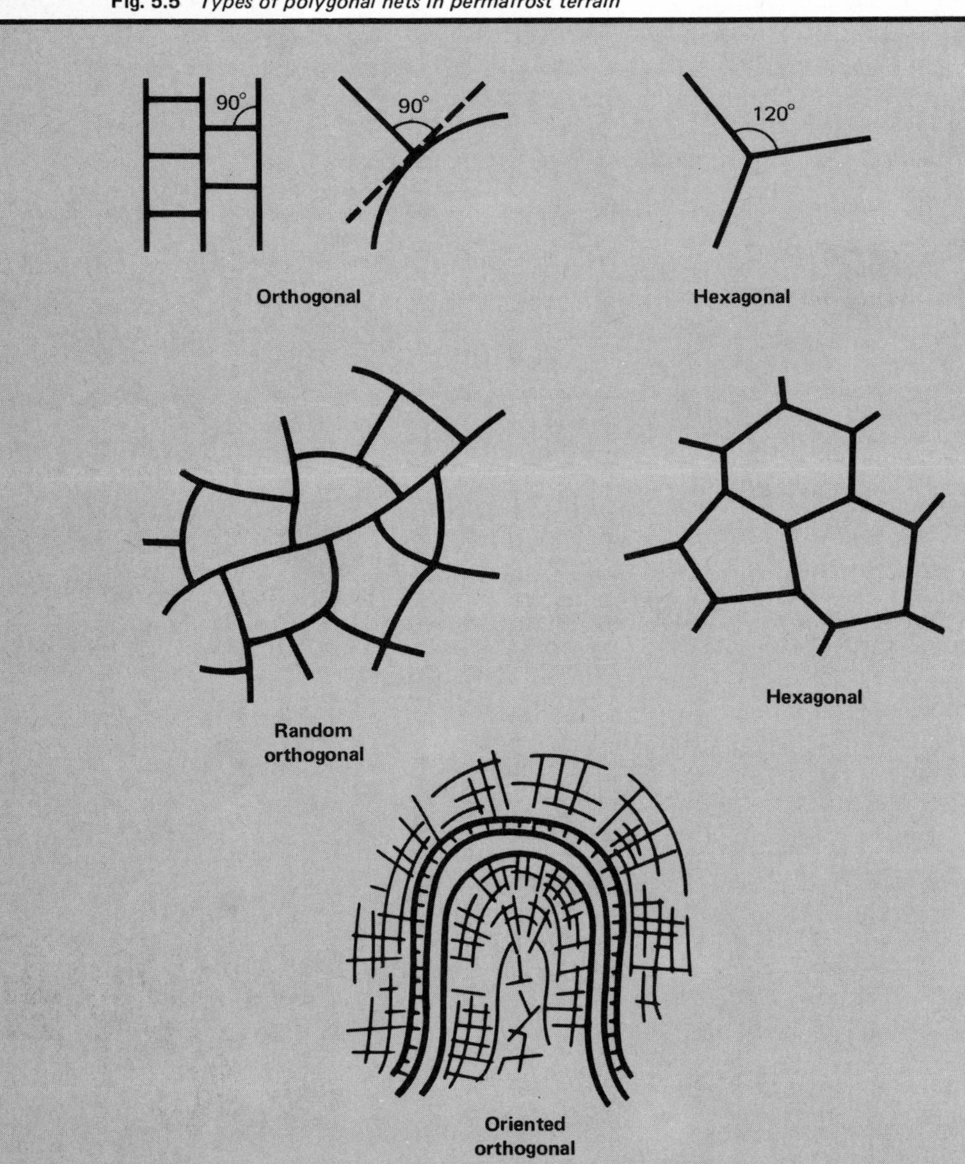

sion contrasts with the many descriptions of polygonal ground in which the authors have described a tendency for hexagonal or angular junctions to dominate (e.g. Leffingwell, 1919; Black, 1952). The implications of the hexagonal pattern, and angular intersections of 120 degrees (see Fig. 5.5), are that the frost cracks develop at a series of points and that each crack develops more or less simultaneously. On the other hand, the orthogonal pattern is thought to indicate an evolutionary sequence in which primary frost cracks develop in an essentially random pattern. These are then followed by secondary frost cracks which progressively divide up the area and which have a tendency for an orthogonal intersection pattern. Lachenbruch (1966) classifies the resultant crude polygonal network as a 'random orthogonal system' (Fig. 5.6) in contrast to an 'oriented orthogonal system' (Fig. 5.7) which develops in the vicinity of large bodies of water. In the case of 'oriented orthogonal systems', one set of cracks develops normal to the water body and the other at right angles. The apparent dichotomy between the theoretical tendency towards an orthogonal pattern and existence of hexagonal patterns in reality is explained by Lachenbruch as being the result of the obscuring of the initial orthogonal intersection pattern by the growth of large ice wedges and the upthrusting of the adjacent sediments. Alternatively, it has been suggested that the hexagonal

Fig. 5.6 *Oblique air view of ice-wedge polygon terrain in central Banks Island, from an elevation of 200 m. The pattern is predominantly a random orthogonal one. The complete range from low-centred polygons with raised rims and depressed centres to high-centred peaty mounds can be observed.*

Fig. 5.7 *Oblique air view of oriented orthogonal ice-wedge pattern in an abandoned channel, Mackenzie Delta, Canada. As the water progressively shrunk in the channel, ice wedges developed normal to the retreating water line, to be followed by wedges parallel to the water line.*

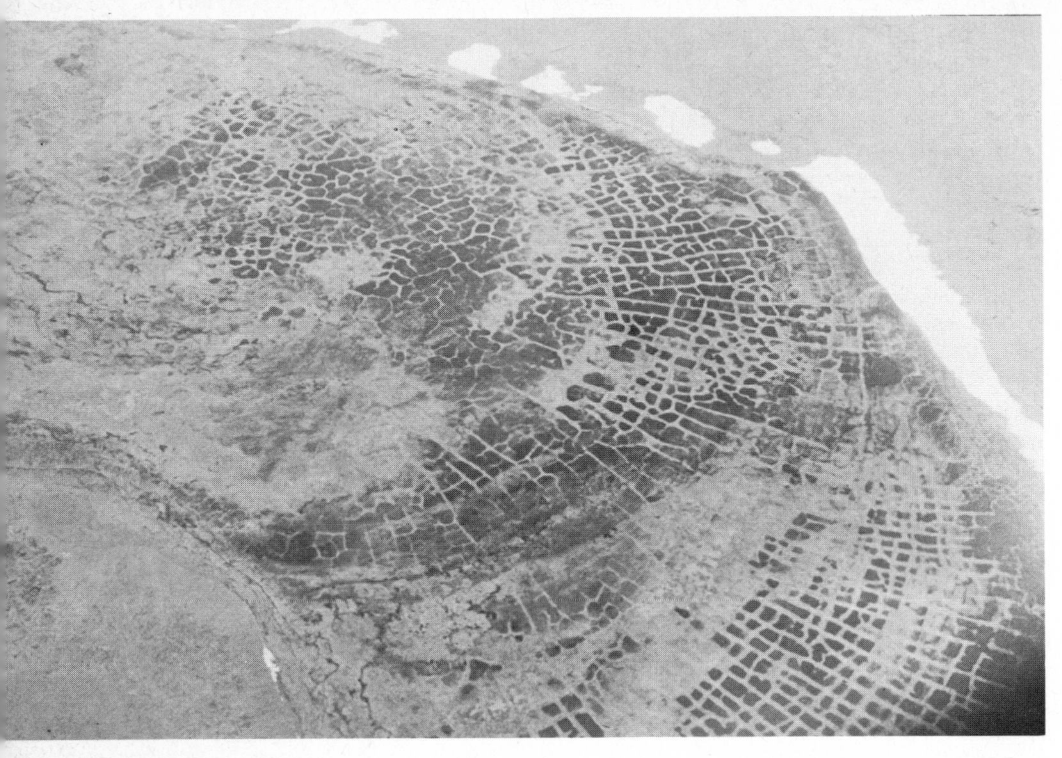

pattern of ice-wedge polygon nets is best developed in homogeneous materials subject to uniform cooling, while the orthogonal pattern develops in heterogeneous materials.

The polygonal network of ice-wedge bodies, particularly syngenetic ice-wedge growth, has also been the subject of much Russian investigation. According to Dostovalov (1960; Dostovalov and Popov, 1966) the size of the polygonal net reflects the severity of the climate. When temperatures are relatively warm large rectangular polygons are formed by primary and secondary fissures. However, with increasingly severe winter temperatures, these rectangles are successively broken up by fissures of consecutively higher orders into smaller and smaller bodies (Fig. 5.8). As such, the probability of repeated cracking is smaller the higher the order of fissure generation; in relatively warm winters, only fissures of low order are formed, but in cold winters, cracking occurs in the higher orders of fissure generation. Therefore, the thickness of ice wedges growing simultaneously but formed by fissures of different orders of generation will be different; the higher the order of fissure, the smaller is the ice-wedge thickness. The advantage of such a hypothesis is its simplicity. On the other hand, until detailed quantitative measurements, similar to those by Mackay on Garry Island, are undertaken in a variety of different climatic environments,

Fig. 5.8 *The successive growth of frost fissures of higher orders and the development of the polygonal net according to B. N. Dostovalov.*

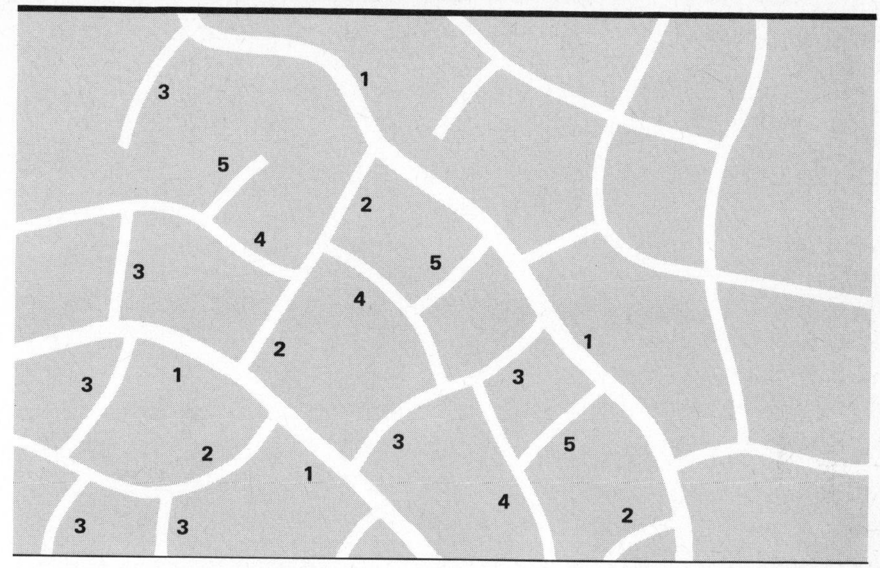

there is no proof that either a hierarchy of fissures exists or that cracking is related solely to the severity of the climate.

Climatic and palaeogeographic significance The formation of ice wedges is of climatic and palaeoclimatic significance: (a) they require the presence of permafrost and (b) they can only form when air temperatures drop well below 0°C. Based upon the distribution of active ice wedges in Alaska and elsewhere, Péwé (1966b) has concluded that ice wedges only form in areas where the mean annual air temperature is −6°C or colder. In Alaska for example, it is clear that active ice wedges only occur in the zone of continuous permafrost. Those ice wedges found in the zone of discontinuous permafrost are generally inactive. Their existence today reflects the present climate which is not mild enough to destroy them and implies the previous existence of colder conditions than at present. It follows that the recognition of ice-wedge casts in temperate latitudes today is one of the few really reliable indicators of past permafrost conditions and low mean annual air temperatures.

Certain Russian investigators believe that the shape and distribution of syngenetic ice-wedge bodies reflect palaeogeographic and geotectonic conditions. To appreciate this, the mechanism of growth of syngenetic ice wedges must be considered (Fig. 5.9). According to Dostovalov and Popov (1966), the growth of an ice wedge and the shape of its cross section will depend upon the ratio of the depth of cracking, the width of repeated fissures, and the rate of sediment accumulation. In syngenetic growth, the ice wedge grows in association with the accumulation of sediments. If the mean depth of the fissure remains constant, the growth in the width of the wedge must cease at some stage (Fig. 5.9, stage III). After that, the wedge grows vertically. In cases of slow sedimentation, the limit of ice-wedge thickness increases and is reached at a rather later stage than

Fig. 5.9 *Schematic diagram to show the mode of formation of syngenetic ice wedges, according to Dostovalov and Popov (1966)*

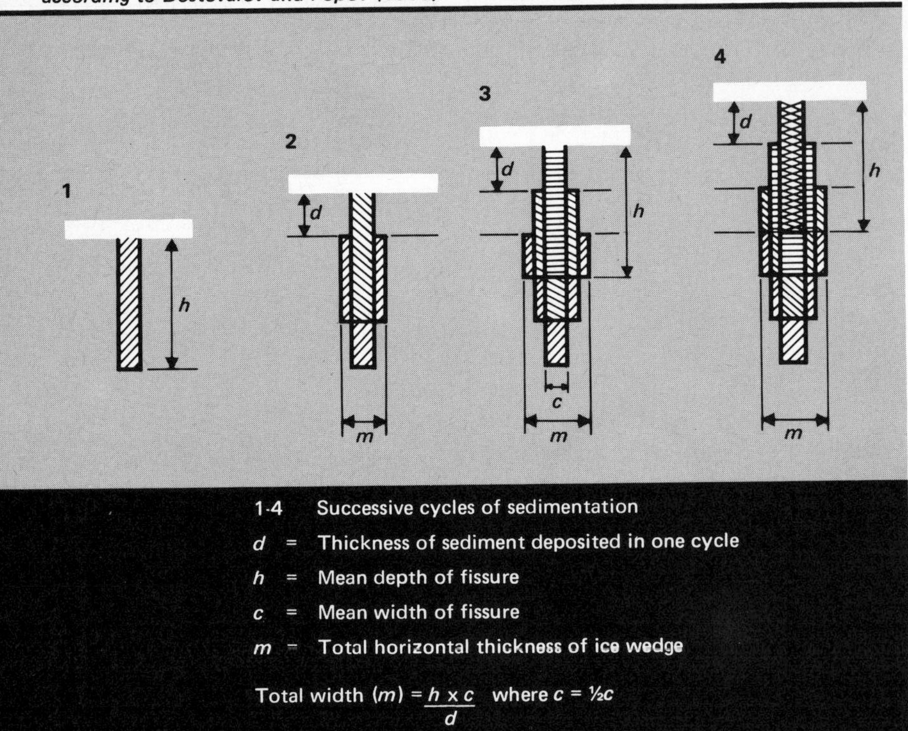

in the previous situation; however, since the growth of the ice-wedge width is limited, the compressive stress developed during its growth is also limited and, in fact, is progressively moved to the upper part of the ice wedge. Thus, in contrast to epigenetic growth, syngenetic growth is capable of producing ice wedges of quite considerable dimensions, especially on flood plains and other aggradational surfaces with well developed sedimentation cycles.

If syngenetic growth takes place during a cooling climatic environment, such as at the onset of a glacial period, the polygonal ice-wedge net must become smaller due to the generation of fissures of higher order (Dostovalov, 1960). With continued surface aggradation, these smaller ice wedges will occur at successively higher elevations within the sediments. During a warming of the climate on the other hand, the reverse happens; the smaller ice wedges in the fissures of higher order cease to grow, and only the larger, first and second order ice-wedge fissures remain active. The latter continue to grow upwards in association with the continued sedimentation. For example, in the Yana—Indigirka lowlands of northern Siberia, the width of many ice wedges decreases in the younger sediments near to the surface (Popov, 1969). This is interpreted to reflect a general amelioration of temperatures following the Middle Pleistocene epoch, together with an increase in the continentality of the area. The latter is thought to be related to the progressive northward migration of the Arctic coast of Siberia consequent upon long continued coastal sedimentation and aggradation. It would appear that the increase in continentality resulted in a decrease in the moisture available for ice-wedge growth, and that warming temperatures concentrated ice-wedge growth only in the fissures of lower order.

It is debatable whether syngenetic ice-wedge growth, as described from the lowlands of Siberia, is applicable to many other areas, particularly the North American Arctic. Few areas have experienced the same long and uninterrupted periods of alluvial sedimentation and aggradation during the Pleistocene as did Siberia. It may be that the Siberian Arctic and subarctic coastal plains and interior lowlands are the only areas where terrain conditions, vast rivers and Quaternary history are favourable for their development. In the western Arctic for example, the majority of the field evidence points to epigenetic ice-wedge growth as being more typical, even in the aggrading alluvial environment of the Mackenzie Delta.

Pingos and other ice-cored mounds

A pingo is an ice-cored hill which has been domed up from beneath by either the intrusion of water under pressure which freezes or by the growth of segregated ice lenses. The word, which in Eskimo means a hill, was first used by A. E. Porsild (1938) in describing the pingos of the Mackenzie Delta of Canada. Subsequently, the term has been used to describe features in east Greenland, Alaska and elsewhere. In Siberia, the term 'Bulgannyakh' is used.

The genetic term 'hydrolaccolith', which applies to all ice intrusions, is not to be confused with the current usage of the term 'pingo', since the latter is restricted to perennial intra-permafrost features which are different from seasonal frost mounds, peat hummocks and other smaller ice-cored mounds (Muller, 1968). Moreover, many workers now assign a greater role to segregation ice, particularly in the middle and later stages of growth of a pingo, than to injection ice (e.g. Mackay, 1973, p. 999).

Pingos vary from a few metres to over 60 m in height and up to 300 m in diameter (Figs 5.10–5.11). They possess a variety of forms ranging from symmetrical conical features, to asymmetric and elongate features. It cannot be assumed that all pingos have a recognisable and typical form. Their one common characteristic, often concealed by 1.0–10.0 m of overburden, is a core of massive ice or icy sediments. This may be remarkably pure and with an absence of internal structures, or it may be layers of icy sediments. Fractures and faulting are sometimes seen within the pingo core. Frequently, the pingo is ruptured at the top to form a small star-like crater, and a small lake may develop in the crater from the melting of the ice core. This is the first stage in the decay of a pingo, the melting of the ice core, the collapse of the updomed sediments, and the ultimate formation of a shallow rimmed depression (Fig. 5.12).

Within a periglacial landscape of low relief dominated by mass wasting, a well developed pingo is a striking geomorphic form. It is probably on account of this that the pingo has received so much attention in the literature, and why pingos which occur within more hilly environments have not received comparable attention. On the other hand, although they represent a classic periglacial phenomenon, they should not be regarded as typical features of all periglacial regions. In fact, their existence and development is usually the result of a number of distinctive and limiting geomorphic and hydrologic conditions. This should be borne in mind in the following discussion. Pingos are also important from a palaeoclimatic viewpoint since they, like ice wedges, clearly demand the presence of frozen ground for their formation.

According to Muller (1968), the distribution of pingos is confined approximately to latitudes 65–75°N, in those areas where permafrost is either discontinuous or alternatively, continuous but thin. However, pingos are known to occur in areas of thick permafrost and, as more information becomes available upon the variety of pingo-like forms that also exist, it is probably unwise to state definitive limits for pingo distribution at present. On the North American continent, the greatest concentration of pingos, over 1 440, is to be found in the western Arctic coastal plains (Mackay, 1962), but significant numbers are also found in Alaska (Holmes *et al.,* 1968), the Yukon Territory (Hughes, 1969), the western Arctic islands (Fyles, 1963; Pissart, 1967b; French, 1975a), and in Greenland (Muller, 1959; Cruickshank and Colhoun, 1965; O'Brien, 1971). In the USSR the various forms of intrusive ice have been mapped by Shumskiy and Vtyurin (1966). Perennially heaved mounds (i.e. bulgannyakhs) are widely distributed throughout the Arctic and subarctic mainland of Siberia. Major concentrations occur in central Yakutia, in the coastal regions adjacent to the deltas of the Ob, Lena and Indigirka Rivers, in the Chukchi Peninsula, and in the upper Amur River basin, in Chitinsk Oblast. No details are available for adjacent parts

Fig. 5.10 *Large bulgannyakh (pingo) in Olong Erien alas, central Yakutia, USSR.*

Fig. 5.11 *Closed system pingo developed in shallow lake basin in the Mackenzie Delta, Canada.*

of northern China and Mongolia, but it is probable that they occur in these regions also since permafrost is known to exist.

The formation and growth of pingos and pingo-like forms are varied and still not clearly understood. As more information upon their distribution becomes available, it is becoming clear that a wider range of conditions exist for pingo growth than was first thought. The presence of obviously relic pingos (e.g. Pissart, 1967b) in present-day periglacial environments complicates attempts to identify the conditions for present-day pingo growth. There is also debate concerning the mechanism of growth and, in particular, as to the origin of the ice which forms the pingo core; traditionally, this has been classified as 'injection ice' frozen from a pool of water under pressure (e.g. Muller, 1959; Shumskiy, 1964). Recent studies by Mackay (1973) suggest that all gradations between pure injection ice at one extreme to icy sediments of a segregated nature at the other may constitute the core of a pingo. In general terms, however, one may distinguish between 'open system' and 'closed system' pingos, a distinction first proposed by Muller (1959) and widely adopted subsequently.

Open system pingos Open system pingos occur in areas of thin or discontinuous permafrost where surface water can penetrate into the ground and continue, as sub- or intra-permafrost water, circulating in unfrozen sediments. Where this water rises towards the surface it freezes to form localised bodies of

Fig. 5.12 *Oblique air view of a collapsed pingo, Sachs River lowlands, southern Banks Island, Canada.*

ice which force the overlying sediments upwards (Fig. 5.13). Thus, open system pingos usually develop in distinctive topographic situations, such as in valley bottoms or in lower valley side slopes, where hydrostatic head or pressure can develop sufficiently to provide a constant and regular water supply to the freezing plane. Gravels and sands and other porous and permeable materials are favourable for this process.

In general, open system pingos occur as isolated features, or as small groups within the same locality. It is not uncommon for a new pingo to develop inside the crater, or on the flank of an older one. Many are oval or oblong in form, and a high proportion of open system pingos examined in the field are ruptured to varying degrees. The largest known concentration of open system pingos occurs in the unglaciated areas of central Alaska and the Yukon where over 700 have been identified (Holmes ***et al.,*** 1968; Hughes, 1969). Open system pingos are also reported from Greenland (Muller, 1959) and they are sometimes referred to as 'East Greenland' pingos. It is probable that many of the perennial frost mounds (bulgannyakhs) described by Soloviev (1973a; 1973b) in central Yakutia are also open system features. The fact that nearly all open system pingos lie in unglaciated terrain is an interesting aspect of their distribution which is not clearly understood. In North America, they are extremely rare in areas glaciated within the last 25 000 years; none are known in Alaska and only two are known in the Yukon Territory (Hughes, 1969). Differences in the extent and thickness of permafrost in glaciated and unglaciated terrain may be a factor.

Hydrologic conditions and the role of artesian pressures in the formation of

Fig. 5.13 *Open system pingos.* **A** *– Schematic diagram illustrating growth of open system pingos. From Muller (1968).* **B** *– Nature of occurrence and geomorphic situations of open system pingos in Alaska. From Holmes et al., (1968)*

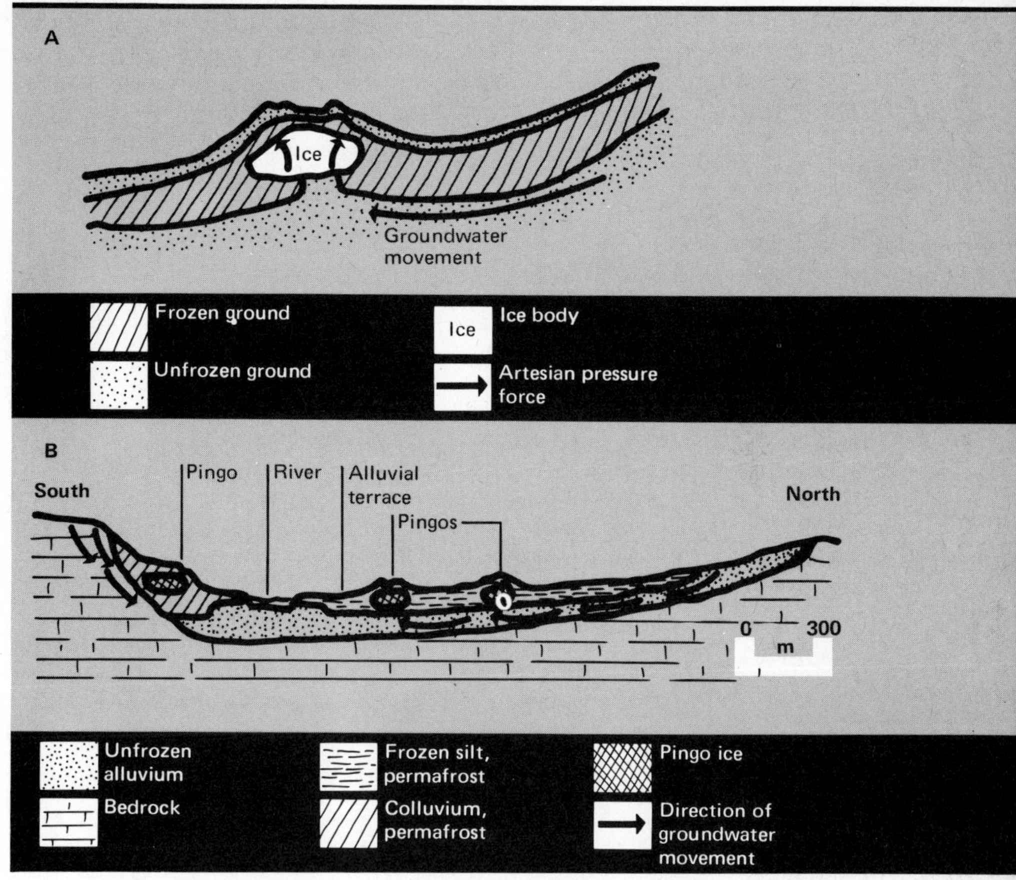

open system pingos has been discussed earlier (Chapter 4, pp. 67–8). Some further comments are in order concerning the growth mechanism of open system pingos. Since the role of artesian pressure is not to force the overlying sediments upwards but merely to ensure a steady but slow supply of groundwater, it follows that the development of open system pingos by 'injection' with the core of the pingo being composed solely of 'injection ice' (Muller, 1959) is not necessarily the only growth mechanism. Mackay (1973, p. 1000) has argued that pingo growth from injection ice, which requires a constantly replenished pool of water beneath the ice core, would represent a very unstable condition. If water were injected faster than it could freeze, pressure would rise until the pingo ruptured and water flowed from the ground as a spring. Equally, if water were injected slower than it could freeze, the unfrozen water pool would freeze and cease to exist. In fact, an unlikely long term balance would be required between three independent variables; water pressure which is determined by conditions external to the pingo, the overburden strength which would vary with time of year, and the rate of freezing which would depend upon temperature. Since all three may change independent of the others, this balance will rarely be maintained for the total growth lifetime of a pingo. This implies that the growth of an open system pingo probably requires a certain amount of segregated as well as injection ice. In Yakutia, where both 'flat' bulgannyakhs (dome-like elevations 2–5 m high) and upstanding large bulgannyakhs (hills 10–50 m high) exist in close juxtaposition (e.g. Olong Erien alas near Abalakh), it is also thought that the ice core is of several origins, produced both by injection from groundwater under pressure below and by segregation (Soloviev, 1973b, pp. 148–51). According to Soloviev, the flat bulgannyakhs are primarily of a segregated nature, composed of icy sediments, while the larger forms possess massive ice cores 5–10 m thick formed through the repeated injections of water. The Yakutian bulgannyakhs occur within alas depressions, 20–30 per cent of which have bulgannyakhs on their floors, and their evolution is directly related to the development of the alas thermokarst relief; this is discussed more fully in Chapter 6 (pp. 111–15).

Closed system pingos Closed system pingos occur almost exclusively in areas of continuous permafrost, often in alluvial lowlands with little vertical relief. For the most part, they occur in specific geomorphic situations and are brought about by the local downward aggradation of permafrost into a previously unfrozen zone. The largest concentration (over 1 350) of closed system pingos occurs in the Pleistocene coastal plain of the Mackenzie Delta of Canada (Mackay, 1962), and closed system pingos are sometimes referred to as 'Mackenzie Delta' type pingos. Others occur in the modern Mackenzie Delta, the Yukon coastal plain, western Victoria Island, and southern Banks Island. Isolated pingos, or small groups have also been reported from the District of Keewatin, Baffin Island and other areas. Conical mounds discovered on the floor of the Beaufort Sea have also been interpreted as submarine pingos, possibly of a closed system origin.

In the Mackenzie Delta, closed system pingos typically occur within small shallow lakes or former lake beds (Fig. 5.11). They usually occur singly and not in groups, although some drained lake beds are known in which three or four small pingos are growing (Mackay, 1973). The specific site of pingo growth is that of a shallow residual pond, usually less than 2 m in depth. The final size and shape of the pingo reflects that of the residual pond in which it grew.

According to Mackay (1962; 1973), a closed system pingo develops when permafrost advances into a shallow lake basin which is either being drained or is infilling (Fig. 5.14). The freezing of the surface layers produces a closed system of supersaturated and unfrozen soil beneath the lake. As the freezing plane advances in the unfrozen zone, pore water is expelled in advance of the freezing plane (Mackay, 1973, p. 982). This expelled pore water provides the continuing water supply and hydrostatic pressure necessary for pingo growth. The growth of the ice then leads to the updoming of the surface sediments and the formation of the pingo.

The growth of several closed system pingos in the Mackenzie Delta has been measured by Mackay (1973) using precise levelling techniques for the 1969–72 period, and air photographic evidence before then. In one locality 20 km southwest of Tuktoyaktuk, rapid coastal retreat has resulted in the drainage of a lake and the subsequent growth of four small pingos in the last 35 years. Air photo-

Fig. 5.14 *Nature and growth of a closed system pingo in the Mackenzie Delta, Canada. From Mackay (1972a)*

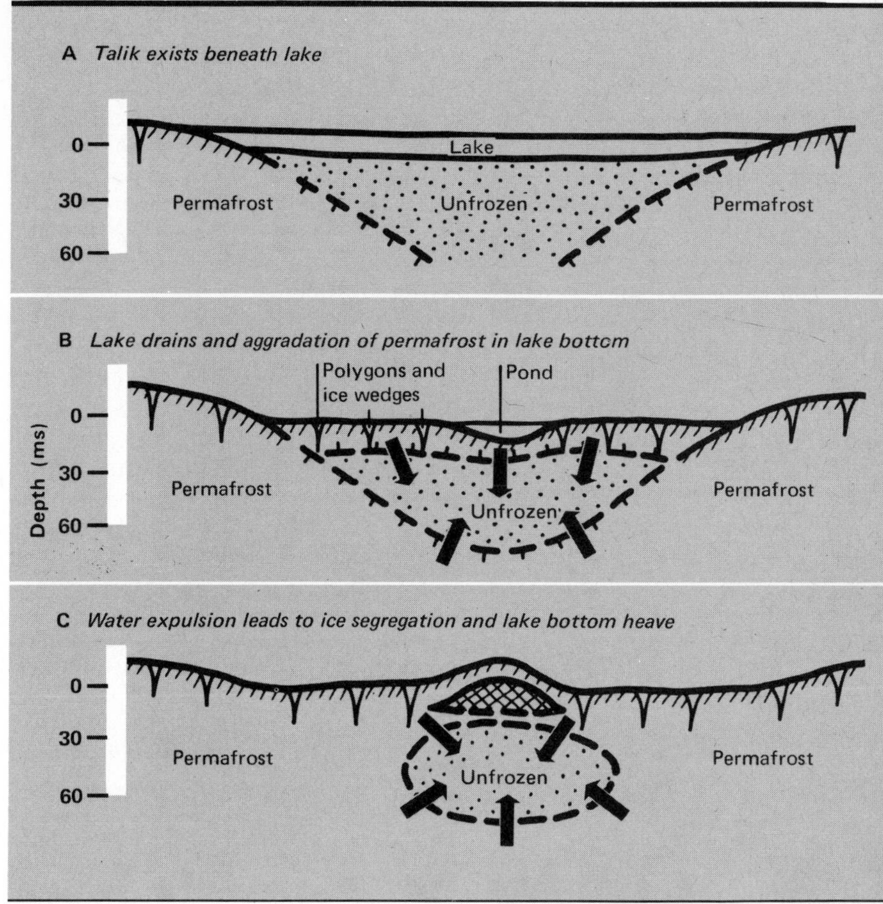

graphs taken in 1935 show the lake to be intact. By 1950, the lake had been drained and four small residual ponds were left on the floor of the lake basin. By 1967 four small pingos had grown precisely in the centres of the residual ponds, and by 1971, fresh looking tension cracks had developed on two of the pingos (Mackay, 1973, pp. 989–90).

Mackay has concluded that a pingo growth rate of 1.5 m/year over the first one to two years of growth may be typical but that the rate decreases rapidly with time. In general, pingos probably attain their maximum dimensions rather early, after which they grow higher rather than wider, with the tops growing faster than the sides, i.e. the growth of excess ice is concentrated beneath the centre of the pingo. In the Mackenzie Delta, the largest pingos probably took over 1 000 years to form. As a rough estimate, Mackay suggests that possibly 15 pingos may commence growth in a century, and that there are probably 50 or more actively growing today. Similar conclusions have been reached by Russian investigators, although quantitative observations on bulgannyakh growth are lacking. In Siberia for example, some of the open system bulgannyakhs are known to have developed in recently drained alas depressions in the last 50 years. Eye witness accounts indicate growth rates in the beginning stages to be of the same order of magnitude, 0.5–2.0 m/year (Soloviev, 1973a).

Significant advances have also been made in the theoretical understanding of the actual growth process of closed system pingos. The ice in a pingo core has been traditionally classified as 'injection ice' frozen from a pool of water injected under pressure. It is now believed that the majority of pingo growth is by ice segregation rather than by the freezing of a pool of water. In the early growth stages, free water may be injected faster than it can freeze, and injection ice can form. However, as Mackay has argued (see above), such a condition is unstable and unlikely to be maintained for the duration of the pingo growth. The source of water and the associated positive pore water pressures which favour ice segregation and lensing, is the result of permafrost aggradation and the expulsion of pore water ahead of the freezing plane. Moreover, it is likely that, at different stages of pingo growth, different types of ice formation will assume greater or lesser importance. For example, Mackay (1973) has illustrated theoretically how, for a constant overburden pressure which is a function of pingo height, an increase in pore water pressure from a low to high value can probably cause a change from the freezing of pore ice, through segregated ice, to injection ice, and finally rupture of the pingo. Several stages of pingo growth can be recognised. In the pore ice stage the ice core does not form. Instead, as the pore ice freezes, the entire lake bottom heaves slowly upwards because of the 9 per cent volumetric expansion of water freezing to ice. When the pore water pressure equals or exceeds the overburden pressure, segregated ice tends to form and pingo growth commences. This is a segregated ice stage. As pore water pressures continue to increase, the overburden will eventually yield and water will be injected faster than it can freeze. This is a third stage or an injected ice stage. Finally, if water continues to be injected faster than it can freeze, the pingo will rupture. It follows that all types of ice may be present in a pingo core, and that it need not be pure ice; at depth, soil and ice lenses may be intermixed. It is also clear that pingo growth is merely one example, although a dramatic one, of frost heaving of the ground.

Pingo-like features In addition to the open and closed system pingos as described above, it is possible that several other types of ice-cored hills may exist. Both Muller (1959) and Mackay (1962, p. 22) mention pingos which belong to neither of the two categories already outlined. In particular, Pissart (1967b) has described pingo-like forms on Prince Patrick Island which are of interest since they clearly demand very different explanations.

In one category are over 100 pingos which exist in a broad north–south zone along the summit of the island. They are far from any lakes or other surface water bodies, and since the island is underlain by thick and continuous permafrost, they are of neither the open or closed system types. Generally the mounds have an approximately circular plan and are of relatively small dimensions, varying in height from between 1 and 13 m, and with average dimensions of about 60 m. Pissart believes their location is related to faults in the bedrock some 50 m below the sands and gravels of the Beaufort Formation which covers much of the island. But, in the absence of detailed geophysical research, the exact mechanism of their formation is unclear. Since the permafrost is probably 300–400 m thick, the hypothesis that sub-permafrost water was able to move through the permafrost by ways of the faults without freezing en route is not convincing. If it is accepted that the thick permafrost does indeed constitute a barrier to the movement of water from depth, it may be that the features developed prior to the appearance of the barrier, perhaps early in the Wisconsinan.

In a second category are approximately 50 pingos or pingo-like features which occur close to sea level, either on low coastal terraces, or in present-day alluvial flats. In the Satellite Bay area, these pingos possess a variety of forms ranging from a classical conical form with crater, approximately 3 m high and 50 m in diameter, to long elongated forms. Of the latter, the most striking is comparable to a small esker, some 1 300 m in length, between 40 and 70 m in width, and less than 9 m high. A distinctive characteristic is a central depression aligned along the major axis of the ridge. This is interpreted as being the line of axial rupture which has allowed the ice core to melt partially. To explain this feature, Pissart suggests that a closed system situation developed at some previous time along the line of an old valley which has been inundated by a rise of sea level and infilled with near shore sands. More recent sea level lowering led to the exposure of the sediments and the progressive freezing of the saturated sands. This lead to the injection of water under pressure towards the surface, the updoming of an elongated ridge, and the rupture of the summit.

A number of similar elongated and partially collapsed ice-cored hills are also found on adjacent Banks Island (Pissart, 1975; French, 1975a). They rarely exceed 15 m in height and occur predominantly on the low fluviatile terraces of the major rivers in the interior lowlands. It is hypothesised that the majority of these features grew, in an unstable fashion, through the freezing of local taliks which had developed beneath the deeper parts of old river channels.

The recognition of pingo-like features in areas other than classic pingo localities emphasises the variety of pingo forms and origins. It is clear that, in certain instances, neither a closed nor open system hypothesis is appropriate. It is also clear that closed system pingos require specific and localised conditions for their growth, and that the variety of such conditions is not yet fully appreciated. It is highly likely that the majority of pingos, particularly closed system pingos, which are to be found today in periglacial environments are inactive and developed at some time in the past. Climatic and terrain conditions at that time may, or may not, have been similar to the prevailing conditions. Finally, the variety of forms and sizes, and the fact that pingos may occur in clusters or in isolation makes the recognition and interpretation of relic Pleistocene pingos in temperate latitudes difficult.

Hydrolaccoliths and frost mounds A large number of smaller mounds with ice cores or ice lenses beneath them exist in present-day periglacial environments. They are usually annual or short-lived perennial features (e.g. see Maarlveld, 1965; Lundquist, 1969). By definition, these features are not regarded as pingos, mainly on account of their size, their transitory nature, and their location primarily within the active layer. The variety of features suggests that they are not all the same and may be of different origins.

For example, in the USSR, the term 'bugor' has been used to describe small gently rising and oval-shaped mounds which occur in the subarctic and tundra regions of Siberia (Dostovalov and Kudryacev, 1967). They occur in groups or scattered, ranging in height from 5 to 10 m, and varying in length from 100 to 500 m and in width from 50 to 80 m. Analogous features may occur in the North American Arctic. For example, Porsild (1955), Pissart (1967b) and the writer have observed low circular mounds scattered over the western coastal plains of both Prince Patrick Island and Banks Island. They rarely exceed 2 m in height, and are usually between 15 and 50 m in diameter. They are frequently used as owl perches and stand out as relatively dry sites, cut by tundra polygons and with a lichen and/or grass surface cover. The origin of these bugor-like features is not clear since they show no apparent relationship to topography. According to Porsild, these mounds are the result of localised ice injection or ice segregation which occurs beneath silty and/or peaty beds occasionally found within the sands and gravels of the underlying Beaufort Formation.

Even smaller hydrolaccoliths are almost certainly of a segregational origin. Bird (1967, p. 203 and Plate 38) has described mounds of clear ice covered with vegetation or peat, less than 4 m in diameter and 1 m high, which occur on Southampton Island, and Sigafoos (1951) has described slightly larger features from the Seward Peninsula of Alaska. In both cases, the peaty soils have favoured the development of a lens of clear ice immediately beneath the

Fig. 5.15 *Small hydrolaccolith associated with low-centred ice-wedge polygons, Masik Valley, southern Banks Island. The mound is composed of a body of pure ice immediately beneath the organic mat.*

Fig. 5.16 *Suggested cycle of growth of seasonal or short-lived perennial ice cored mound. From French (1971b)*

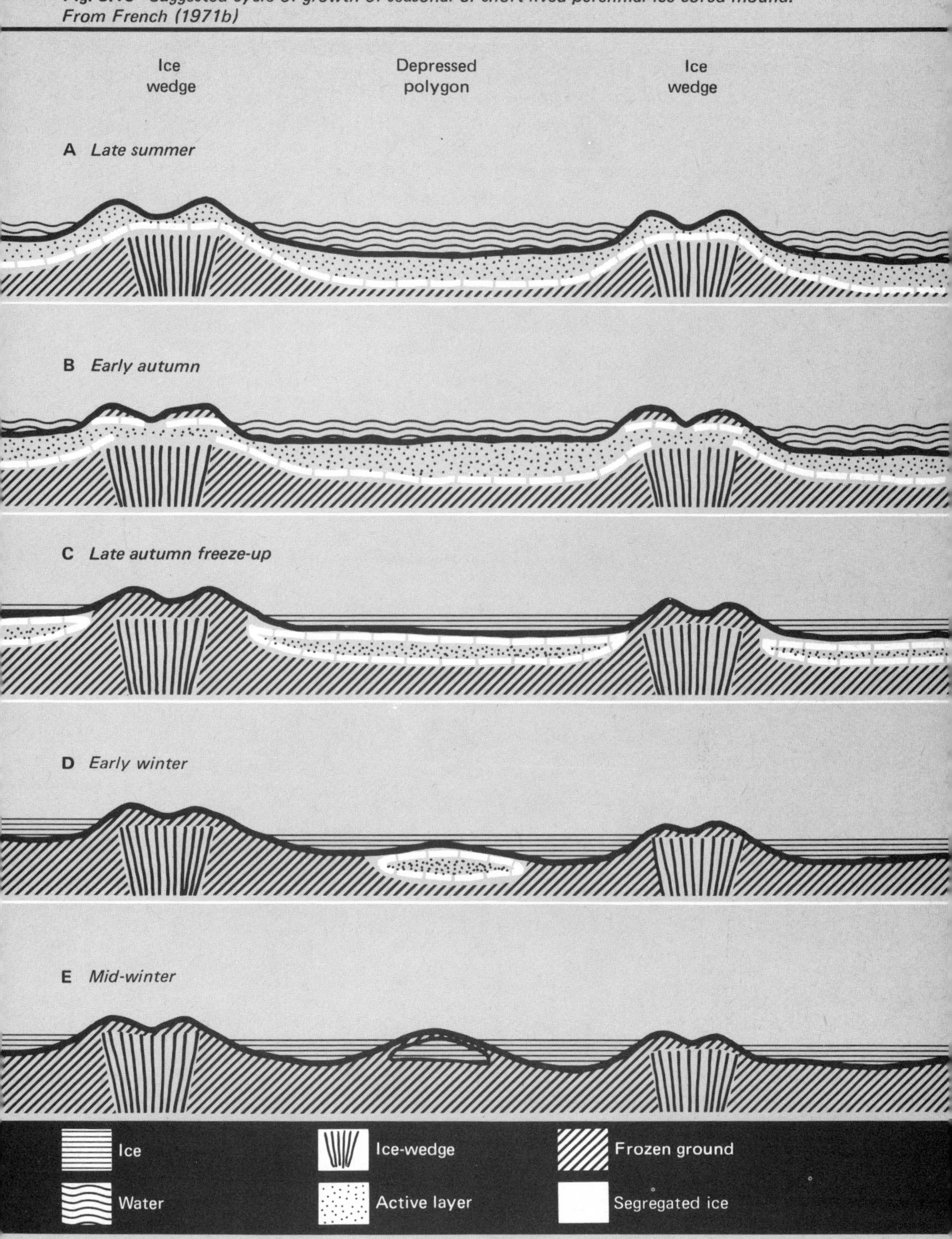

vegetation sod. Slightly different features have been observed by the writer in the Masik Valley of southern Banks Island (French, 1971b). Small hummocks, composed of almost pure ice lenses, occur within the centres of well developed low centred ice-wedge polygons (Fig. 5.15). The surrounding terrain is poorly drained and in the summer months water covers the centres of the polygons. The hummocks are thought to be either seasonal or short-lived perennial features resulting from localised ice segregation. This occurs during freeze-up each year when there is a contraction of the water saturated active layer under cryostatic pressure. The upstanding ice wedges which band the polygons create a closed system as soon as the downward freezing of the ground at the ice-wedge surface meets the permafrost table (Stage C, Fig. 5.16). There is some similarity, therefore, between these features and the closed system pingo, as regards origin and mode of development.

6 Thermokarst

The term 'thermokarst' was first proposed by the Russian, M. M. Ermolaev, in 1932, to describe irregular, hummocky terrain due to the melting of ground ice. This terrain appeared similar to the sinkhole and karst plain topography of limestone regions. Subsequently, the term has been applied specifically to the process of ground ice melting accompanied by local collapse or subsidence of the ground surface (Popov, 1956; Kachurin, 1962). Although Dylik (1968) argued initially to reserve the term for the melting of underground ice as opposed to buried glacier or surface ice, it is now accepted that the word 'thermokarst' applies to the melting process of all ground ice bodies, irrespective of origin. Moreover, the meaning of the word is rapidly being enlarged to include not only the process of subsidence and collapse, but also a large number of more complex activities. For example, the variety of slope processes called cryoplanation or thermoplanation, and the mechanisms of thermal abrasion, thermal erosion, and thermo-erosional wash are all considered to be thermokarst controlled processes (Kachurin, 1962, p. 52; Dylik, 1971, 1972).

Despite the apparent similarity of topography, it should be made quite clear that thermokarst is not a variety of karst. The latter is a term which is applicable to limestone areas where the dominant process – solution – is a chemical one. Underlying the development of thermokarst, however, is a physical, i.e. thermal, process – ground ice melting, which is peculiar to regions underlain by permafrost.

Thermokarst processes are believed to be some of the most important processes fashioning the periglacial landscape. They achieve their greatest activity in lowland such as the western Arctic and northern Siberia where alluvial sediments with high ice contents are widespread. Some types of thermokarst are probably the most rapid erosional agents presently operating in the tundra and Arctic regions. In alpine regions, in much of the eastern Canadian Arctic, and in the Canadian Shield and other areas underlain by consolidated and resistant rocks, soils with high ice contents are relatively rare and localised. In such regions, thermokarst development is probably less important in landscape development.

In the USSR thermokarst phenomena are well known and there is extensive literature available. Thermokarst phenomena have been reported from many parts of the Soviet Union ranging from the subarctic and Arctic regions of northern Siberia in the north (Dostovalov and Kudryacev, 1967) to the mountainous areas of the central Tien-Shan in the south (Aleshinskaya, 1972). In parts of central Yakutia and eastern Siberia, it is thought that over 40 per cent of the land surface has been affected, at some time or another, by thermokarst processes (Czudek and Demek, 1970a). By contrast, in the periglacial regions of North America, there are few documented studies dealing specifically with naturally occurring thermokarst and its importance may have been underestimated. Likewise, in the interpretation of Pleistocene periglacial features, thermokarst processes do not seem to have been fully considered (e.g. Dylik, 1964b).

Within this context, this chapter outlines the various causes of thermokarst and the controls over its subsequent development. Emphasis is placed upon landscape modification both by thermokarst subsidence and thermal erosion, in the belief that this will facilitate the better understanding of slope and fluvial processes, which are examined in Chapters 7 and 8.

Causes of thermokarst

The development of thermokarst is due primarily to the disruption of the thermal equilibrium of the permafrost and an increase in the depth of the active layer. Thermokarst subsidence is then a function of the new equilibrium depth of the active layer and the supersaturation of the degrading permafrost. This can be illustrated with a simple example (Fig. 6.1). Consider an undisturbed tundra soil which has an active layer of 45 cm. Assume also that the soil beneath 45 cm is supersaturated and yields on a volume basis upon thawing, 50 per cent excess water and 50 per cent saturated soil. If the top 15 cm were removed, the equilibrium thickness of the active layer, under the bare ground conditions, might increase to 60 cm. As only 30 cm of the original active layer remained, 60 cm of the permafrost must thaw before the active layer can thicken to 60 cm, since 30 cm of supernatant water will be released. Thus, the surface subsides 30 cm because of the thermal melting associated with the degrading permafrost, to give an overall depression of 45 cm.

There are many reasons why thermal disequilibrium and permafrost degradation may take place. These are summarised in Fig. 6.2. On a large scale, regional climatic conditions might change, while on a small scale, there are an

Fig. 6.1 *Diagram illustrating how disturbance of high ice content terrain can lead to permanent ground subsidence. From Mackay (1970)*

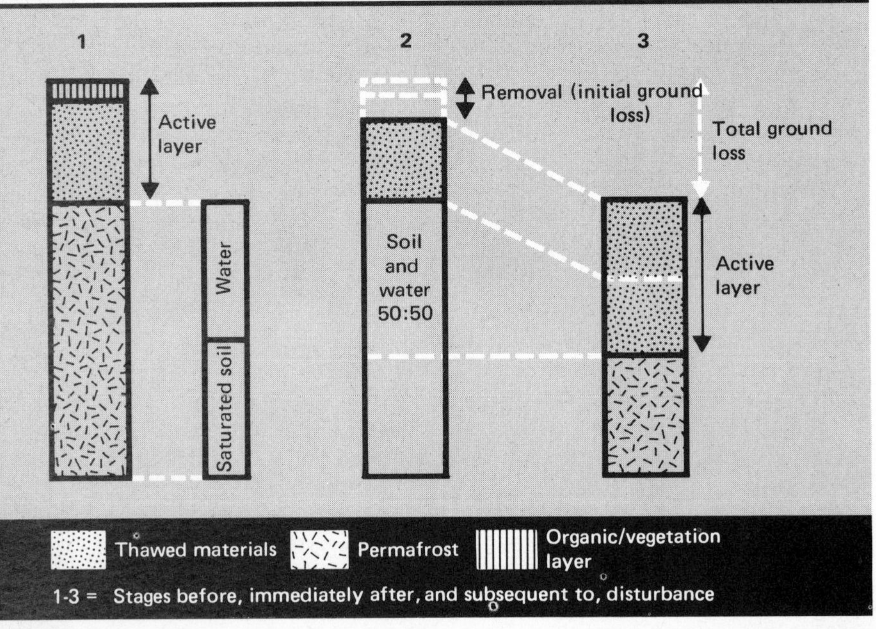

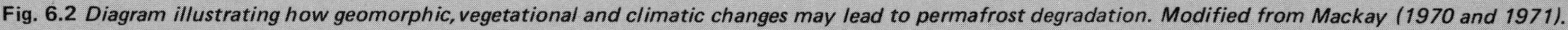

Fig. 6.2 *Diagram illustrating how geomorphic, vegetational and climatic changes may lead to permafrost degradation. Modified from Mackay (1970 and 1971).*

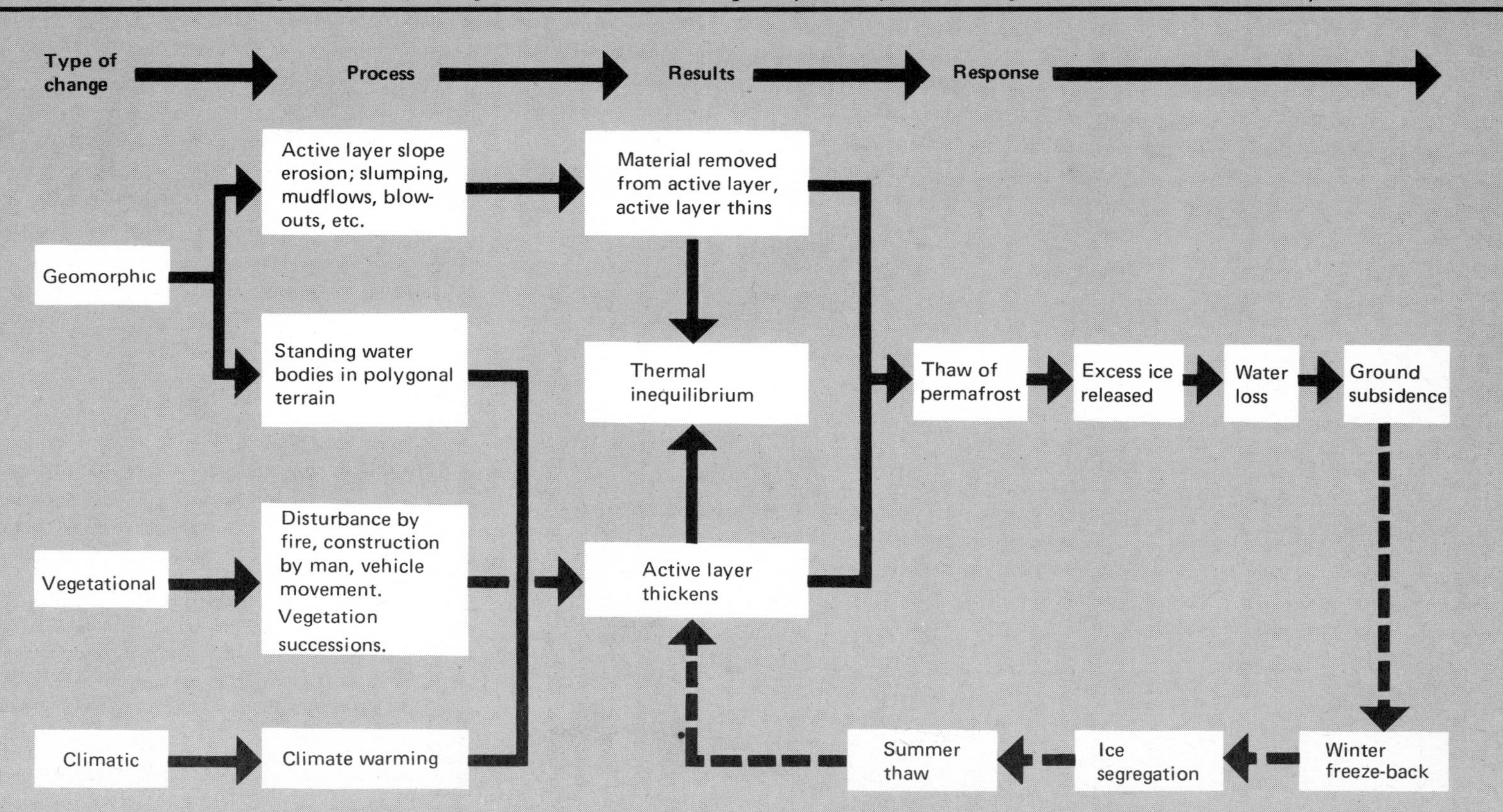

infinite number of local conditions. With respect to the regional climatic changes, it can be theorised that thermokarst development should be most intense under the following natural conditions: (1) when associated with rises in the mean annual temperature of the soil, i.e. an amelioration of climate, and (2) when associated with an increased amplitude of temperature, i.e. an increase in continentality. A combination of these two conditions should lead to a maximum growth in the depth of the seasonally thawed layer. However, in reality, an increase in the mean annual temperatures of present-day permafrost regions would probably lead to a diminution of heat exchange in the soil and a lower summer soil temperature, since it would most likely be associated with an increase in precipitation and cloud cover. As a result, there would be a decrease in the depth of thaw, which is the exact opposite to what is required for thermokarst development. Therefore, the most likely natural climatic cause of thermokarst would be a progressive increase in continentality resulting in a greater range of soil temperature values and summer thaw depths.

In central Siberia, where active thermokarst occurs on a regional scale, it probably reflects a long term regional climatic amelioration. In the western Arctic of North America however, any active thermokarst is probably the result of local, non-climatic factors. Palynological evidence suggests climatic conditions progressively ameliorated during the period 11 000–8 500 years B.P. and between 8 500 and 4 000 years B.P. may have been significantly warmer. This could have induced a period of thermokarst development. After 4 000 years B.P. however, the climate cooled to that of today and recent thermokarst activity due to regional climatic change has probably been negligible.

Nevertheless, in a stable climate, it is clear that thermokarst can develop in response to a variety of geomorphic and/or vegetational conditions (see Fig. 6.2). These may be either natural or man-induced. One natural cause for example, is the presence of polygonal ice-wedge systems. In summer, water accumulation occurs in the central depression along the actual thermal contraction crack, or at the junction of ice wedges, or within low centred polygons. These shallow bodies of standing water invariably favour more intensive thawing during summer and impede the winter freeze-up. The concentration of water also has a chemical effect upon the underlying ice vein since the water is often extremely pure. Once initiated, the concentration of water in the summer, and of snow during the winter, increases as the depression grows larger, thus promoting the further melting of the ground ice without any supplementary agents. This phenomenon has been called 'self-developing thermokarst' (Aleshinskaya, 1972). As such, it can occur within exceptionally severe periglacial and arctic environments. For example, the thickness of the active layer in the Indigirka lowlands of Siberia, where the permafrost temperature is –9°C or less, is twice as great beneath 20–25 cm of water than in adjacent ground (Table 6.1). Another natural cause of thermokarst development is associated with lateral stream erosion developing a thermo-erosional niche (Walker and Arnborg, 1966) which causes the undermining and slumping of the overlying material. The thermal regime of the terrain is disrupted as a consequence. Often, this process is associated with asymmetrical slope and valley development. For example, on eastern Banks Island, retrogressive semicircular ground ice slumps occur predominantly upon the steeper west and southwest facing slopes of asymmetrical valleys and were initially triggered by the lateral migration of the stream

Table 6.1 *Depth of the active layer in ice-wedge polygon terrain in the Yana–Indigirka lowlands of northern Siberia, illustrating thermal effects of standing water bodies. From Czudek and Demek (1970a)*

Moisture conditions in central part of polygon	Depth of active layer (cm)	
	Latitude 70°N	Latitude 71°N
Relatively dry	30–34	23–25
Moist	38–41	31–33
Water layer, 7–10 cm deep in polygon centre	52–56	39–42
Water layer, 20–25 cm deep in polygon centre	62–69	51–53
Lake, 30 x 40 m with depth 25 cm or more	75+	

to the base of that slope (French and Egginton, 1973). A third natural cause of thermokarst development in the boreal forest zone is a local forest fire; for example, in a part of the Siberian taiga, the active layer has increased from 40 to 80 cm in 12 years following a fire in 1953 (Czudek and Demek, 1970a), and at Inuvik, in the North West Territories of Canada, where a forest fire occurred in 1968, Heginbottom (1973) has observed an increase in the depth of the active layer in the burned over area of approximately 40 per cent in four years (Table 6.2). Other local causes for various types of thermokarst include ice-push and scour along coastlines, cyclical changes in vegetation, local slope instability, and deforestation or the disruption of the surface vegetation by man.

Table 6.2 *Depth of thaw measurements at the site of the 1968 forest fire at Inuvik, NWT, Canada. Data from Heginbottom (1973)*

Terrain type	Undisturbed	Burned	Bulldozed
Depth of thaw (cm)			
September 1969	75+	75+	75+
September 1971	77	100	138
September 1972	85	110	140
Surface subsidence (cm)			
September 1969 to September 1971	6	14	21
September 1971 to September 1972	14	12	8

The amount or extent of morphological change associated with the thermokarst process depends upon (a) the magnitude of the increase in depth of the active layer, (b) the ice content in the soil, and (c) the tectonic regime of the region. Factor (a) has already been dealt with in the previous paragraph. With respect to factors (b) and (c), two points need to be stressed. First, thermokarst forms can only usually develop in lowland areas experiencing stable or rising tectonic histories. In subsiding regions, the developing thermokarst forms are quickly filled with deposits. Second, although the actual amount of morphological change will depend upon the amount of excess ice present in the ground, the distribution of the ground ice will determine, to a large extent, the pattern and nature of the ground subsidence. For example, vein ice, which is distributed both vertically and horizontally throughout the soil in polygonal systems, will lead to mound-like thermokarst forms. Segregated and soil ice by contrast, being distributed relatively uniformly in the horizontal plane and predominantly in the upper horizons, lead to widespread surface subsidence.

It is difficult to generalise about the distribution of thermokarst phenomena since the climatic, permafrost and ground ice controls vary not only with latitude but also between each other. In general, however, thermokarst processes are favoured in the tundra and forest regions, and in particular, along the southern margins of the permafrost zone. In the latter, permafrost degradation is currently taking place, summer temperatures are high, the period of summer thaw is at its longest, and the depth of the active layer is at its deepest (e.g. Zoltai, 1971; Thie, 1974). Moreover, in these regions the surface vegetation has a most important control over the subsurface temperatures and thaw depths, and if disturbed for any reason will radically alter the thermal regime. In general, the more luxurious the vegetation, the less is the depth of thaw. Thus, in such areas, the potential for man-induced thermokarst, induced by the modification or destruction of the surface vegetation will be extremely high. North of the treeline and in the high Arctic, the situation is slightly different. In some respects, thermokarst becomes less important in the far north because the active layer is shallow, and the period of summer thaw is increasingly restricted. The absence of vegetation further reduces the complexity of the thermal regime of the soil, and the potential for man-induced thermokarst is greatly reduced. On the other hand, natural thermokarst processes, particularly ground ice slumping and self-developing thermokarst, may assume greater relative importance. The existence of ice-cored terrain and the presence of large amounts of ground ice near the surface (ice wedges, icy sediments, massive ice bodies) are important factors.

Thermokarst subsidence and thermal erosion

Of the various processes included within the term 'thermokarst', a basic distinction should be made between thermokarst subsidence and thermal erosion. Thermal erosion is a dynamic process involving the 'wearing away' by thermal means, i.e. the melting of ice. The easily identifiable characteristics of thermal erosion are flowing water, a slope, and ice which can be melted. By contrast, thermokarst subsidence is 'thermal solution' or more precisely 'thermal melting'. The loss of water then results in the subsidence. Thermal melting depends upon heat conduction, for example, from a pool of water directly overlying icy soil, or from conduction through an intervening layer of unfrozen soil. Therefore, quite unlike thermal erosion, no flowing water or surface gradient is required. Thermokarst subsidence can operate just as efficiently upon a flat and well drained area as in a poorly drained valley bottom. The only difference might be the amount of morphological change and this would depend upon the amount of ground ice present at the two localities.

From a geomorphic point of view, thermal erosion generally results in lateral permafrost degradation or backwearing while thermokarst subsidence results in permafrost degradation from above or downwearing. This distinction, which has been proposed by Czudek and Demek (1970a) to describe the thermokarst features of Siberia, appears useful and enables one to identify a number of distinctive and separate thermokarst phenomena. In essence, lateral permafrost degradation takes place as the result of cliff retreat, lateral river erosion, and marine or lacustrine abrasion. Ground ice slumps, thermo-erosional niches, and thermokarst (thaw) lakes are some of the more distinctive features developed in this way. Permafrost degradation from above occurs in flat undissected terrain mainly on watersheds, and is predominantly a process of subsidence and collapse of the ground surface. It may operate over extensive areas, and the ultimate end result of thermokarst downwearing is the destruction of the original surface relief and the creation of a new thermokarst relief at a lower elevation. The development of closed depressions, collapse features, and hummocky irregular terrain are typical of this process. In the mature stages of thermokarst development, it is not uncommon to have both thermokarst subsidence and thermal erosion operating together.

Thermokarst subsidence

The simplest downwearing thermokarst process is that of straightforward subsidence of the ground, the amount of ground loss being a function of the new equilibrium depth of the active layer and the amount of water (ice) lost from the soil (see earlier).

The most distinct thermokarst forms develop in areas of considerable ground ice, particularly where well developed ice wedges and ice-wedge polygons are present. Since such conditions occur widely throughout the periglacial regions, the resulting thermokarst forms are of more than casual importance. They have been described in detail from central Siberia where Soloviev (1962) and others believe thermokarst phenomena develop in a predictable and sequential fashion. The type area for this kind of relief is upon terraces of the Lena and Aldan rivers in the central part of the Yakutian ASSR.

The alas thermokarst relief of central Yakutia As with most 'type areas', a number of relatively unique factors aid in the development of the distinct topography. In central Yakutia, one may distinguish at least three controlling factors in the development of the thermokarst relief. First, the terraces are underlain by considerable thicknesses of alluvial silty loams in which segregated ice lenses may constitute up to 80 per cent ice content. In addition, thick syngenetic ice wedges reported to exceed 50–60 m in vertical extent in some areas underlie between 30–60 per cent of the surface of the terraces. In terms of ground ice conditions therefore, the area is exceptionally well suited for thermokarst development. Second, the geomorphic history which enabled such conditions to occur are rather uncommon. The central Yakutian lowland appears to have remained unglaciated for much, if not all, of the Quaternary. It acted as a stable aggradational environment in which alluvial sediments were deposited over a long period of time by the ancient Lena and Aldan rivers. Third, the present climatic conditions are continental and among the most extreme in the world. For example, the annual temperature range at Yakutsk, at latitude 62°N, is 62°C while at Sachs Harbour, at latitude 72°N in the western Canadian Arctic, the equivalent value is only 36°C. As a result, active layer depths in excess of 2.0 m are not uncommon in Yakutia where air temperatures in August often reach 30°C. Since the greater the amplitude of the mean annual temperature, the greater is the thawing problem, it follows that thermokarst development in Yakutia will be particularly favoured. In more general terms, therefore, one must regard the thermokarst terrain of central Yakutia as being rather unique, and it is unlikely that exactly similar conditions will be found elsewhere.

Bearing these considerations in mind, it is instructive to examine the nature and development of the various thermokarst relief forms found in Yakutia. The sequence of alas thermokarst relief, as suggested by Soloviev (1962; 1973b) is outlined in Fig. 6.3. In the first stage following the increase in the depth of thaw, the polygonal system of ice wedges begins to thaw. High centred polygons or 'degradation polygons' (Katasonov and Ivanov, 1973) begin to develop with trough-like depressions along the ice wedges. An undulating, convex surface quickly develops but the vegetation cover is not destroyed. However, as soon as the preferential thawing and subsidence along the line of the ice wedges exceeds 1–2 m in depth, slumping starts, the vegetation cover is broken, and the polygon centres become distinct conical mounds. These mounds are called 'baydjarakhs', a Yakutian term used to describe silty or peaty mounds. They are commonly between 3 and 4 m high, 3 and 15 m wide and up to 20 m long. As such, they closely resemble the dimensions of the polygons from which they have developed. Young baydjarakhs have a clearly truncated conical form but old baydjarakhs have a more convex form. From the air, baydjarakhs resemble a cobblestoned surface, distributed in a checkerboard or rectilinear pattern. They are found not only in Yakutia but also in the Tamyr Peninsula, near Dickson, and in other parts of Siberia adjacent to the Laptev Sea. In the discontinuous permafrost zones where the active layer is at its thickest, or in areas where the ice wedges have almost disappeared, similar features are called 'grave-yard mounds' (Popov, Kachurin and Grave, 1966).

The second stage of thermokarst downwearing is characterised by the progressive collapse and decay of the baydjarakhs as a depression develops in the centre of the baydjarakh field, often with a central hollow or sinkhole. With the linkage of these sinkholes, continuous depressions form with steep slopes and uneven bottoms which are called 'dujodas' in Yakutia.

By stage III (Fig. 6.3), a distinct depression with steep sides and a flat bottom has developed through the continued collapse of the baydjarakhs on the sides of the dujodas. This depression is called an 'alas' in Yakutia, denoting a circular or oval depression with steep sides and a flat floor with no trees but with a thermokarst (thaw) lake (Fig. 6.4). The latter, upon obtaining a minimum depth at which it does not freeze to the bottom during the winter, promotes the development of a talik zone beneath the lake. Thus, as the ground ice melts and the sediments beneath the lake consolidate, the thaw lake deepens and subsurface thawing is further enhanced. Ultimately, alas development proceeds to the complete thawing of the permafrost or to the formation of a stabilised talik. In the fourth stage, the alas lake disappears either by infilling with alas deposits or by draining to a lower alas level or to a river valley. Permafrost aggradation takes place as the talik progressively freezes. Ice segregation results in a slight elevation of the alas floor and new epigenetic ice wedges begin to develop. Sometimes, bulgannyakhs (see Chapter 5) are formed either by the intrusion of water

Fig. 6.3 *The sequence of development of alas thermokarst relief in central Yakutia, according to P.A. Soloviev (1962; 1973b)*

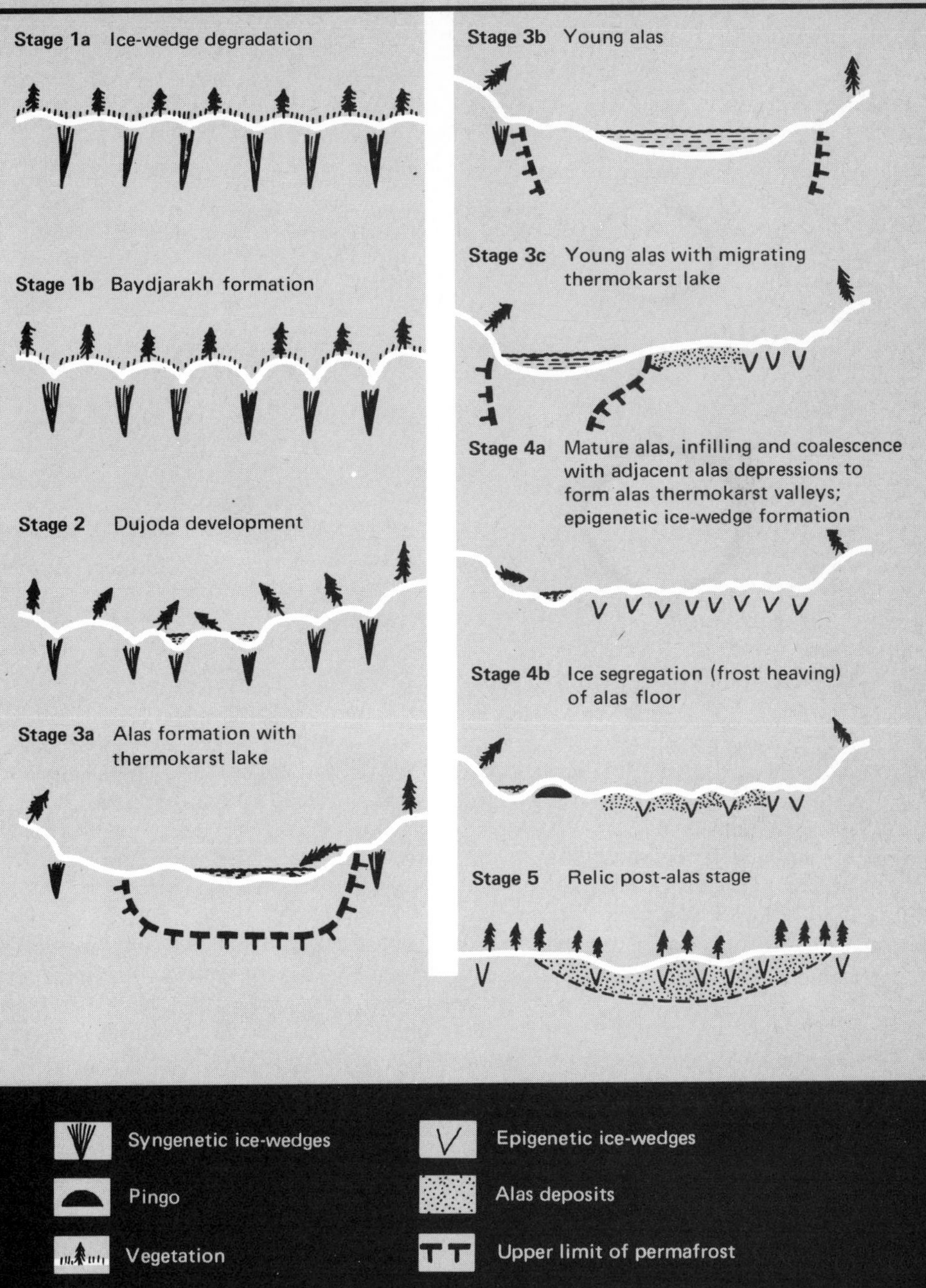

Fig. 6.4 *A small alas depression which has developed in historic time near Maya village, central Yakutia, USSR.*

under pressure into the frozen ground (open system) or by the freezing of local taliks (closed system). The majority of the larger bulgannyakhs in the major alasses are probably of the open system type since the permafrost beneath the floor of the alas depression is often thin or discontinuous. The end result of this sequence of alas relief is the development of a flat depression with gentle slopes and an undulating bottom.

In central Siberia, alas formation has greatly modified the lowland areas. In one region near to Yakutsk nearly 40 per cent of the initial land surface has been destroyed by alas formation (Fig. 6.5). The coalescence of adjacent alasses has led to the formation of large depressions often in excess of 25 km^2 and to the development of complex thermokarst valleys. The latter are irregular in plan consisting of wide sections (alas depressions) connected by narrow reaches cutting through the intervening watersheds. Other characteristics of these valleys are right angle turns, blind spurs, and a general misfit relationship with the overall topography and drainage. The long profiles are also irregular due to the variable ice content and resulting differential settlement of thawed sediments within the various alas depressions. However, with time, the deepest alasses are filled and the higher floors are lowered as erosion progresses along thermokarst gullies originating along epigenetic ice wedges. A graded long profile develops and there is an overall general tendency towards the development of exceptionally smooth and broad depressions.

The rate of alas formation may vary considerably since there are reports of

Fig. 6.5 *Map showing extent of alas thermokarst terrain in the vicinity of the Abalakh settlement, central Yakutia, USSR. From Soloviev (1973a). The depressions give the appearance of pitted outwash (kettle) topography, yet the area has not been glaciated.*

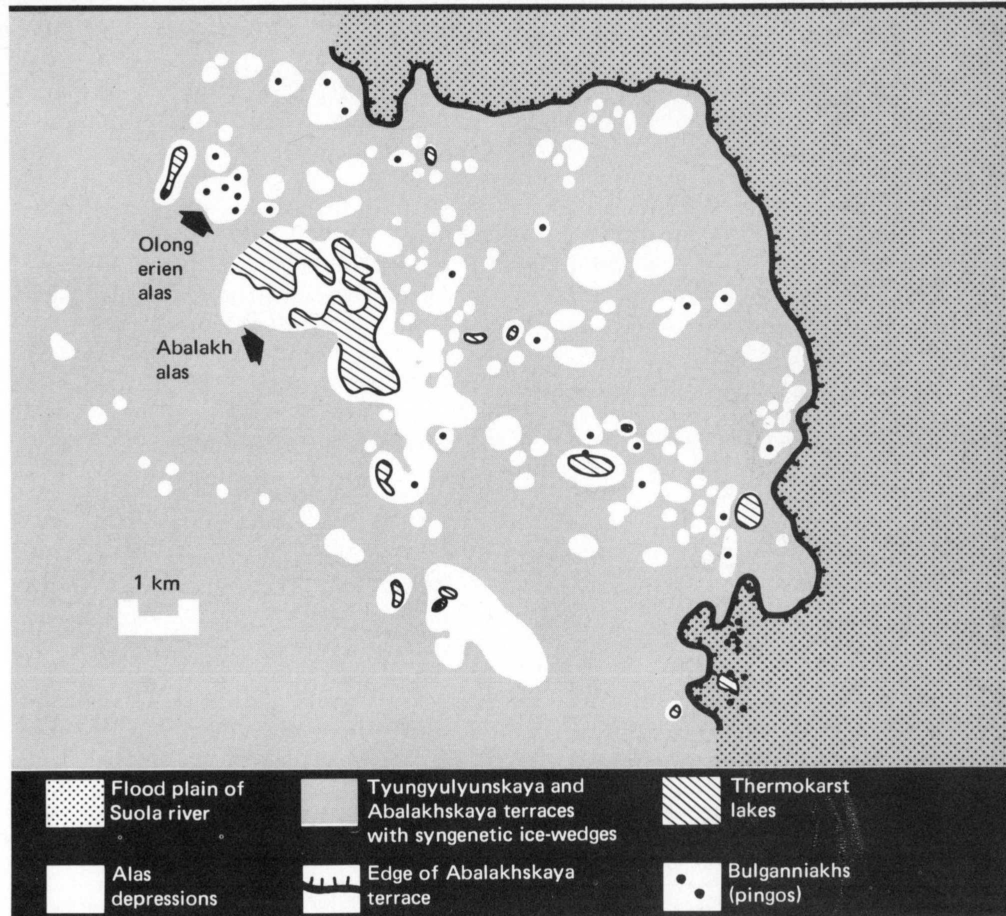

some alasses developing within historic time, while others are obviously very old features. Kachurin (1962) regards the beginning of thermokarst development in the USSR to coincide with the warmer climate of the Atlantic Period (8 200–5 300 years B.P. in Western Europe) but that thermokarst was then inhibited in the following Sub-Atlantic Period when temperatures dropped. In recent years, it has recommenced under the present climatic regime. According to Grigoryev (1966), many of the alasses of the Yakutia region developed during the Holocene climatic optimum which, in that region, was approximately 4 000–5 000 years B.P. and Soloviev (1962) has concluded that many of the thermokarst forms are not active today; in fact, probably only 10 per cent of the terrain in the vicinity of Yakutsk is undergoing active thermokarst modification today.

Alas relief in North America In North America, the full range of alas thermokarst features has not been identified. Given the rather different geomorphic history and climatic conditions of much of North America, it is unlikely that similar relief exists. On the other hand, certain aspects of the alas thermokarst relief find striking analogies with certain North American phenomena.

For example, in Alaska Rockie (1942) first drew attention to the pitting and settling of ground caused by the thawing of ice wedges in recently cleared field systems near Fairbanks. Péwé (1954) and others have subsequently referred to the hummocks as 'thermokarst mounds'. In many respects, these features are analogous to the baydjarakhs or 'grave-yard mounds' of the USSR. The 'dujoda' stage also has analogies in Alaska with the thaw sinks, funnels and 'cave-in' lakes reported from the Seward Peninsula and other areas by many workers (e.g. Wallace, 1948; Péwé, 1948; Hopkins, 1949). However, the circular alas depression, and in particular with bulgannyakh or pingo growth within the depression has not been reported. The closest comparable features are the thermokarst depressions and lakes of the Mackenzie Delta in which closed system pingos are developed (see Chapter 5). In general though, few of the thermokarst depressions of the Mackenzie Delta have the same striking relief as the Yakutian alasses and they are more the result of lateral growth rather than subsidence. Probably the most analogous terrain to the alas thermokarst relief is the ice-cored topography described by Rampton (see Chapter 5, pp. 82–4) from the Mackenzie Delta of Canada. In particular, thermokarst development along the course of small creeks has produced 'macro-beaded' drainage systems where a series of depressions, partly lake filled, are connected to each other by narrower sections or shallow interfluves. In total, the depressions form what appears to be a former stream course.

In both form and origin, these macro-beaded drainage systems are clearly analogous to the thermokarst valleys in Siberia. On a smaller scale, enclosed depressions and small sinks of a thermokarst origin have been described (e.g. Zoltai, 1971; Thie, 1974) from the southern fringe of the discontinuous permafrost zone in northern Manitoba and Saskatchewan. There, the melting out and collapse of peat plateaus, often as the result of fire, can lead to the development of a shallow enclosed or semi-enclosed grassy depression or 'collapse scar', some 2–5 m deep and anything up to 200–300 m in diameter. Since these features occur in the forested zone and are surrounded by locally forested terrain, they clearly resemble the appearance of the typical alas. According to Thie (1974), melting is exceeding aggradation of permafrost and has done so for the past 150 years. Only approximately 25 per cent of the once occurring permafrost (60 per cent of land area) is now present.

Ice-wedge thermokarst relief Although ice wedges and ice-wedge polygon relief has been studied in detail in the western North American Arctic (e.g. Carson and Hussey, 1962; Hussey and Michelson, 1966; Mackay, 1963a), the genetic relationship between the various wedge polygon relief forms and the thermokarst process has not been emphasised.

In North America, two types of tundra ice-wedge polygon relief forms are commonly recognised: a high centred type and a low centred type (Fig. 6.6). Invariably, the high centred type develops through the low centred type. The latter are characteristic of low marshy areas and commonly possess upstanding rims, often in excess of 50 cm in height, which correspond to the upthrust sediments adjacent to the ice wedge, and a low wet centre composed of sedges and

Fig. 6.6 *Typical relief and vegetation of low- and high-centred ice wedge polygons*

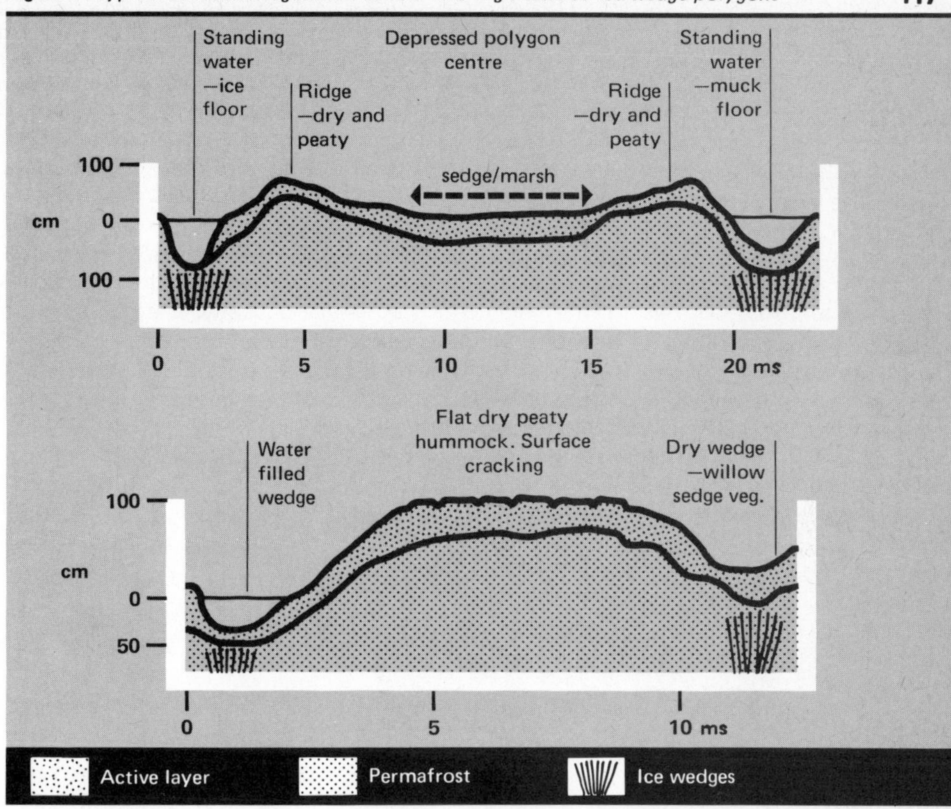

tussock tundra (see Fig. 5.3). In exceptionally poorly drained areas, the centre of the polygon together with the central depression running along the line of the thermal contraction crack and between the two upthrust rims, are filled with standing water. As such, the low centred tundra polygon is a natural relief form for the initiation of self-developing thermokarst. The most favoured location for thermokarst melting of the underlying ice wedge is at the junction of two or more ice wedges and small irregularly shaped lakes often persist in such localities throughout the summer. If integrated drainage occurs in such terrain, it assumes a 'beaded' pattern. At the same time, water accumulating within the central depression often results in the beginning of a small tundra (thaw) pond which, through amalgamation with adjacent polygons, will increase in size. The change from low centred to high centred polygons is usually brought about by a widening of the upthrust ridges, either by continued ice-wedge growth or by vegetation growth, such that the ridges begin to encroach upon the central sedge and tussock area. An elevated peaty area finally emerges surrounded by troughs along the underlying ice wedges (see Fig. 5.6).

Once the high centred polygon terrain has developed, it inevitably favours thermokarst degradation by concentrating surface water and thermal erosion along the lines of the underlying ice wedges. Shumskiy and Vtyurin (1966) have summarised this process in a number of stages. In the initial destruction stage, the polygons (i.e. degradation polygons) may be either plane, with raised edges, or with a convex form. Then, in the mature destruction stage, the polygon form develops to be either convex with deep thawing furrows or to assume a domed, 'grave-yard mound' type of appearance (Fig. 6.7). Ultimately, it may proceed to a thermokarst depression or alas. To complete the classification, relic convex forms or mounds are included in which the ice wedge has thawed completely and the permafrost has disappeared.

Rather dramatic relief which has developed as the result of thermal melting and erosion operating preferentially along ice wedges has been described by the writer from eastern Banks Island (French, 1974b, pp. 791–3). In areas of ice-rich silts of glacio-lacustrine origin which possess large ice wedges, the result is striking baydjarakh or badland terrain. In one area drained by two small streams, a large amphitheatre-like hollow has been eroded nearly 0.5 km in diameter, and several hectares in area (Fig. 6.8). Drilling to depths of 6 m revealed the silts to possess 20–30 per cent excess ice on average, and ice wedges exceeding 6 m in depth. Fluvial erosion has operated preferentially along the ice wedges to isolate sharp conical mounds on the floor of the depression, and to give a serrated edge to the amphitheatre. The presence of shallow lakes together with a number of localised 'active layer failures' (see pp. 145–7) within the amphitheatre suggest development of the amphitheatre has proceeded in two stages: (1) fluvio-thermal

Fig. 6.7 *Oblique air view of degradation polygons or 'grave-yard mounds' occurring on sloping terrain, eastern Banks Island, Canada. The relative relief of the mounds is 2–3 m; the size of the polygons is indicated by the tents.*

Fig. 6.8 *Oblique air view from 50 m elevation of terrain undergoing rapid thermal melting and erosion operating preferentially along ice wedges, eastern Banks Island, Canada. The conical mounds are over 8 m high. Remnants of the original polygon surface can be seen in both the foreground and mid-distance.*

erosion along ice wedges and the development of the baydjarakh-like mounds, and (2) the subsequent degradation of the mounds brought about by the high ice content of the silts, and their exposure to thawing from all sides. Since the natural water (ice) content of the silts exceeds their liquid limits, the silts are extremely mobile upon thawing.

Backwearing thermokarst

In contrast to thermokarst subsidence or downwearing thermokarst, thermokarst processes also operate laterally to produce distinctive relief features through backwearing. Ground ice slumps and thaw lakes or depressions are two features which develop in this regressive manner. For simplicity, they are discussed separately.

Ground ice slumps Ground ice slumps (Fig. 6.9) result from the melting of supersaturated frozen ground (i.e. ground which possesses excess ice). They attain their largest dimensions and most frequent occurrence in areas underlain by massive icy bodies, or silts of high ice content. They are known to occur in forested (Czudek and Demek, 1970a; Zoltai and Pettapiece, 1973), tundra (Mackay, 1966), and high Arctic (French, 1974b) environments. In general, they are short-lived but rapidly developing features, the majority of which become

Fig. 6.9 *An active ground ice slump approximately 35 km west of Johnson Point, eastern Banks Island, Canada. Ground ice is exposed in the headwall and there is active slumping and sliding of material across the face.*

stabilised within 30–50 summers after their initiation. They are important in that they represent one of the most rapid erosive processes currently operating in present-day periglacial environments.

Ground ice slumps resemble earthflows of more temperate regions but because there is no surface of rupture, features characteristic of rotational slides are largely absent. The crown has no lunate tension cracks paralleling the main scarp but, instead, there are minor cracks in the active layer along hummocks, created as the hummocks become detached and fall down the face. From a process point of view, ground ice slumps may best be described as 'retrogressive thaw flowslides'. The term 'thermocirque', proposed by Czudek and Demek (1970a) to describe analogous features in Siberia, is graphic but fails to identify the nature of the process or processes involved.

In plan, the slump face can be divided into three sections: (a) the soil active layer which is a vertical or near vertical face, (b) the ice face which is a smooth thermal erosion belt, with gradients varying from between 70° and 20° downslope, and (c) a scarp base composed of slumped debris. Erosion takes place through the thermal melting of the ground ice which causes a crumbling and slipping of the overlying sediments into the hollow, which then liquify into mud through the addition of the melted ice. Excavation is then by mudflow or solifluction, or, if there is sufficient moisture, by fluvial erosion.

Mackay (1966) has identified three main processes involved in scarp retreat:

(a) free fall or sliding of the active layer, (b) erosion, both thermal or mechanical, on the exposed and frozen scarp face, and (c) slumping of thawed debris from above the frozen scarp face. The first process is usually related to the thawing of the active layer. The overlying material at the lip of the scarp dries out and cracks along the outline of small hummocks or desiccation cracks. Ultimately, these blocks break off and fall down the ice face. The second process is a reflection of the insolation melting of the ice face, which is accentuated by the abrasive effects of the sliding earth clods from the vertical overhang above. The third process is the most significant in terms of material removed and involves the progressive thawing of the scarp face, with negligible movement, until the overburden is 1–2 m thick. Then, rapid slippage along the frozen surface exposes a surface of failure not unlike that of a rotational landslide. According to Mackay, the majority of scarp retreat is accomplished by this third process.

The rate of retreat of the slump scar will depend primarily upon the ice content of the underlying materials. Mackay (1966, pp. 70–1) has demonstrated how, if the ice content is at least 200–300 per cent, a scarp face of up to 6 m in height may be able to retreat with little transport of thawed soil beyond the foot of the scarp. This is because the 'thaw hole' developed at the foot of the retreating scarp would be large enough to accommodate all of the volume of thawed material from the overlying scarp, assuming that the meltwaters were allowed to escape.

Thus, if a slump were active for several successive years, the slump floor could be almost flat and the retreat process could continue indefinitely assuming that the ice content remained the same. Field measurements of scarp retreat reveal this process to be extremely active; Mackay (1966) concluded that average retreat rates are probably between 2 and 5 m p.a. varying with the ice content. However, rates in excess of this have been observed, especially in localities where continual wave action removes material from the base of the slump as quickly as it accumulates.

The size of ground ice slumps may vary considerably depending upon the initial relief of the terrain, and the amount and distribution of the ground ice. In the Mackenzie Delta, an average sized feature may be 50–100 m in diameter and possess a scarp face of between 2 and 7 m in height. Some slumps however are polycyclic features and several hectares in extent. In the high Arctic, the slumps are exceedingly broad and shallow, since the height of the headwall including the ice exposure rarely exceeds 2 m, while the maximum horizontal dimensions may often exceed 200 m (French, 1974b; Lamothe and St-Onge, 1961). Thus, they are slightly different to those which occur in the tundra and forest tundra zones, and probably reflect differences in the annual rate, depth and duration of thaw in high Arctic regions. Ground ice slumps appear to be widespread throughout the Canadian high Arctic, and occur wherever unconsolidated ice-rich sediments are found.

On the silty morainic terrain of eastern Banks Island, ground ice slumps attain a regional density of 0.5/km^2 (French and Egginton, 1973). Both active and stabilised features can be recognised. Maximum rates of headwall retreat of between 6.0 and 8.0 m year^{-1} appear typical (French, 1974b). Since the maximum size of the slumps as revealed on air photographs does not usually exceed 200–300 m in diameter, French and Egginton (1973) have concluded that the majority of slumps become stabilised within 30–50 summers of their initiation. Several causes of the initial slope failure or disturbance that first revealed the icy sediments have also been suggested. They are illustrative of some of the more common trigger mechanisms. For example, many of the slumps occur on the steeper slopes of asymmetrical valleys and have been triggered by lateral stream erosion undercutting the west-facing bank and causing collapse later in the summer. Others are found adjacent to lakes and probably owe their initiation to lacustrine effects such as wave action or ice pushing (see Chapter 10). Finally, some slumps, unrelated to any obvious trigger mechanism and occurring randomly, may have developed initially from simple active layer slope failures (Fig. 7.5) which expose the surface of the permafrost to thawing. Such failures are not uncommon in permafrost regions, and have been widely reported in the high Arctic (e.g. Bird, 1967, pp. 217–20). In the western Arctic, where over 50 per cent of the total precipitation falls as summer rain, the development of high pore water pressures is not uncommon after prolonged rain. Similar conditions are favoured in years of exceptionally high winter snowfall and/or rapid and late thawing (see Chapter 7, pp. 145–7).

The ultimate stabilisation of the headwall is related to the balance between the rate at which debris is supplied to, and removed from, the base of the headwall. This will be a function of the relative height of the icy sediments in relation to the base of the headwall. Assuming that the icy sediments occur as a body parallel to the regional slope of the ground, as the ground slope gradually decreases towards the top of the slope, less of the icy sediments are exposed above the base of the headwall. Ultimately, when the icy sediments are lower than the base of the headwall, the feature becomes inactive. Given the generally subdued relief of slopes developed in unconsolidated sediments under mass wasting processes in the high Arctic, it is possible, therefore, for extensive headwall retreat to occur and for complete slopes to be consumed by regressive slumping.

Thermo-erosional cirques are a form of ground ice slump associated with ice wedges and reflect a combination of processes, including both subsidence and backwearing. They have been described from several parts of Siberia (Czudek and Demek, 1970a; Dylik, 1971), where they occur along the banks of the larger rivers. The first stage in their development is the thawing of ice wedges exposed in the bank leading to the formation of a number of small gullies. These gullies are distributed regularly along the bank, and their spacing correspond to the dimensions of the polygons on the surface above. As ice wedges parallel to the river also begin to thaw, a number of 'baydjarakh' forms develop. These eventually thaw and an irregular hollow develops. The thermocirque is then bounded by an overhanging scarp or vertical headwall which is, in fact, an ice wedge in oblique section. The headwall then retreats causing the collapse of the overlying sediments and vegetation into the hollow. Since massive syngenetic ice wedges may underly as much as 50 per cent of the terrain in certain areas of Siberia, the potential for continued retreat of the headwall is considerable. In North America, thermocirques have not been described as such. In general, as was pointed out earlier (pp. 85–93), there does not appear to be the same syngenetic ice-wedge development in North America as there is in parts of Siberia.

Thaw lakes and depressions Perhaps the most common thermokarst phenom-

Fig. 6.10 *Oblique air view of oriented thaw lakes, Baillie Island, Arctic coastal plain, District of Mackenzie, Canada.*

ena which occur in the lowland periglacial environments are shallow rounded depressions, often with semicircular or circular ponds or lakes within them (Fig. 6.10). In the North American literature, they have variously been called thaw lakes, thaw depressions, thermokarst lakes and tundra ponds (Black, 1969a). They have attracted considerable attention particularly in Alaska (e.g. Wallace, 1948; Hopkins, 1949; Black and Barksdale, 1949) and in the Mackenzie Delta (Mackay, 1963a). Similar features have been reported widely from the Russian and Siberian Arctic coastal plains (e.g. Dostovalov and Kudryacev, 1967), and, in fact, are uniquitous in flat lowland areas wherever silty alluviums with high ice contents are present. Fluviatile terraces, outwash plains and Arctic coastal areas are the most favoured localities for their widespread development.

The initial cause of the thaw lake or depression may be the quite random melting of ground ice, the subsidence of the ground, and the accumulation of water in the depression. For example, in the Seward Peninsula of Alaska, in the boundary zone between the discontinuous and continuous permafrost, Hopkins (1949) has described thaw depressions and some of the small lakes that occupy these depressions. The depressions appear to have been initiated by locally deep thawing, the result of (1) disruption of vegetation cover by frost heave, (2) accelerated thaw beneath pools occupying intersections of ice-wedge polygons, and (3) accelerated thaw beneath pools in small streams. In some instances, where the permafrost was particularly thin, the lakes have pierced or thawed the permafrost and subterranean drainage into the underlying thawed sediments has occurred. This has created 'cave-in' lakes and thaw 'sinks' (Wallace, 1948). In areas of more continuous permafrost, however, the lakes often originate in poorly drained areas characterised by low centred ice-wedge polygons. In such areas, the ground ice content is exceptionally high and the depressed polygon centre favours standing water bodies. Differential thawing is thereby promoted and, through coalescence of adjacent polygon lakes, larger tundra ponds are created.

Thaw lakes vary considerably in size, frequency and shape. Some attain diameters of 1–2 km but the majority are smaller features rarely exceeding 300 m in diameter. They are also extremely shallow, in all cases less than 3–4 m deep, and often less than 0.3 m deep in the littoral margins.

The development and growth of thaw lakes proceeds in a distinctive manner. To begin with, a thaw lake may be of an irregular shape; however, it rapidly assumes a circular and smooth form through the melting of the permafrost and an undercutting of the vegetation causing collapse and retreat of the bank. As such, thaw ponds are dynamic features of the tundra landscape since they are constantly migrating, changing shape and coalescing with adjacent depressions and lakes. In Alaska, average annual rates of bank retreat vary from 15 to 20 cm per year according to Wallace (1948), depending upon the strength and persistence of wave (wind) action, and the temperature of the water, among others. Hopkins (1949) has outlined an idealised growth cycle from youth to old age. In essence, the thaw lakes grow in size to begin with, coalescing with adjacent lakes. As the thaw lake migrates, the old lake bed emerges as a shallow depression. Tundra vegetation begins to develop upon the newly exposed and silty lake floor and organic matter begins to accumulate. At the same time the lake itself begins to slowly infill with silts and organic matter. Alternatively, the lake basin is emptied catastrophically by the expansion of an adjacent depression. Subsequently, ice segregation and heave takes place on the floor of the depression, and solifluction and mass-wasting will take place upon the banks, moving material into the depression. Ultimately, the depression is infilled and all traces of the former lake and lake basin may be obliterated from the landscape.

The growth, drainage and rebirth of thaw lakes and depressions is probably an extremely rapid geomorphic process, and, in the case of small lakes, may complete itself within 2 000 to 3 000 years. In the Barrow region of Alaska, ^{14}C dating of organic matter in two drained lake basins suggest ages of several thousand years and other data suggest that most surface features in the vicinity of Barrow are not older than 8 000 years (Black, 1969a). In the Umiat region, Tedrow (1969) has dated material from a level within a thaw depression at approximately 4 500 years old and to be some 4 000 years younger than the initial surface in which the thaw depression first developed. These dates give some idea as to the relative magnitude and rapidity of the thaw lake cycle.

One of the more perplexing aspects of thaw lakes is that they are often elongate in shape, with a common and systematic orientation of the long axis (e.g. Fig. 6.10). The thaw lakes of the Barrow Point region are exceptionally good examples of this phenomenon and possess long axis orientations approximately N9W–N21W (Carson and Hussey, 1962). Other oriented lakes have been described all along the Arctic coastal plain of Alaska, Yukon Territory and the Mackenzie District (e.g. Black and Barksdale, 1949; Mackay, 1963a) as well as in the Old Crow Flats area of interior Yukon (Price, 1968), and in other parts of Arctic Canada (e.g. Dunbar and Greenway, 1956, pp. 132–4; Bird, 1967, pp. 212–16).

A variety of oriented forms can be recognised. For example, in northern Alaska, the lakes are commonly elliptical or rectangular in shape, ranging in size from small ponds to large lakes 15 km long and 6 km wide (Black and Barksdale, 1949). The ratio of length to breadth varies from 1 : 1 to 1 : 5. The depths of the lakes suggest two varieties (a) with a shallow shelf surrounding a deeper central part, which may be 6–10 m deep, and (b) with a uniform saucer shaped cross profile with depths of less than 2 m. In the Tuktoyaktuk Peninsula of the Mackenzie Delta, there are very similar oriented lakes (Mackay, 1963a) and their shape has been classified as being either lemniscate, oval, triangular or elliptical. In the Cape Bathurst area, many of the lakes are 1–2 km long and with a length–breadth ratio of 2: 1. In western Baffin Island, however, in the eastern Arctic, on the Great Plain of Kaukdjuak, many thaw lakes are clam-shaped with one straight edge and with length–breadth ratios varying between 1.5–2.5 to 1 (Bird, 1967, p. 215).

The cause of the orientation of thaw lakes has attracted considerable controversy, particularly in Alaska (e.g. Carson and Hussey, 1963; Price, 1963). In most cases, the long axes of the lakes are at right angles to the present-day prevailing winds, and some relationship between the two variables is clearly apparent. However, it must be stressed that oriented lakes are not solely a feature of the periglacial environment since many are found in other morpho-climatic regions (Price, 1968). At the same time, it is clear that oriented lakes are of a contemporary nature and are forming today under present periglacial climatic conditions since they are sometimes found upon plains and terraces which have only recently emerged from beneath the sea. Mackay (1963a) has concluded in a general fashion that the oriented lakes of the Mackenzie Delta represent an equilibrium condition in response to the processes currently operating. Others have focused attention upon the role of the present winds which either produce wave current systems which scour at right angles or which deposit sediment on the east and west shores insulating them from further thaw (e.g. Carson and Hussey, 1962). One suggestion is that wind-induced littoral drift reaches a maximum at the corner ends of lakes, and that the eroded materials are then distributed uniformly along the long axis shorelines. This explanation does not account for either the orientation of very small lakes where such circulation systems are not so well developed or for the lack of erosive currents at the ends until after the basin is elongate. At present, it is difficult to distinguish between cause and effect in the circulation pattern of thaw lakes. More detailed studies concerning wave and current effects upon thaw versus transportation and deposition of sediments within arctic lakes are required before any firm conclusion can be made.

Fluvio-thermal erosion processes

A rather specific type of lateral or backwearing thermokarst process occurs along river banks and is the combined effect of both the thermal and fluvial erosive capacities of running water. This is termed 'fluvio-thermal' erosion.

An important characteristic of the fluvial regime of many periglacial areas is the marked peak of surface run-off which occurs in the early weeks of summer

Fig. 6.11 *A thermo-erosional niche developed in sands and gravels, Ballast Brook, northwest Banks Island, Canada. The niche extends inwards for over 4 m. There is sloughing of material from the face and the building of a ridge of material at the foot which will ultimately cover and protect the niche.*

(see Chapter 8, pp. 168–74). It reflects the rapid snowmelt at that time. A consequence is that floodwaters are capable of eroding river banks not only by normal, mechanical or abrasive means, but also by the thermal melting of the permafrost. In unconsolidated sediments, the result is the development of a 'thermo-erosional niche' at the flood water level (Fig. 6.11), which may extend beneath the banks for upwards of 10 m. Subsequently, the undercut frozen sediments may collapse along a line of weakness, often an ice wedge, destroying the thermo-erosional niche. This process is common in many present-day periglacial regions and has been described by Péwé (1948) along the lower Yukon River near Galena in Alaska, by Walker and Arnborg (1966) in the Colville River Delta of northern Alaska, and by Czudek and Demek (1970a) who give several illustrations of how this process operates along some of the major rivers in Siberia (Fig. 6.12). Even in Pleistocene periglacial environments, where, in all probability, the fluvial regime was rather different, fossil thermo-erosional niches and collapsed blocks have been identified (Dylik, 1969a). In coastal regions too, the development of a thermo-erosional niche and subsequent collapse is an important mechanism of coastal erosion in permafrost. For example, Grigoryev (1966) has described thermo-abrasion niches some 3 m high and 20 m deep occurring at the base of sea cliffs along the Laptev Sea and other areas of

Fig. 6.12 *Examples of lateral and thermal river erosion in Siberia. Modified from Czudek and Demek (1970a)*

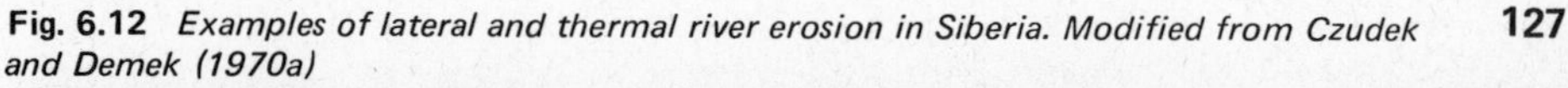

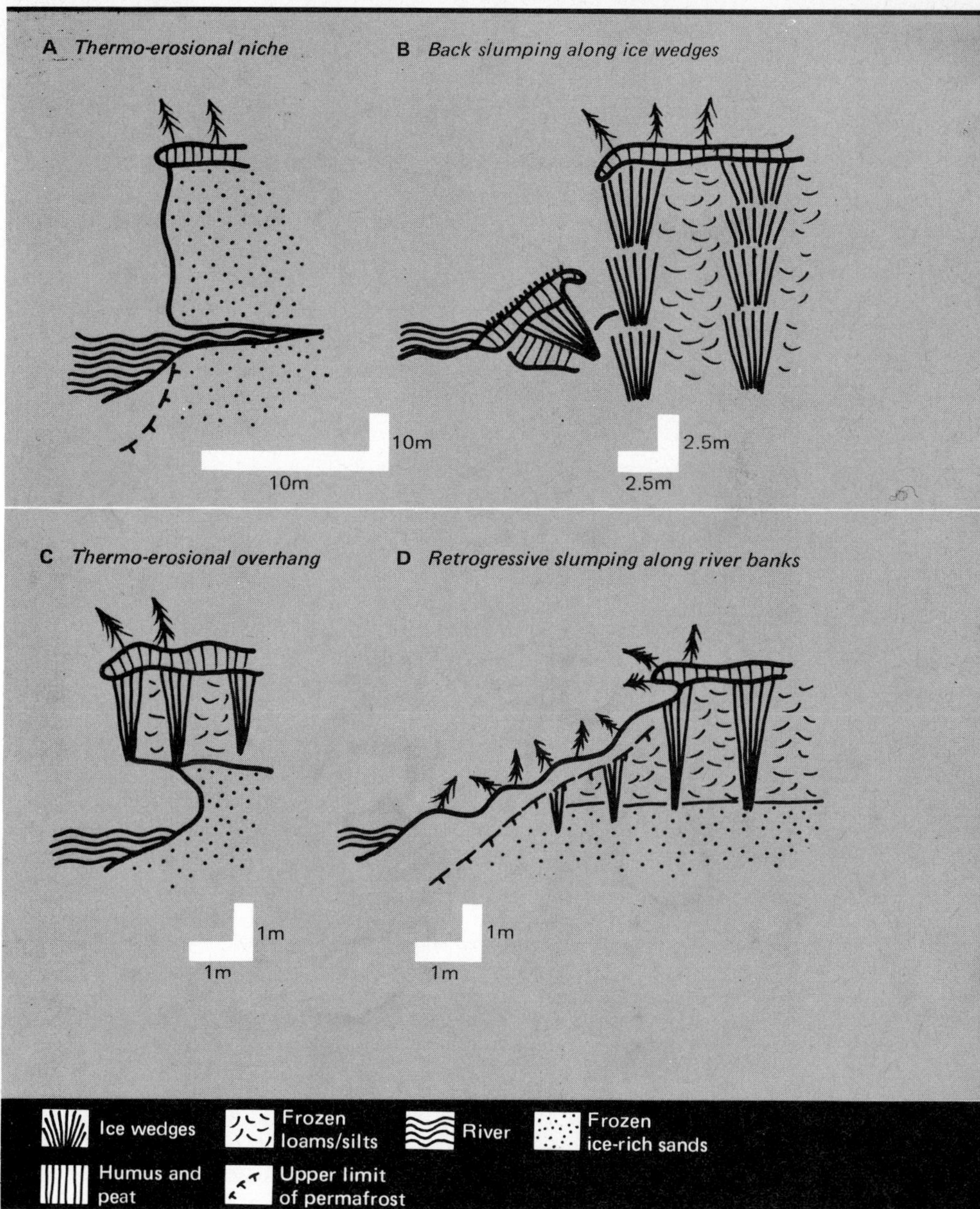

northern Siberia, and near Barrow, Alaska, a number of coastal erosion studies have emphasised this process (e.g. Hume ***et al.,*** 1972).

Two types of river bank modification were recognised by Walker and Arnborg (1966). The first, river bank sloughing, is the less important. In the first few weeks after the recession of the floodwaters, air temperatures rise and there is a thawing and sloughing from an oversteepened bank face. Sediment accumulates, therefore, in front of the niche, and may seal it until the following year. According to Walker and Arnborg, bank sloughing does not account for more than 1 m of bank recession per year. By contrast, very rapid and drastic morphological change can result from the collapse of the overlying frozen sediments along the line of ice wedges which parallel the river. In one instance, total bank retreat of over 10 m was observed in a two-day period.

Any discussion of thermal erosion within the periglacial environment should not restrict itself solely to the relationship of the thermo-erosional niche to ice wedges along river banks. For example, thermal erosion is but one aspect of fluvial erosion, and this topic is dealt with in Chapter 8. Also, thermal erosion is a factor influencing the nature of periglacial slopewash ('ruissellement'). The nature of slopes is discussed in the next chapter. In fact, thermal erosion is intimately linked to the whole question of slope and valley development in periglacial environments.

Man-induced thermokarst development

As the population and permanent settlement of areas underlain by permafrost continues to expand, the amount and extent of man-induced disturbances to the ground surface might be expected to increase also. The Soviet Union, by virtue of its long history of northern settlement and exploration, has by far the greatest experience of these problems of any northern country. As early as 1925, experiments were being undertaken to determine the effects of vegetation changes upon the underlying permafrost, brought about by either deforestation or by ploughing (Table 6.3).

In Alaska, a number of similar experimental studies have also been undertaken. One of the earliest, initiated by the United States Corps of Engineers in 1946, was near Fairbanks, Alaska, and involved the cutting and/or clearing of the surface vegetation. In the completely stripped area, the permafrost active layer increased from 1 m to over 3 m in thickness over a 10-year period. Subsequently, other experimental studies have emphasised the importance of the insulating qualities of the surface layers of organic material (e.g. Kallio and Reiger, 1969; Brown ***et al.,*** 1969). For example, in an alpine tundra environment in central Alaska, plots which were sheared of living vegetation, or sheared and covered with a mulch of tundra vegetation showed a two-time increase in the depth of thaw within a few years (Brown ***et al.,*** 1969). If this increase in thawing occurs in sediments with reasonably high ice contents, then thermokarst subsidence and/or erosion may take place. Even the smallest disturbances to the tundra may induce thermokarst. Mackay (1970) has described two examples which illustrate this point. Both are from the Mackenzie Delta region of Canada. In the first example, an Eskimo dog was tied to a stake with a chain 1.5 m long in an area of earth hummocks. In the 10 days of tether, the dog had trampled and destroyed the vegetation of that area. Within two years, the site had subsided like a pie plate to a depth of 18–23 cm and the depth to permafrost had increased 10–13 cm in the depression. In the second example, a number of walking trails were established along the coast of Garry Island between 1964 and 1966. Very slight seepage down the trails resulted in extensive thaw of the ice

Table 6.3 *Some examples of the effects of man induced vegetational changes upon permafrost conditions in the USSR, according to P. I. Koloskov, 1925. Quoted in Tyrtikov (1964)*

A – *Increase in average July soil temperature at 40 cm after ploughing, Yenesie region, Siberia*

Soil	Previous vegetation cover	Temperature increase (°C)
Semi-bog	Forest	+14
Peat-bog	Grass	+12
Semi-bog	Grass	+9
Light sod, gravelly	Grass	+3

B – *Soil temperature changes as a result of deforestation, Amur Province*

Depth of measurement (m)	Thickness of snow (cm)	Soil temperature changes within one year (°C)
0.2	20	+0.5
	10	−1.6
0.3	20	+0.7
	10	−1.2
0.4	20	+0.6
	10	−0.6

wedges with a collapse of the unsupported vegetation and soil overburden. In one case, where the trail cut inland from the coast, a gully had developed through thermal erosion along the ice-wedge system, and had progressed 30 m inland in two years.

These examples illustrate the extreme sensitivity of permafrost to man-induced surface modification. Disturbance almost invariably leads to an increase in thawing and this results in subsidence, slumping and sometimes gullying. The major variables are, of course, the ice content of the soil and second, the thickness and insulating qualities of the surface vegetation.

Perhaps the most common cause of man-induced thermokarst on a large scale is the clearance of the surface vegetation for agricultural or construction purposes. If this takes place in areas underlain by polygonal ice wedges, a distinctive hummocky ground develops since there is preferential settlement along the thawing ice wedges. A classic example has been described from the Fairbanks region of Alaska where ground was cleared in the 1920s for agricultural purposes. The following 30 years saw the development of a pattern of mounds varying from 3 to 15 m in diameter and 0.3 to 2.4 m in height in the cleared area (Rockie, 1942). Péwé (1954) has interpreted this topography to reflect the thawing of underlying ice wedges. The so-called 'mima mounds' of Washington State are also thought to be similar, but fossil, thermokarst mounds (Péwé, 1948; Richie, 1953). Another example is described by Haugen and Brown (1969) from Umiat, Alaska, where a trail was scraped of surface vegetation in the 1940s. Subsequently, a polygonal micro-relief has developed, some of the mounds being 20–30 m in diameter. In all cases, there is a certain similarity

Fig. 6.13 *Oblique air view of the disturbed terrain adjacent to the Sachs Harbour airstrip, Banks Island, Canada. The raised sections were used to transport material from the depressed areas (old borrow pits) to the airstrip.*

between these man-induced features and the baydjarakh terrain already described.

In the high Arctic islands, an excellent example of active man-induced thermokarst can be seen adjacent to the Sachs Harbour airstrip on southern Banks Island (Fig. 6.13). The underlying sands and gravels are ice rich with approximately 25–30 per cent excess ice and natural water (ice) contents of between 50 and 150 per cent (French, 1975b). The surrounding surface terrain is characterised by a well developed ice-wedge polygon pattern and there is every reason to believe that similar ice wedges underlie the area of the airstrip. When the airstrip was constructed between 1959 and 1962, it was necessary to eliminate a natural depression in the proposed airstrip site. Using bulldozers, the thawed surface sediments were stripped from the adjacent terrain and transported via ramparts to the airstrip. Material was removed over a period of three summers and an area of approximately 5 hectares (50 000 m^2) was disturbed. In places, 1.5–2.0 m of material was ultimately removed.

Since their development in 1962, the borrow pits on both sides of the airstrip have undergone progressive subsidence and thermokarst modification. Today, a hummocky type terrain has been created, composed of small mounds and depressions (Fig. 6.14). In places, the hummocky terrain is dissected by linear troughs or depressions commonly 1.0–3.0 m in width and with a maximum relative relief of 100—150 cm. Numerous bodies of standing water exist in many of the depressions, and in the spring water drains across the airstrip at its point of lowest elevation causing gullying and maintenance problems.

The examination of air photographs taken in 1964 indicate that this terrain developed within three years of the initial disturbance. By 1968, the linear depressions had begun to assume a broad polygonal alignment, probably reflecting the larger ice wedges within the underlying polygonal net. Detailed levelling and field observations in 1972 and 1973 suggest that subsidence and permafrost degradation is still active, over 10 years after the initial disturbance. Thaw depths in disturbed terrain where the Dryas tundra vegetation had been removed were between 10 and 20 cm greater on average than in undisturbed terrain, and the mounds were subsiding at a rate of between 5.0 and 10.0 cm per year. Given the present relief and the fact that it had developed from flat floored borrow pits in 10—12 years, this suggests that thermokarst is beginning to slow down. It is probable, therefore, that following the initial disturbance and the thickening of the active layer, maximum subsidence quickly occurred along the ice wedges to produce the linear troughs. Subsequently, the intertrough mounds have undergone further degradation through increased thawing from all sides, the release of excess water, and the settling of the thawed sediments under gravity

Fig. 6.14 *Thermokarst topography developed in the borrow pits adjacent to the Sachs Harbour airstrip. Standing water bodies indicate active melting of the permafrost. The underlying silts, sands and gravels are ice-rich with 25—30 per cent 'excess ice' content. Photograph taken in 1973.*

processes. Water, collecting within the depressions and troughs, has further accentuated the thawing and subsidence and has promoted the gullying of the airstrip.

A second common cause of man-induced thermokarst is related to the movement of vehicles and the associated vegetation destruction. In Canada, a classic example of this type of terrain disturbance was the bulldozing of seismic lines in the Mackenzie Valley in the middle 1960s. Seismic survey constitutes one of the methods by which oil exploration companies obtain geophysical data. Most seismograph lines are run during winter when the ground is frozen and when bulldozers can be used to consolidate or remove snow to obtain a smooth compact surface upon which the geophones are placed. Canadian government regulations now restrict such activities to the winter months. However, in late summer 1965, some 300 km of seismic lines were bulldozed and long strips of vegetation and soil, approximately 4.25 m wide and 25 cm thick were removed. This debris was piled in irregular ridges along the edge of the seismic line. Thermokarst subsidence and erosion by running water subsequently transformed many of these lines into prominent trenches and canals over much of their length (Kerfoot, 1974). In one instance, where the seismic line depression had a relative relief of 1.55 m, it was calculated that 13 per cent of the relief change was the result of material removed by the bulldozer, 45 per cent was due to the redistribution of debris along the edge of the seismic line, and 42 per cent due to the subsequent settling of the ground surface as a result of thermokarst subsidence. In other areas, where the seismic lines passed over an area underlain by ice wedges, the most pronounced thermokarst subsidence occurred along the position of the ice wedge and the floor of the settled area developed a distinctive corrugated appearance, with several closed depressions as much as 3 m deep (Kerfoot, 1974).

The movement of tracked vehicles across the tundra in the summer months has been particularly destructive of the surface vegetation. Large areas of Alaska are now permanently scarred by these tracks, since they will not disappear for many decades. Once present, they initiate thawing of the underlying ice wedges or, if upon a steep slope, they channel snowmelt and surface run-off within them, to promote thermal erosion and gully development. In general however, gullying is relatively rare and its importance has been over-estimated. Usually, the prevalence of low surface gradients over much of the tundra, the paucity of winter snowfall and spring run-off in many Arctic regions, and the differential settling of the ground by thermokarst subsidence creates an irregular surface which precludes the development of integrated drainage and extensive gully erosion. For the most part, the problem of vehicle track disturbance is an aesthetic one.

These examples of man-induced thermokarst represent one aspect of a wide range of engineering problems which are unique to periglacial regions. From a geotechnical point of view, they all emphasise the need for thorough investigation of ground ice conditions before construction or movement, and the desirability of minimising surface terrain disturbance once activity is underway. In keeping with recent environmental conservation attitudes, Canada and the United States are now developing regulations designed to ensure minimal terrain disturbance in northern regions. The off-road movement of tracked vehicles in the summer months is now prohibited in both Canada and Alaska. Oil company exploration activities and the hauling of heavy equipment is now limited to winter months. In Canada, a significant government achievement was the production of a series of 'Terrain Classification and Sensitivity' maps at a scale of 1:250 000 for the proposed Mackenzie Valley pipeline and transportation corridor in 1972. They illustrate land capability and performance in relation to superficial and bedrock geology and terrain conditions. Criteria used include grain size, active layer conditions, excess ice and natural water (ice) contents, Atterberg limits, and the nature of organic surface material. An evaluation of the terrain units as a source of construction materials was also included. In 1973, a number of 'Terrain Disturbance Susceptibility Maps' at a scale of 1:50 000 were produced for selected areas (e.g. Norman Wells, Map 22, P. J. Kurfurst, 1973) in which terrain was mapped and ranked mainly according to its ground ice and vegetation cover, and its susceptibility to thermokarst. For example, in the highest category was terrain underlain by organic and inorganic clays and silts, possessing low permeability and high plasticity, on slopes in excess of 5°, with moderate to high ice contents and irregular patches of surface organic cover. According to the legend, such terrain is a poor source of fill material, large detachment slides and retrogressive flowslides are common, and there is a high susceptibility to major thermokarst slumps and rapid gullying due to disturbance. Hopefully, such maps will assist in minimising terrain disturbance and man-induced thermokarst in the future, as well as provide a base upon which engineering decisions can be made.

7 Hillslope forms and processes

Hillslopes constitute that part of the periglacial landscape which occurs between the summits of hills and interfluves and their drainage channels. In both form and process, hillslopes are interdependent with stream channels and the geometry of drainage basins. Stream channels provide the means by which weathered and transported debris is removed from the base of the slope. The geometry of the drainage basin determines the basic length and height parameters of the slope.

In detail, however, our understanding of the course of slope development in periglacial environments is limited. There are several reasons for this. First, there is a lack of quantitative measurements and field observations. Second, slope changes through time are poorly documented because of their general slowness. Third, it is often argued that, because of climatic change, the periglacial cycle of slope development rarely runs its full course and there is the inevitable problem of distinguishing between the effects of past and present climates. Finally, although significant advances have been made towards the understanding of certain slope processes, the link between hillslope form and process is still difficult to make. For these reasons, the present chapter serves only as an introduction to future investigations. It is organised around a discussion of first, the mass wasting process as applied to periglacial environments and second, the various slope form assemblages that occur in present-day periglacial environments. Finally, a short section considers the nature of periglacial slope evolution.

Mass wasting processes

Mass wasting is the term applied to the downslope movement of debris under the influence of gravity. It is a process which is not unique to periglacial environments. On the other hand, mass wasting processes probably reach their greatest intensity and efficacy under periglacial conditions. There are at least three reasons for this. First, the dominance of frost action in periglacial regions is itself a cause of mass wasting through the frost creep mechanism. Second, the high moisture content of the active layer is particularly favourable for the operation of mass wasting processes. This situation is brought about by the melting of ground ice as the active layer thaws each spring, and the downward percolation of moisture into the soil as the result of snowmelt and/or summer rain. Third, the presence of permafrost directly aids mass wasting in several ways. Not only does permafrost effectively limit the downwards infiltration of water into the ground inducing high pore water pressures, but the permafrost surface acts as a water lubricated slip plane for movement of the overlying active layer.

A number of processes which operate separately or in association are usually included within the term mass wasting. It is not unreasonable to assume that in any one area a specific process or group of processes will dominate, depending upon such factors as climate, lithology, and local topography. The different types of mass wasting may be summarised as being of either a slip, flow, or fall nature. Flow involves gelifluction, creep and slopewash; slip involves active layer failures and ground ice slumps; fall involves avalanches and rockfalls. The various mass wasting processes may be summarised under the headings of solifluction, slopewash, solution, nivation, and rapid mass movement.

Solifluction Solifluction is regarded as one of the most widespread processes of soil movement in periglacial regions. The term was first used by J. G. Andersson to describe the 'slow flowing from higher to lower ground of masses of waste saturated with water' which he observed in the Falkland Islands (Andersson, 1906, p. 95). Since solifluction, so defined, is not necessarily confined to cold climates, the term 'gelifluction' has been proposed to describe solifluction associated with frozen ground (e.g. Washburn, 1973). Intimately associated with gelifluction is a second process, 'frost creep'. This is defined as 'the net downward displacement that occurs when the soil, during a freeze–thaw cycle, expands normal to the surface and settles in a more nearly vertical direction' (Benedict, 1970, p. 170). When operating together the two processes of gelifluction and frost creep constitute the movement which is generally termed solifluction in the modern sense.

Suitable conditions for gelifluction occur in areas where the downward percolation of water through the soil is limited, and where the melting of segregated ice lenses provides excess water which reduces internal friction and cohesion in the soil. It has been shown that significant gelifluction occurs usually when moisture values nearly reach or exceed the Atterberg liquid limits (e.g. Washburn, 1967; Harris, 1972). In contrast to gelifluction, frost creep has traditionally been regarded as a much slower process, incapable of producing the distinct lobes and stripes associated with gelifluction. Displacement of particles at the ground surface due to the upheaving of fine particles by needle ice (see p. 33) constitutes one of the few instances of rapid and noticeable creep. The actual amount of frost creep will clearly decrease with depth and depend upon such variables as the frequency of freeze–thaw cycles, the angle of slope, the moisture available for heave, and the frost susceptibility of the soil.

The nature of the creep and gelifluction components, and their relation to each other, is illustrated in Fig. 7.1. There are three types of movement that must be considered. First, the Potential Frost Creep (***PFC***) is the horizontal movement associated with the movement from P_1 to P_2, consequent upon the freezing and heaving of the ground. Since the mechanics of the frost heave process have been discussed in Chapter 4 they need not be repeated here. What is important to note is that because heave takes place at right angles to the slope, the amount of ***PFC*** will increase in direct proportion to the angle of slope. Gelifluction, ***G***, is the second major component of horizontal movement. The majority of gelifluction movement occurs in the spring and subsequently decreases later in the summer as the active layer thickens, snowpatches disappear, and the ground is subject to desiccation by wind. Only in favoured localities, such as immediately below perennial or long-lasting snowpatches does gelifluction continue throughout the summer. The third component of movement is that of Retrograde Movement (***R***). This is a negative or upslope movement acting in the opposite direction to gelifluction. As the ground settles upon thawing, there is a tendency for the thawed material to settle back against the slope rather than purely vertically. This movement only becomes apparent if it exceeds gelifluction.

Fig. 7.1 *Diagram illustrating the components of solifluction movement. A - the movement of a surface particle due to frost creep, gelifluction and retrograde movement. B - the frost creep mechanism. From Washburn (1967) and Dylik (1969a)*

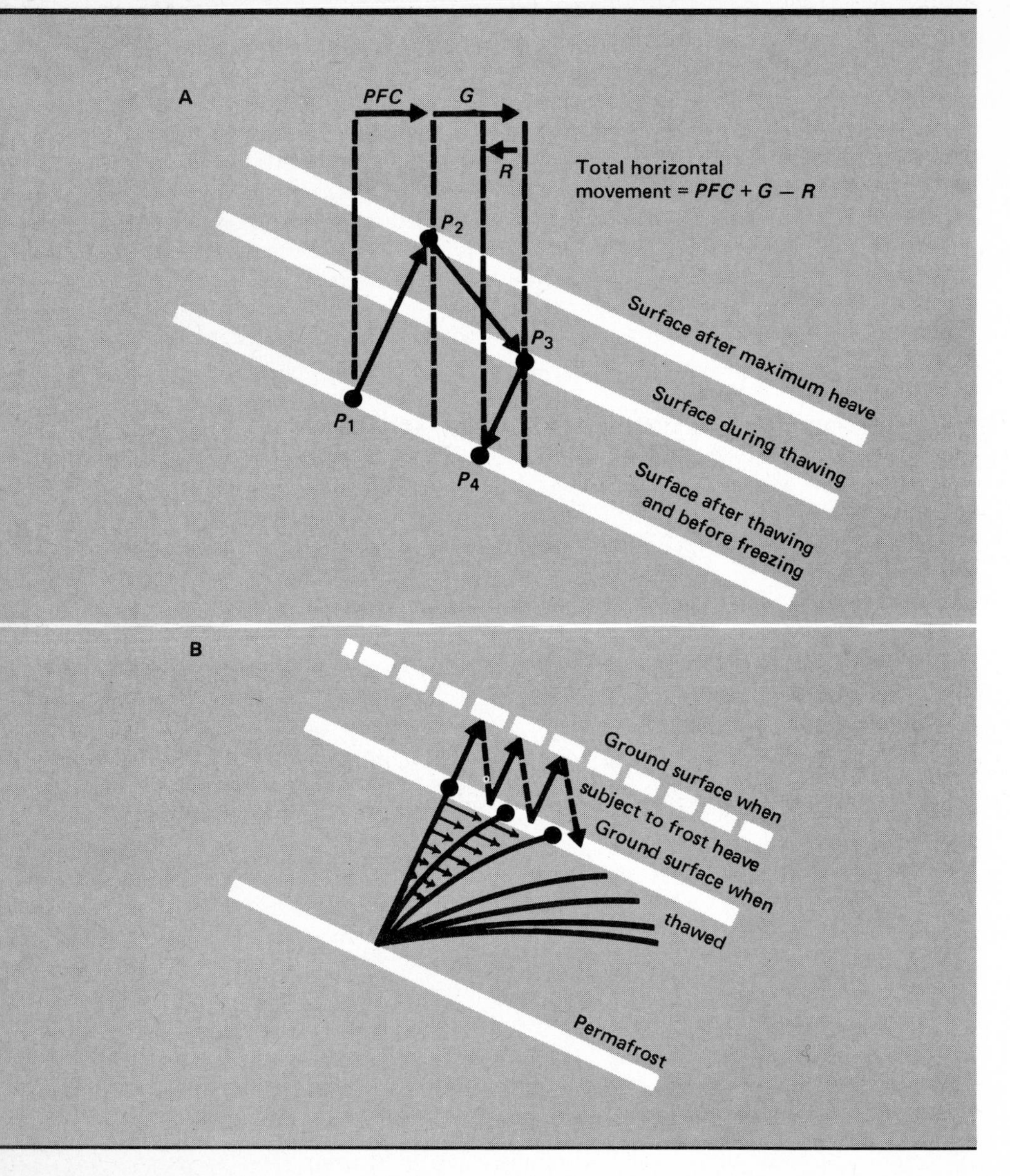

The relative importance of each component of movement will vary from locality to locality, depending upon site factors such as moisture availability or regional factors such as the seasonal rhythm. In an alpine region with strong diurnal and seasonal rhythms, such as the Colorado Front Range, frost creep is the dominant process operating for much of the year (Benedict, 1970). Only in specialised micro-environments such as the axial positions of wet gelifluction lobes, do the effects of gelifluction exceed the effects of frost creep. Retrograde movement and gelifluction both occur in the spring and minimise the effect of the other. On the other hand, observations from a high latitude environment, such as East Greenland, indicate that either process may dominate in a given year, and in different sectors of the same slope depending upon moisture conditions. According to Washburn (1967), frost creep tended to exceed gelifluction by not more, and probably less, than 3:1 in most years, and frost creep commonly resulted in 30–50 per cent of the total movement. In East Greenland, it would appear that frost creep due to the annual cycle is more important than creep due to short term cycles.

Numerous studies have investigated the magnitude and rate of movement of solifluction by direct field measurement. Surface movement has been monitored by marked stones and pegs placed on the surface or protruding through the surface (e.g. Pissart, 1964; Washburn, 1967), while variations of movement with depth have been studied by the insertion of plastic cylinders which subsequently deform (Rudberg, 1962; Harris, 1972), by rectangular tin foil markers (French, 1974a), by probes attached to strain gages (Williams, 1962; Price, 1973), and by linear motion transducers (Everett, 1966). Some of the various rates recorded and published in the literature are summarised in Table 7.1.

Table 7.1 *Some recorded rates of solifluction movement*

Locality	Reference	Gradient (degrees)	Rate (cm year^{-1})
Spitsbergen	Jahn (1960)	3–4	1.0–3.0
Spitsbergen	Jahn (1961)	7–15	5.0–12.0
Karkevagge, Sweden	Rapp (1960a)	15	4.0
Tarna area, Sweden	Rudberg (1962)	5	0.9–1.8
Norra Storfjell, Sweden	Rudberg (1964)	5	0.9–3.8
French Alps	Pissart (1964)		1.0
East Greenland,	Washburn (1967)		
'Wet' sites			3.7
'Dry' sites			0.9
Colorado Rockies	Benedict (1970)		0.4–4.3
Okstindan, Norway	Harris (1972)	5–17	1.0–6.0
Ruby Range, YT, Canada	Price (1973)	14–18	0.6–3.5
Sachs Harbour, Banks Island, NWT, Canada	French (1974a)	3	1.5–2.0

In general, most investigations indicate there is a progressive decrease of movement with depth and that solifluction movement is laminar in nature. Usually, movement is restricted to the uppermost 50 cm of the active layer. Slope angle appears to be of little importance in influencing solifluction movement, although there is certainly a significant difference in solifluction movement between

Table 7.2 *Average subsurface movement at five sites on slopes of 2–4° at Sachs Harbour, Banks Island, western Canadian Arctic. From French (1974a). All data in centimetres*

	Depth	Site 1	Site 2	Site 3	Site 4	Site 5	Total average movement, 1969–72
	3.0	5.8	4.9	4.0	5.9	5.6	5.3
	7.0	4.9	4.6	4.2	5.7	5.6	5.0
	15.0	4.7	4.9	4.2	5.4	3.9	4.6
	30.0	4.2	4.5	4.5	4.3	2.1	3.9
Total average movement, 1969–72		4.9	4.7	4.2	5.3	4.3	4.7
Average annual movement, 1969–72		1.63	1.56	1.40	1.76	1.43	1.56

slopes of less than 5° and those over 15°. It is difficult to generalise about rates, but Table 7.1 suggests that average rates of surface movement of between 0.5 and 4.0 cm $year^{-1}$ are typical for most environments. On very low angle slopes, rates in excess of 2.0–2.5 cm $year^{-1}$ are probably unusual; equally, rates in excess of 6.0–10.0 cm $year^{-1}$ probably involve some other gravity controlled process in addition to solifluction and are usually only encountered on steep slopes. The decrease of movement with depth is usually of a linear nature. For example, Table 7.2 gives annual average rates of movement over a three-year period for depths of 3, 7, 15, and 30 cm at five sites on Banks Island, Canada. In all sites except No. 3, there is a progressive decrease in movement with depth. Other investigations by Rudberg (1964) and Price (1973) suggest that the depth of movement does not usually exceed 50–60 cm, and is often less. In general, therefore, and if one assumes an average surface movement rate of between 0.5 and 5.0 cm $year^{-1}$, and that movement in the upper 50 cm is on average one-quarter of this amount, it can be calculated that solifluction transports between approximately 6.0–60.0 cm^3 cm^{-1} $year^{-1}$ of material downslope (Young, 1972, p. 60).

Vegetation has a marked influence upon movement rates since, by restricting surface movement, it may cause greater subsurface movement. This probably explains the subsurface movement pattern of site No. 3 in Table 7.2. More detailed information is given by Price (1973) who has investigated solifluction movement on slopes of different aspect in the Ruby Range of the Yukon Territory. Certain sites showed a convex profile of subsurface movement (i.e. greater subsurface movement) while others showed the normal decrease in movement with depth. Movement also varied with aspect and micro-climate. On the north- and east-facing slopes where vegetation was most poorly developed, the greatest rates of movement occurred (2.4–2.7 cm $year^{-1}$); lower values were recorded on southeast-facing slopes (1.6 cm $year^{-1}$), where the vegetation cover was particularly well developed. Movement was least (0.7 cm $year^{-1}$) on southwest-facing slopes which, exposed to the prevailing winds, were drier than the other three slopes. With respect to subsurface movement, only the southeast-facing slope showed greater subsurface movement than surface movement. According to

Fig. 7.2 *Solifluction sheet on very gentle slopes, northwest Banks Island, Canada.*

Price, this reflected the inhibiting influence of a thick organic turf layer combined with a shallow active layer and abundant moisture just beneath the surface.

Solifluction activity produces a number of distinct morphological forms. Perhaps the least studied but most widespread solifluction form is the solifluction sheet. Usually the result of extensive gelifluction and frost creep, solifluction sheets produce uniform expanses of smooth terrain, often at angles as low as 1–3° (Fig. 7.2). The downslope edge of such sheets is usually characterised by numerous small lobate forms with Dryas-banked risers, often only a few centimetres high. Not only are these solifluction sheets capable of movement on extremely low angled slopes, but they are capable of transporting large erratic boulders. The latter, which are sometimes referred to as 'ploughing blocks', are rafted on the surface, their undersides resting at or near the permafrost table. Sometimes, they leave a shallow trough upslope indicating the path of their movement. Solifluction sheets give rise to extensive rectilinear slopes of between 1 and 3° in angle. Solifluction sheets are probably best developed in high Arctic regions where the absence of vegetation enables solifluction to operate uniformly, and where there is extensive lowland terrain underlain by unconsolidated sediments. In the tundra and forest tundra areas further south, vegetation hinders the sheet-like extension of solifluction, and instead, favours more localised lobate movement.

Lobes and terraces are relatively well known solifluction phenomena. They have been reported to occur in most periglacial environments, but are probably best developed in alpine regions or areas of significant local relative relief. The lobes and terraces commonly give rise to a stepped, tread-like slope which may range in angle from 3–5° upwards to 15–20°. The micro-relief of such slopes varies from vertical turf risers in excess of 2–3 m in height, down to small dryas-backed risers only a few centimetres high. Each riser is separated by treads of low angle and varying extent. Both creep and gelifluction contribute significantly to the movement of lobes, with gelifluction being more important in the wetter axial parts and creep more important in the drier peripheral parts. Obviously, for lobes to develop, the gelifluction movement must be greater than frost creep and be concentrated in well defined linear paths. Where movement is more uniform, terraces develop. Particularly favoured locations for the development of lobes are immediately below snowpatches while terraces often develop on the lower parts of valley slopes. Given time, solifluction lobes and terraces will advance downslope progressively burying other solifluction lobes developed at lower elevations.

Sections excavated through solifluction features usually show one or more buried organic layers over which the lobe or terrace has advanced. Radiometric dating of this material is one method of calculating past rates of movement. The fabric of solifluction material can also be studied in sections. Besides its heterogeneous and non-sorted nature, the next most obvious characteristic of many solifluction deposits is the preferred alignment and inclinations of pebbles. In most cases, pebbles have their long axes oriented in the direction of movement (Fig. 7.3), a condition also found in Pleistocene deposits (see p. 235). Rather more problematic is the common tendency for the downslope end of pebbles to be inclined upwards at angles between 20 and 45° (Fig. 7.3). It probably reflects the combined effects of stone tilting due to freezing (see Chapter 3) and material

Fig. 7.3 *Fabric diagram of colluvium on the gentle slope of an asymmetrical valley, Beaufort Plain, northwest Banks Island, Canada. From French (1971a)*

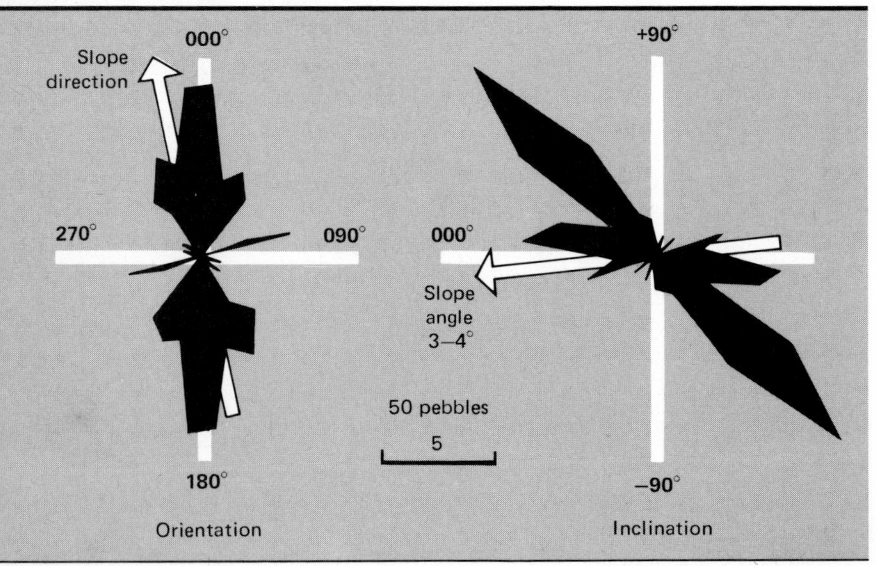

from upslope which progressively overrides the upslope end of the stone first so as to slightly increase overburden pressure at one end of the pebble.

Some lobes and terraces are marked by a concentration of large stones or boulders at their downslope ends. These have been variously called 'stone garlands', 'block banked terraces' or 'boulder steps'. The stones are commonly tilted upwards and appear to be 'emerging' from the terrace or lobe. Their existence is usually interpreted as indicating that subsurface movement is greater than surface movement. Turf-banked features on the other hand, indicate surface movement to be greater and pebbles at the downslope ends of these features often possess a more horizontal or even a 'submerged' character (i.e. downslope inclination).

Slopewash The importance of slopewash ('ruissellement') as both an erosive and transportational process in periglacial environments has been neglected in the past. This reflects the fact that (a) slopewash was commonly considered an azonal process and therefore not unique to the periglacial environment and (b) early field workers were attracted by the importance of the rather more recognisable solifluction and frost-action phenomena. This neglect is unfortunate since slopewash is probably a significant process of erosion and transportation operating on unvegetated slopes in periglacial regions, and in several instances, may exceed the effect of solifluction. However, at present there is an almost complete lack of quantitative data to substantiate this claim, and detailed field observations and measurements are desperately needed.

Slopewash processes are well known in non-periglacial regions of the world. A number of different processes contribute to slopewash. According to Carson and Kirkby (1972, pp. 188–230) erosion may be by (1) rainsplash impact, (2) unconcentrated surface wash, commonly termed sheetwash, (3) concentrated surface wash, commonly termed rillwash, and (4) subsurface wash, in which particles are transported through the soil pores. In the case of the surface wash processes, particles are carried along in an overland flow. Where the flow is thin, individual particles are removed from the surface without any marked concentration of erosion to give uniform denudation. If the flow varies in thickness, or if it is turbulent, flow becomes progressively concentrated into smaller channels of higher intensities of erosion to give differential denudation. It has also been found that vegetation is the major influence upon the rate of transport of sediment. Vegetation intercepts precipitation, breaks the impact of raindrops, improves soil structure and infiltration around roots, binds the soil, and subdivides overland flow around plant stems and litter. According to Carson and Kirkby (1972, p. 219), rates of surface soil transport are greatest in arid and semi-arid areas of the world which receive less than 1 000 mm of precipitation per annum and least in humid temperate areas with continuous vegetation. For example, in the UK several hundreds of measurements for a variety of sites give a median value of only 0.09 $cm^3\ cm^{-1}\ year^{-1}$. On the other hand, studies in the arid and sparsely vegetated regions of the western United States have recorded values ranging between 200 to over 1 000 $cm^3\ cm^{-1}\ year^{-1}$ (e.g. Leopold ***et al.***, 1966; Schumm, 1964; Carson and Kirkby, 1972, p. 209).

Since many high Arctic periglacial regions are also arid or semi-arid, and since vegetation is often sparse or totally lacking, it is not unreasonable to assume that high rates of surface soil transport may be typical. If correct, and bearing in mind average solifluction movement of between 6.0 and 50.0 cm^3 cm^{-1} $year^{-1}$ (see previous section), it follows that surface wash may be as important a transporting agent as solifluction in certain high Arctic areas. Such a conclusion must be modified for areas where the vegetation cover is greater. For example, in the boreal or taiga forest zones, surface wash is probably insignificant. Other areas probably lie between these two extremes. As a generalisation, one might assume that slopewash progressively increases in importance northwards from the treeline, and bears an inverse relationship to the percentage of ground covered by vegetation. As precipitation amounts decrease in the extreme high latitudes, a zone of maximum slopewash effectiveness might be expected to occur in the transition from tundra to polar desert.

Over and above these general considerations, there are additional reasons for regarding slopewash processes as being particularly effective in periglacial regions. First, the presence of permafrost prevents infiltration and favours surface run-off even on very low angled slopes. Second, periglacial areas have a high effectiveness of run-off since evaporation rates are low and run-off is concentrated in the short summer season. Third, moisture for slopewash in non-periglacial regions is usually considered to be derived from rainfall and/or snow-melt. In periglacial regions however, the most important source is the melting of the pore ice and/or small segregated ice lenses as the active layer thaws each spring. As has been outlined earlier (p. 77), pore ice or ice cement is the most widely distributed ice type in permafrost, and Dylik (1972, p. 172) has stressed that the melting of such ice is '. . . the essential requirement for periglacial downwash'.

There is a further factor which makes periglacial slopewash distinct from normal, azonal, slopewash. Because of the greater thermal heat capacity of water than of mineral soil, the water present in the soil promotes the slow melting of permafrost. The melting of the pore ice in particular results in the liberation of individual mineral particles small enough to be transported downslope. This process is called 'thermo-erosional wash' and is an erosive process associated with essentially unconcentrated wash and subsurface wash. According to Soviet authorities, thermo-erosional wash is capable of operating upon slopes of very low angle indeed, and is the major process leading to the general flattening of periglacial terrain. This process of relief reduction has been called 'thermo-planation' (Kachurin, 1962; Soloviev, 1962; Dylik, 1972). However, its operation has yet to be adequately demonstrated.

The effectiveness of the various types of slopewash will depend on the nature of the terrain, the angle of slope, the nature and amount of ground ice, and the height of the permafrost table. For example, in continental areas with deep active layers and well developed boreal or taiga forests, conditions are best suited to subsurface wash. If throughflow is concentrated at any one locality, and if lithological conditions are favourable, local collapse hollows may develop which are similar in appearance to limestone solution holes. This process is termed 'suffosion' by Soviet investigators (e.g. Kachurin, 1962; Anisimova ***et al.,*** 1973). It is well known in parts of central Yakutia where springs, emerging from the foot of terraces and fed by supra- and subpermafrost waters, excavate large amounts of fine sands (Anisimova ***et al.,*** 1973). North of the treeline and where the depth of thaw is less, sheetwash and rillwash operate widely on unvegetated slopes (e.g. Jahn, 1961; Bird, 1967, p. 246), particularly beneath snowpatches and on nivation or cryoplanation benches (e.g. St-Onge, 1965, 1969; Demek, 1968, 1969a). One of the few quantitative measurements of slopewash is provided by Jahn (1961) from Spitsbergen; approximately 12–18 g m^2 $year^{-1}$ of sediment was being washed downslope beneath large perennial snowbanks.

Very few phenomena can be attributed solely to the action of slopewash, since slopewash is but one of a complex of processes operating upon slopes. It is believed that slopewash and intense freeze–thaw activity during the Pleistocene resulted in the deposition of rhythmically bedded slope deposits (Dylik, 1960). Such deposits have been widely recognised in present-day temperate latitudes, especially in Europe where they are termed 'grèzes litées' (see pp. 234–5). However, similar deposits have not been widely reported from high latitudes and it is probable that high latitudes do not experience the repeated freeze–thaw oscillations that favour the frost shattering and slurrying which the stratification of 'grèzes litées' suggests. The more typical slopewash deposits currently being formed in high latitudes are the colluvium layers and other fine unconsolidated sediments that accumulate at the foot of slopes and below snowbanks as the result of the continual percolation of meltwater downslope.

Solution The downslope percolation of rainwater and snowmelt may effect considerable denudation by solution if the underlying sediments are of a calcareous nature. This is on account of the solubility of calcium carbonate ($CaCO_3$) in water. In areas underlain by relatively pure limestones, this process assumes particular importance. Furthermore, since the solubility of $CaCO_3$ in water increases with a decrease in temperature, it has been suggested that solution in periglacial and polar regions is considerably greater than in other regions (e.g. Corbel, 1959). In his study of denudation in northern Sweden, Rapp (1960a) concluded that solutional loss was by far the most important agent of denudation. However, other studies in different periglacial environments suggest that solutional activity is no greater under periglacial conditions than under other conditions. For example, using data obtained from the central Canadian Arctic, Smith (1972) has shown that (a) solution rates are actually smaller than in lower latitudes, (b) carbonate concentrations in standing water bodies are of the same order of magnitude as those of temperate regions, and (c) that concentrations of carbon dioxide in snowbanks, as suggested by Williams (1949), do not necessarily exist. Some typical limestone solution values are listed in Table 7.3. It would appear that a wide range of solutional values occur in periglacial environments and that, other factors being equal, precipitation controls the solution rate. The generally low rates of limestone denudation which characterises the polar and continental localities reflects the relative aridity of these areas while areas of greater moisture and more maritime situations have higher rates.

The significance of limestone solution is usually considered within the context of limestone or 'karstic' relief. In periglacial environments, the presence of permafrost in many areas and the inability of water to percolate downwards suggests that limestone solution is incapable of producing a karst-like relief similar to that which is found in other environments. Thus, St-Onge (1959) has remarked upon the absence of solutional effects in terrain developed in gypsum on Ellef Ringnes Island, and Bird (1967, pp. 257–70), in a detailed account of the limestone scenery of the central Canadian Arctic, has concluded that

Table 7.3 *Some rates of limestone denudation by solution under periglacial and non-periglacial conditions*[1]

Location	Mean annual precipitation (mm)	Net rate of denudation (mm/1 000 yrs)	Total carbonate hardness (ppm)[2]	
			$CaCO_3$	$MgCO_3$
Periglacial				
Somerset Island, NWT, Canada	130	2	84	22
Spitsbergen	300	27		
Tanana River, central Alaska	450	40		
Svartisen, northern Norway	740–4 000	400		
Non-periglacial				
Western Ireland	1 000–1 250	51–53	66	11
Southern Algeria	60	6		
Indonesia	200–300	83		

(1) Data from J. N. Jennings, ***Karst***, M.I.T. Press (1971), Table 6; M. M. Sweeting, ***Karst Landforms***, Macmillan Press Ltd, Tables X and XI; and Smith (1972).
(2) Standing water bodies.

solutional effects are weakly developed and, in most instances, overwhelmed by those of physical weathering. Nevertheless, the fact that solutional activity is concentrated in the active layer is of significance. It may be that the generally subdued nature of many periglacial regions, whether or not they are underlain by limestone, is in part the result of solutional activity associated with thermo-erosional wash.

Nivation Nivation is the term given to the combined action of frost shattering, gelifluction and slopewash processes which operate in the vicinity of snowbanks. It is an erosive process involving localised and intense physical weathering, the result of an abundant moisture supply percolating into the rock beneath and around the snowbank. Lithology, in determining debris size, controls the agents of removal and consequently, the nature of the landforms which result. In its simplest form, nivation processes erode shallow hollows or cirque-like basins (Fig. 7.4) which form on slopes and upland surfaces (Cook and Raiche, 1962b). If active, they are occupied by semi-permanent snowbanks; if inactive, the snowbanks are lacking. Debris is removed from the hollow by sheetwash, rillwash and gelifluction.

The initial growth of nivation hollows is poorly understood, and there is a lack of quantitative data available to describe the nature and rapidity of the nivation process. Once the hollow has developed, however, the process is self-generating since snow, remaining in the hollow, prolongs the action of the freeze–thaw, gelifluction and slopewash processes. Nivation hollows have been categorised as being either transverse (i.e. linear) with the major axis lying trans-

Fig. 7.4 Oblique air view from 300 m elevation of large transverse nivation hollow with snowbank and rill and sheetwash zone below snowbank, central Banks Island.

verse to the drainage, longitudinal (elongated downslope) or circular (a transition between the other two types). The most common type of nivation hollow is undoubtedly the transverse hollow. Prevailing winds are an important control over their location and distribution. They often occur on the upper parts of slopes which are in the lee of the prevailing winter winds. Structural benches and other slope irregularities, if located in a lee slope position, are particularly suited for snowbank accumulation and nivation processes. If the lithological conditions are favourable (see pp. 159–61), nivation processes can result in a variety of erosional landforms (e.g. St-Onge, 1969), and are important in the initial stages of cryoplanation (see p. 162) and the development of irregular, faceted slope forms.

Rapid mass movement The final category of mass wasting processes are those involving the relatively rapid or even catastrophic movement of material downslope. Slope failures of various sorts, rockfalls and avalanches are the more important of these processes. All are rather more localised and/or periodic processes than the other mass wasting processes previously described.

Localised and small-scale slope failures which are confined to the active layer are relatively common in Arctic lowlands underlain by fine grained unconsolidated sediments. Although no systematic study of these features has yet been undertaken in high Arctic regions, they appear to occur quite randomly and

Fig. 7.5 *An active layer slope failure, Masik Valley, southern Banks Island. It occurred on terrain underlain by shales of Cretaceous age, between 1 and 5 August 1969, following summer snowfall.*

been undertaken in high arctic regions, they appear to occur quite randomly and are usually restricted to the middle or upper parts of slopes, with the entrance and exit of the failure plane well within the slope (Fig. 7.5). Casual observations indicate that they occur on slope angles which are usually greater than 10–15°. The cause of such failures is attributed to local conditions of soil moisture saturation and resulting high pore water pressures. This leads to a point where the shear strength is exceeded and the slope becomes unstable. The surface of the permafrost table acts as a lubricated slip plane for movement and controls the depth of the failure plane. The majority of active layer failures in high Arctic regions are small, the slump scar or hollow often being no more than 2–5 m in diameter. In particularly fine grained sediments possessing high liquid limits, failure takes the form of a mudflow which may extend downslope for distances in excess of 100 m in certain cases. If icy sediments are exposed at the permafrost table, these will then melt and the slump scar will develop in a regressive manner. The extreme case of regressive slumping is that of ground ice slumping (see pp. 119–22). In some instances, a simple active layer failure may be the initial 'trigger' for extensive ground ice slumping.

In the Canadian Arctic islands, conditions are particularly suited for frequent active layer failures and they have been widely reported (e.g. Rudberg, 1963; Bird, 1967). Not only is the active layer thin and fine grained unconsolidated sediments widespread, but over 50 per cent of the total precipitation falls as summer rain. While not appreciable in absolute quantity (often less than 75 mm), it is usually concentrated into two or three periods of prolonged light rain, at which time saturated soil conditions can prevail. Years of excessive rainfall and/or cooler than normal summers are years in which frequent failures occur; in drier years, instability is less frequent. Favourable conditions also occur in those years when the spring thaw follows a winter of above average snowfall or when the spring thaw is late and therefore rapid. In both cases, high pore water pressures develop at a time when the active layer is not fully developed.

Similar slope failures are also known to occur widely in the boreal forest zone of the Mackenzie Valley lowlands where they have been termed 'detachment failures' (Hughes, 1972). They may be of either a simple rotational nature, or of a flowslide or solifluction nature. They are particularly common along river banks and on ice-rich colluvial slopes developed on shales, fine grained tills and glacio-lacustrine sediments. In other areas of the boreal forest/tundra transition zone, active layer failures may develop following the destruction of the surface vegetation by forest fires (e.g. Zoltai and Pettapiece, 1973) or man-induced terrain disturbance (e.g. Heginbottom, 1973). In both instances, the destruction of vegetation results in a greater penetration of rainwater into the active layer. Furthermore, there is a deepening of the active layer and the possibility of thawing of substantial bodies of ground ice, which leads to flowslides if the liquid limits are exceeded.

Where resistant rocks outcrop in the form of vertical or near vertical free faces, rockfails assume local importance. Extensive scree slopes build up below the free face. Theoretically, if the scree is not renewed, the free face will be progressively eliminated by the accumulation and upward growth of the scree slope. Often, as in parts of the eastern Canadian Arctic, extensive and imposing rock faces have been inherited from the previous glacial period; in other areas, these slope forms occur in association with specific geologic structures, with sea cliffs, or with deeply incised stream valleys. The weathering and recession of these free faces occurs primarily through the melting of interstitial ice and the loosening of rock particles by frost wedging in the small joints and fractures that are inevitably present. Eventually, the sheer strength of the material is reduced below the level at which it is capable of countering the stresses imposed by gravity and a rockfall results (Carson and Kirkby, 1972, pp. 125–8).

The rockfall process is a sporadic one and may vary in importance from year to year, as well as spatially across the free face. It is difficult, therefore, to establish average rates of debris movement without long term and extensive observations. In northern Sweden, rockfalls have been studied in some detail by Rapp (1960a). It has been shown that rockfalls occur most frequently in the spring and autumn when the air temperature fluctuates around the freezing level. Rockfalls may also be accelerated by undercutting at the base of the slope, such as in the case of marine cliffs, or by gullying below the resistant rock outcrop if softer rocks lie immediately beneath.

A number of studies have attempted to measure the rate of rockwall retreat under periglacial conditions. The values recorded are listed in Table 7.4. The rates, which range from 0.007 to 1.30 mm year^{-1} should be treated with caution since all are gross approximations. For example, R. Souchez computed the volume of scree material lying downslope of a raised beach dated at approximately 8 000 years B.P. to arrive at his figures. This involved assumptions

Table 7.4 *Some rates of cliff recession under periglacial and non-periglacial conditions*

Location	Lithology	Recession (mm year^{-1})	Source
Periglacial			
Spitsbergen, Mt Templet	Limestones and sandstones	0.34–0.50	Rapp (1960b)
Spitsbergen, Mt Langtunafjell	Limestones and sandstones	0.05–0.50	Rapp (1960b)
Northern Lapland, Karkevagge	Schists	0.04–0.15	Rapp (1960a)
Ellesmere Island, NWT, Canada	Dolomitic limestones	(1) 0.30–0.80 (2) 0.50–1.30	Souchez; personal communication (1971)
Yukon, Canada	Quartzite, dolomite, shales	0.02–0.17	Grey (1971)
Yukon, Canada	Syenite, diabase	0.007–0.03	Grey (1971)
Austrian Alps	Gneiss, schists	0.7–1.0	Poser, in Rapp (1960*a*)
Non-periglacial			
Mt St Hilaire, PQ, Canada	Gabbro, breccias	0.02–0.04	Pearce and Elson (1973)
Brazil	Granite	2.0	Quoted in Young (1972)
South Africa	Granite	1.5	Quoted in Young (1972)
Southwest USA	Shale	2–13	Quoted in Young (1972)

concerning the shape of the scree, its thickness, and the particle size and porosity (i.e. pore space) within the scree. His computed volumes may be too large therefore, and his retreat rates might need to be reduced in magnitude. On the other hand, Rapp (1960a) estimated the volume of material in fresh rockfalls and then averaged that amount over the whole area of the vertical rockwall in the Karkevagge Valley. This probably led to an underestimate of retreat values since rockfalls do not necessarily occur uniformly over the whole of the free face. Bearing these considerations in mind, the available evidence suggests that maximum rates of rockwall retreat of between 0.3 and 0.6 mm year^{-1} are probably typical for most lithologies. If correct, there is little support for the conventional view that weathering and slope retreat is appreciably faster in periglacial climates than in non-periglacial climates. In fact, there is evidence that exactly the reverse may be the case, especially in some of the more severe periglacial climates where extreme aridity and intense cold prevent effective frost wedging. For example, Souchez (1967) has commented upon the scarcity of scree material and the general geomorphic inertness of slopes in the Sor-Rondane and Victoria Mountains of Antarctica. Furthermore, when the rates of recession as listed in Table 7.4 are compared to recession rates in humid temperate and subtropical semi-arid regions, it is clear that retreat rates in periglacial regions are, in general, at least one order of magnitude less.

In certain periglacial areas characterised by abundant snowfall and high mountain relief, rapid mass movement may occur through various types of snow and debris avalanches. The most suited areas for such activity are middle latitude alpine areas where glacially oversteepened slopes are developed in hard resistant rock. Avalanches are much rarer in high Arctic and continental periglacial regions, even in mountain areas, because snowfall amounts are limited. Most avalanches start as snow avalanches which then pick up varying amounts of rock debris en route, ultimately becoming debris avalanches or slides. More liquid forms such as slush avalanches and mudflows may also occur where excessively wet snow is subject to rapid thaw.

The importance of avalanches and rockfalls as opposed to the other mass wasting processes will obviously depend upon such factors as relief, climate and lithology. In unconsolidated and/or colluvial sediments with low angled slopes, solifluction and slopewash will clearly be more important than rockfalls. Slope failures, rockfalls and avalanches, on the other hand, will be more common on steeper slopes, especially if they exceed the angle of repose. One of the few studies to quantitatively distinguish between the various processes is that of Rapp (1960a), in the Karkevagge Valley, a formerly glaciated trough valley in northern Lapland. Over a period of eight years, the various forms of mass wasting were measured. The most important conclusion was that solution loss, or the transport of dissolved salts, was the largest transporter of material, with a net movement nearly equal to that of all the other processes combined. Next in importance, however, were movements of a rapid flow or fall nature. Surprisingly, solifluction was of only minor significance.

Until comparable data from other areas is available, the Karkevagge data cannot be properly evaluated. Quantitative studies by Everett (1967) from West Greenland, under conditions of lower snowfall and colder temperatures, suggest solifluction to be more important than suggested by Rapp. Furthermore, until additional studies are made of solutional loss in other non-limestone periglacial areas, Rapp's claim for dominance of that process must go unchallenged. Finally, slopewash processes were assumed by Rapp to be of negligible importance yet, as Jahn (1961) has illustrated from Spitsbergen, this process may have the same effectiveness in certain periglacial localities as it has been shown to possess in subtropical semi-arid regions. Given our present state of knowledge, it is difficult, therefore, to make an assessment of the relative importance of the various mass wasting processes.

Periglacial slope forms

There is no one slope form or assemblage of slope forms which may be regarded as unique to the periglacial environment or distinctly 'periglacial' in nature. Although quantitative information upon slope forms is limited, all available evidence suggests that all kinds of hillslope forms are to be found in all kinds of climates. This led Leopold, Wolman and Miller (1964, p. 383) to conclude that '. . . the similarities of form in diverse climatic regions and the differences of form in similar climatic environments emphasises the need not of classification but of understanding the interrelation of climate, lithology, and process'. In this section, an attempt is made to follow such an approach. As a starting point, a number of slope form assemblages are recognised as being

Fig. 7.6 *Typical slope forms found in present day periglacial environments. A - The free face/talus slope; B - Smooth debris mantled slopes; C - Stepped profiles; D - Pediments.*

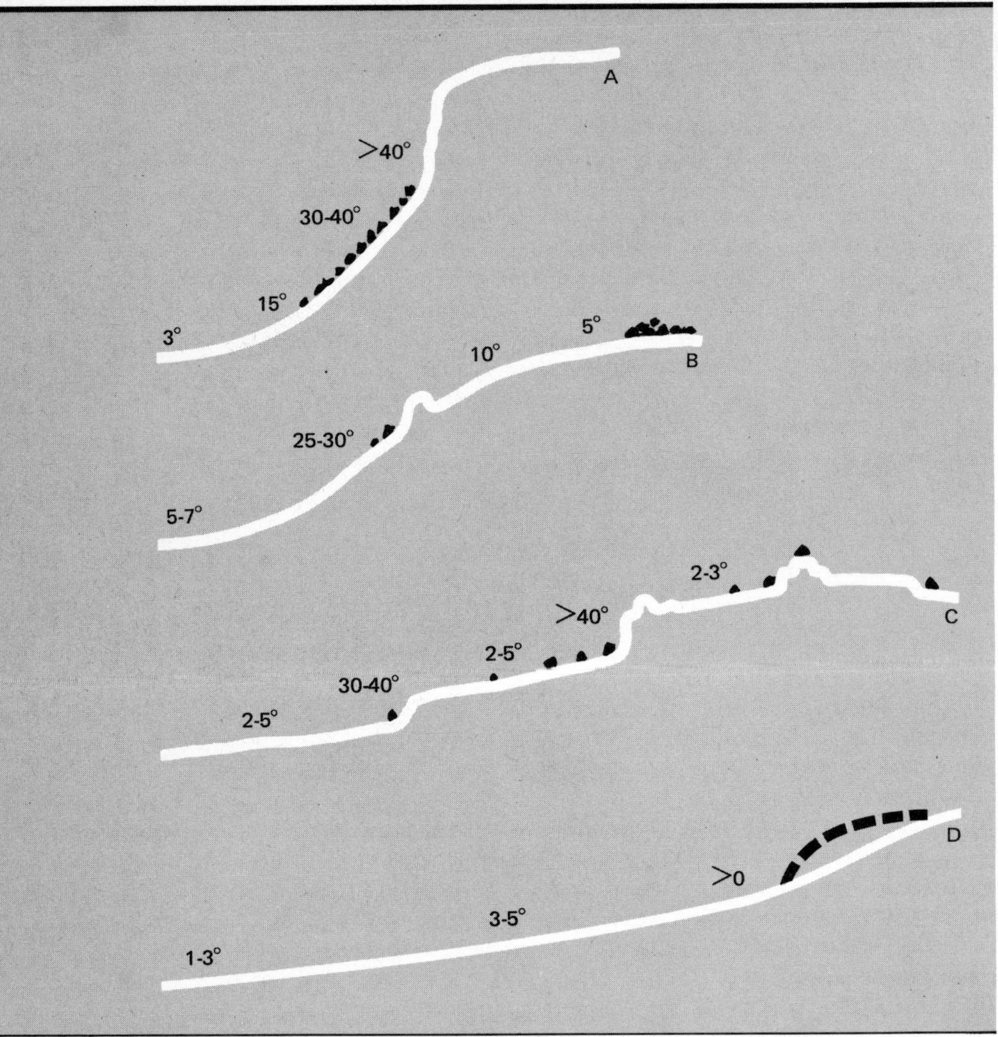

present in periglacial regions (Fig. 7.6). They are identified largely on the basis of the writer's personal experience of the Canadian Arctic. It should be emphasised that all gradations between all types of slope forms exist, and the classification adopted here is purely one of convenience of presentation.

The free face and debris slope profile One of the most widely publicised slope forms of periglacial regions is that composed of a vertical or near vertical rockwall below which is first, a talus or scree slope and second, a basal or footslope complex. According to Jahn (1960), four distinct zones can be recognised on slopes of this type in Spitsbergen: (a) the rock wall, (b) the talus slope, (c) a zone of solifluction activity, and (d) a wash slope. As such, this sort of hillslope form is very similar to the 'standard' hillslope in which there is a crest (waxing slope), a scarp (free face), a debris slope (constant slope), and a pediment (waning slope).

The free face usually stands at angles in excess of 40° and is subject to weathering and cliff recession through frost wedging and rockfalls. The scree slope below is at the angle of repose of the coarser material, usually between 30 and 40° and is progressively built up by rock falling from the free face. The junction between the scree slope and the zone of solifluction is usually marked by an abrupt junction although in some places the two zones merge when moist taluvial slopes develop adjacent to gullies to produce a smooth concave profile. The solifluction zone varies in angle from 25 to 10° and sometimes extends to as low as 3°. It is characterised by a micro-relief of solifluction lobes, terraces and block streams. Downslope, the solifluction zone merges into a low angled wash slope with inclinations of between 2 and 5° in which the effects of solifluction are much reduced and slopewash processes are more important. This is a zone of net accumulation of sediment by wash from upslope, but equally, thermo-erosional wash may contribute to a flattening of the slope through thermo-planation.

Essentially similar slope forms can be recognised in many regions, and they appear to occur wherever resistant rock outcrops. They are particularly common in formerly glaciated areas where the oversteepening of valley side slopes produced rockwalls in excess of the angle of repose. In this sense therefore, this particular slope form assemblage can be regarded as an inherited, glacial form, which is being progressively modified by periglacial processes. In formerly un-glaciated areas, this slope form is rare except where some form of basal under-cutting prevents the free face from being consumed by the talus slope, such as adjacent to rivers or along sea cliffs.

At Karkevagge in northern Sweden, the free face and debris slope assem-blage is the typical slope form (Rapp, 1960a). However, no extensive wash slopes were reported to occur beneath the solifluction zone. This contrasts with the extensive development of wash slopes in Spitsbergen, as reported by Jahn (1960), and probably reflects the difference in intensity of slopewash processes in the two areas. In spite of their approximately similar precipitation amounts, the apparently greater efficacy of slopewash in Spitsbergen than at Karkevagge might be explained by (a) the sparser vegetation cover in Spitsbergen, and (b) the greater susceptibility of sandstones and shales in Spitsbergen to produce sandy and silty particles upon weathering than the granites and gneisses at Karkevagge.

The scree or talus slope lying beneath the free face has been the subject of much investigation. The visual impression of great thickness is misleading since, except for talus cones, the scree is usually less than 5 m in thickness and mantles a bedrock surface. Scree profiles are usually concave and not rectilinear as is often thought, with the steeper section of the scree occurring towards the top of the slope. The reason for this is not clear, but Caine (1969) has suggested that this form reflects a balance between accumulation by rockfall on the one hand, which is greatest near to the foot of the free face, and the spreading of material outwards from the base of the scree by slush avalanches and debris flows. A third process which is important in determining scree profile is the presence or absence of basal erosion. The relationship between these various processes and the form of scree slopes was investigated by Rapp (1960b) at Tempelfjorden, Spitsbergen, and more recently by Howarth and Bones (1972) in the Radstock Bay area of Devon Island, Canada. The Radstock Bay area is a particularly good location for such studies since in some areas, the scree forms part of an active sea cliff and is affected by all three processes to varying degrees, while in others, the scree experiences only meltwater processes (slush avalanches and debris flows) and rockfalls. There are significant differences in the mean angles of scree developed on slopes dominated by different processes. Highest angles of slope occurred on slopes which were subject to periodic basal erosion while lowest angles were associated with scree slopes subject to slush avalanches and debris flows with no basal undercutting. In terms of slope form, Howarth and Bones found that between 80 and 95 per cent of scree slopes dominated by rockfalls or meltwater processes were concave in form. Convexities were rare and only occurred on basally eroded slope forms where they constituted approximately 40 per cent of the profiles.

The movement of material down the scree slope is largely the result of slush avalanches and debris slides, but talus shift is also important. The latter is the slow downslope movement of rock fragments initiated by the melting of the ice binding the surface particles to those beneath and the slippage of the overlying particles on the moistened surface. Talus shift may also be locally initiated by the impact of falling rock and movement of one part of the scree is often transmitted by gravity through the scree. Movement by talus shift is, by nature, extremely variable across and down the slope. Greatest movement usually occurs in mid afternoons and/or during periods of strong solar insolation and surface heating.

Smooth debris mantled slopes A second type of slope form is characterised by a relatively smooth profile with no abrupt breaks of slope and with a continuous or near continuous veneer of frost shattered and solifluction debris. There is no widely developed free face or bedrock outcrop and maximum slope angles are highly variable, ranging from 10° to as high as 25–30°, depending upon lithology. Under certain conditions, and in areas of relatively resistant rock, valley side tors may be present in the upper sections and indistinct mounds of coarser, less weathered debris may constitute the summits of shallow interfluves (Fig. 7.7). In form, the slope profiles extend over the complete range of convexo–concavo forms, from dominantly convex to dominantly concave.

Although the presence of convex and concave slopes is implicit in the periglacial cycle of erosion as proposed by Peltier (1950), they have not been widely reported in the literature as occurring in present-day periglacial regions. This is in contrast to the numerous reports of convexo–concavo slopes in now temperate regions which have been interpreted as being Pleistocene periglacial forms (e.g. Dylik, 1956). In present-day periglacial regions, these slope forms have only been reported from Spitsbergen (e.g. Budel, 1960; Jahn, 1970, pp. 157–9), and from the western high Arctic (Rudberg, 1963; Pissart, 1966b). However, in all probability, convex and concave slopes are more widespread than is thought. Certainly, in central and western Banks Island, where the underlying sediments are sands and gravels of the Beaufort Formation or Cretaceous shales and sands

Fig. 7.7 *Smooth debris mantled slope with bedrock outcrop (tor) on interfluve, 5 km north of Mould Bay, Prince Patrick Island, Canada.*

overlain by a thin veneer of glacial sediments, smooth convexo–concavo slopes are undoubtedly the most common.

Our understanding of these slope forms is limited and the lack of detailed process observations makes their interpretation difficult. It would appear that convex and concave slopes develop in areas of predominantly soft or unconsolidated rocks, usually Cretaceous or Tertiary in age, or in surficial deposits. For the most part, these are lowland, as opposed to upland areas. On such terrain, solifluction and slopewash processes are the dominant agents of landscape sculpture. According to Budel (1960), the convexities and concavities reflect these processes while the steeper middle section of the slope is a debris slope subject to weathering and gravitational processes which lead to backwearing. The maximum angle of the slope, which may range from 10° to 25–30° is essentially an angle of repose and reflects lithological factors. The lower concavity is generally dominated by slopewash and solifluction processes. The slopewash zone lies immediately downslope of the debris slope and is the result not only of the progressive downslope increase in wash but also of the accumulation of snow-patches at that position on the slope. In these locations, the depth of thaw is restricted and vegetation growth is limited because of the lateness of the thaw. As the vegetation cover progressively increases lower down the concavity and as the active layer assumes normal thicknesses, solifluction activity increases at the expense of rill and sheetwash. The solifluction movement is commonly in the form of lobes. According to Pissart (1966b) slopes with inclinations of 5–7°

appear to be the lower limit for the development of such lobes. Below this angle solifluction continues to operate but predominantly in sheets. The lower concavity represents, therefore, the minimum slope across which material brought from upslope is transported, first by sheetwash and then by solifluction. The interpretation of the summit convexity is equally conjectural in view of the absence of quantitative investigations. Numerous theories have been advanced to explain the presence of summit convexities (e.g. see Young, 1972, pp. 93–5). The most probable explanation is that the convexity of periglacial slopes is related to the dominance of slopewash, the volume of which increases downslope from the crest leading to an increase in the power of detachment of particles and of transport. If correct, the convexity is a denudation slope subject to control by removal.

Two further characteristics of this particular slope form need to be mentioned. First, an interesting feature of many convexo–concavo slopes is the tendency for abrupt breaks of slope to develop at the junction between the debris slope and the lower concavity. This phenomenon is reminiscent of the pediment junction of hot arid regions and has been frequently commented upon (e.g. Rudberg, 1963; Pissart, 1966b). A probable explanation is that snowpatches remain longest in depressions and at the foot of lee slopes. Thus, in relation to the total slope profile, there is a sharp increase in the intensity of slopewash below the snowpatch. In particular, the slow but continuous action of sheetwash and thermo-erosional wash leads to the progressive removal of fine particles and the lowering of the surface. The increased removal of material at that point leads to the development of the knickpoint and the steepening of the debris slope above. According to Pissart, these conditions result in steeper slopes evolving less rapidly than gentler slopes in periglacial regions, and explains the frequency of maximum angles of 5–7° and 30–32°. These two groups represent the limiting angles for the operation of gelifluction and slopewash, and gravity processes, respectively.

A second characteristic sometimes associated with this slope form is the presence of rock stacks or tors. These features stand out from the debris covered slope, and are usually angular, frost shattered rock protuberances bounded by two sets of near vertical joints. The most suited lithologies for their development are bedded and highly fissile sandstones, shales and dolomites, all of which are particularly susceptible to frost wedging but which are sufficiently resistant to form tor-like structures. Hillslope tors have been described from Victoria Land, Antarctica (Derbyshire, 1972), where they are developed in sandstones and dolerite, and have been observed by the writer on the Cretaceous shales of Prince Patrick Island (Fig. 7.8) in the Canadian Arctic. They probably occur in other Arctic and Antarctic localities wherever lithological conditions are favourable.

In view of the complexity of the origin of tors and their climatic significance, it is useful to make the distinction between hillslope tors and hilltop or summit tors. Hillslope tors are located in valley side locations and are surrounded by debris covered slopes commonly between 20 and 30° in angle, depending upon lithology. Summit tors are surrounded by slopes of much lower angle, often less than 5–7° and are located on interfluves or points of high elevation relative to the surrounding terrain. Summit tors are examined in the following section dealing with cryopediments and cryoplanation surfaces. The majority of hillslope tors develop through slope retreat and frost wedging. They

Fig. 7.8 Hillslope tor developed in Upper Jurassic sandstones and shales, surrounded by frost shattered angular debris. Prince Patrick Island.

occur on the middle or upper parts of slopes, and are surrounded by coarse angular debris, which rests at the angle of repose and moves downslope under gravity processes. Local site differences, affecting micro-climate, appear to be important. For example, Derbyshire (1972) observed that angular hillslope tors were only developed on west- and northwest-facing slopes. He attributed this to the intense solar radiation received by these slopes which leads to strong surface heating. Rock surface temperatures may exceed 30°C at times in the interior of the Antarctic continent, and thawing of snow lying on dark surfaces has been observed when the air temperature is –20°C (see p. 17). However, frost shattering and slope retreat may not be the only origin of tor-like features. Derbyshire (1972) has shown that chemical weathering and exfoliation may be important processes in the formation of rounded tors in certain localities, and St-Onge (1965) has interpreted tor-like features ('relief ruiniforme') occurring on slopes developed in sandstones on Ellef Ringnes Island in the Canadian Arctic as being the result of wind erosion.

Cryopediment forms Cryopediments are defined as . . . gently inclined erosional surfaces developed at the foot of valley sides or marginal slopes of geomorphological units developed by cryogenic processes in periglacial conditions' (Czudek and Demek, 1970b, p. 101). They were first recognised by J. Dylik as planation surfaces which bevelled slope foots and truncated both glacial and fluvioglacial sediments in central Poland (Dylik, 1957). Dylik concluded that these forms were analogous with the pediments of tropical and subtropical regions, and that they developed through a combination of periglacial processes. The dominant characteristics of cryopediments are (a) their low angles of inclination, (b) their extremely shallow concave, nearly rectilinear, profile, and (c) their geomorphic location at the foot of valley side slopes.

Cryopediments are essentially slopes of transportation and they develop through the parallel retreat of the lower valley side slopes. They can be distinguished from cryoplanation terraces since the latter occur in the middle or upper sections of slopes or on summit and other elevated areas. The processes operating upon cryopediments are the various forms of slopewash, in particular sheetwash, aided by solifluction, while the retreat of the valley side slope at the upper limit of the cryopediment is usually the result of frost action and rillwash processes. In form, cryopediments vary in angle from as much as 8–10° in their upper sections to as low as 1° in their lower sections. The upper limit of the cryopediments is commonly marked by an abrupt break of slope or knickpoint, and the pediment may encroach upon the inter-valley area in the form of embayments. Since cryopediments are slopes of transportation, only a thin veneer of surficial material mantles the slope. Sometimes however, a zone of accumulation of stratified slope deposits develops downslope from the cryopediment proper and extends it in length without any visible change in gradient.

Valley cryopediments have been described from the taiga zone of eastern Siberia (Czudek and Demek, 1973) where they occur in wide deep valleys incised in the various upland plateaus and mountain ranges. The smoothly concave erosional surfaces range from several hundreds of metres to over 3 km in length with inclinations of between 3 and 10°. In places, the pediment surface extends directly to the valley bottom; in others, it grades into fluviatile terraces. In many cases, the valley pediments penetrate from the main valleys into the side valleys and, under favourable conditions, the cryopediments of two opposite valleys merge together to form 'pediment passes'. In the Stanovoj Range, for example, a relief of isolated mountain groups and inselbergs with at least two pediment levels are known to exist.

It is thought that the cryopediments of eastern Siberia developed in the Pleistocene and are continuing to develop under the present-day climatic conditions. In some instances, the veneer of waste material which covers the pediment surface extends downslope without any gradient change to overlap terrace deposits of Middle and Late Pleistocene age. This indicates recent and relatively rapid evolution of the pediment surfaces. According to Czudek and Demek (1973b), the present-day retreat of the steep slope is accomplished in two ways. First, intense frost weathering and creep are concentrated in shallow gullies or 'dells' which act as lines of concentrated erosion and denudation. Second, sapping at the foot of the scarp by nivation processes undercuts and steepens the profile. Present-day transportation across the pediment surface is also indicated by a variety of phenomena. For example, the majority of pediment surfaces are overgrown with taiga (***Larix dahurica***), many of which are inclined or tilted, forming what are termed 'drunken forests'. Russian authorities interpret this phenomenon as reflecting subsurface wash and suffosion processes (see p. 142) currently operating upon the pediments. While this may be true in eastern Siberia, where the active layer often attains thicknesses of 2 m or more, similar tilting of trees in the boreal forest of North America is interpreted as reflecting cryoturbation activity and/or a forest fire vegetational climax, rather than downslope movement (see p. 34). Other phenomena, such as small solifluction terraces and non-sorted stripes on the slopes of the shallow dells which dissect the pediment surfaces, are less equivocal and indicate active gelifluction and surface wash processes.

Cryopediments have not been widely reported in the literature to date. However, it is likely that subsequent investigations will reveal cryopediments to be an important element of the periglacial landscape, and many aspects of pedimentation to be applicable to the arid continental and high Arctic periglacial environments.

Cryoplanation terraces and stepped profiles Many slopes in periglacial regions do not possess any of the typical forms previously described. Instead, they are characterised by an irregular, faceted profile with occasional bedrock outcrops or bluffs separated by gently inclined benches (Fig. 7.9). Furthermore, in many upland areas where such profiles are well developed, the hilltop surfaces are often remarkably flat with occasional tors or frost riven bedrock remnants rising above the summit surface.

These benches are termed cryoplanation terraces. They are defined as erosional surfaces cut into bedrock at the foot of frost riven cliffs and scarps, or in the surrounding of tors (Demek, 1968; Czudek and Demek, 1970b). Each terrace consists of a terrace surface of low angle delimited, at its upslope end, by a steeper riser. In contrast to cryopediments, cryoplanation terraces occur in the middle and upper sections of slopes, and on upland terrain. The terraces may

Fig. 7.9 *The summit and southwestern slope of Shapka Monomakha peak in the Aldanskaya nagorye Mountains, showing tors, cryoplanation terraces, and blockfields. From Demek (1969a).*

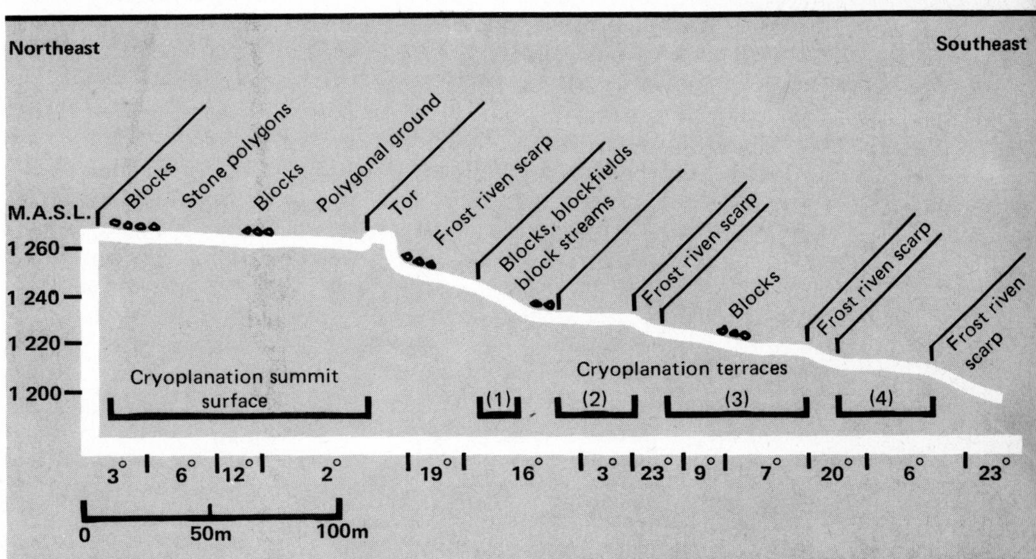

occur individually, but more commonly, they occur in groups, distributed randomly over the slope one above the other, to give a distinct bevelled appearance to the slope. Cryoplanation terraces are thought to develop in essentially the same way as cryopediments through the parallel retreat of the steeper slope section under the combined action of nivation and frost weathering. The terrace flat is a rock surface, thinly veneered with debris, across which material is transported by solifluction and sheetwash processes.

Erosional terraces of the kind described were first reported from Siberia as early as 1890 and subsequent investigations have shown them to be widespread in the northern regions of Eurasia and North America (e.g. Demek, 1969a; Wahrhaftig, 1965; Waters, 1962; St-Onge, 1969). There is an extensive literature dealing with these features and they are well known periglacial slope forms. Cryoplanation terraces are also referred to by a variety of other names. In the Soviet literature, for example, they are known as golcovaya terrasa (golets terraces) after the Siberian word 'golec' for an angular hill or mountain with a flat summit rising above timberline. In the English language, the terms 'altiplanation terrace' (e.g. Eakin, 1915) or 'nivation terrace' (e.g. St-Onge, 1969) are sometimes used.

Cryoplanation terraces are of varying sizes and forms depending upon lithologic conditions and geomorphic location. In form, terraces developed on the sides of isolated hills are often of a sickle-like shape, while those bordering longitudinal upland ridges are elongate and relatively narrow. The dimensions of known cryoplanation terraces vary widely too. The smallest are often less than 50 m in maximum dimensions. Others may have a length of 400–600 m and a width of 150–200 m (Demek, 1969a, p. 42; Wahrhaftig, 1965, p. 16). Somc cryoplanation terraces in Siberia reach even greater dimensions however, being over 1 km in width and several kilometres in length. Summit cryoplanation flats are also of considerable size; the summit flat on the top of the Turku Hill in the Kular Range of Siberia is more than 400 m wide (Demek, 1969a, p. 42). The height of the frost-riven scarps, limiting the cryoplanation terraces, varies not only with the resistance of the rock and the degree to which the riser (free face) is preserved, but also with the inclination of the original slope. On flat summits and on gentle slopes underlain by unconsolidated sediments the height of the risers may only be between 1 and 2 m high (Fig. 7.10) but in other areas with greater relief and overall dimensions of the terraces, the height of the frost-riven scarps may be as much as 10–20 m (Demek, 1969a). The angle of the terrace flat varies between 1 and 12°; usually the larger is the flat and the lower is the inclination of the original slope, the smaller is the gradient. According to Demek, inclinations often fluctuate around 7° on well developed terraces. The flats are usually covered with a thin veneer of colluvial materials formed through the transport and redeposition of the frost shattered debris from the free face. The nature of the deposits closely relates, therefore, to the properties of the bedrock in which the terrace is cut. Occasionally, polygonal ground, block streams, and solifluction stripes develop on the terrace flat.

Several stages in the development of cryoplanation terraces can be recognised (Fig. 7.11). According to Demek (1969b) the intensity and importance of the various processes acting on the terraces is thought to change at each stage. In the beginning, cryoplanation terraces develop in relation to the nivation process which, in the low precipitation regions of the Arctic and Siberia, limits intense

Fig. 7.10 *Cryoplanation terrace developed in unconsolidated sands and gravels of glacial origin, eastern Banks Island.*

physical weathering to the vicinity of snowbanks. As a rule therefore, cryoplanation terraces develop on slopes in the lee of the prevailing winds, or below structural benches or slope irregularities which allow the accumulation of snowbanks. In this initial nivation stage, frost shattered debris is removed by sheetwash and gelifluction. The second stage of cryoplanation is the formation of the cryoplanation terrace when both the tread (flat) and the riser (free face) are developed. Nivation and frost action on the riser result in its steepening and retreat. On the expanding terrace below, a combination of solifluction and wash transports the material away from the frost-riven scarps. The third stage is that of the development of the summit flat or cryoplanation surface, when the frost-riven scarp is ultimately consumed through intersection with an adjacent cryoplanation terrace. In particularly coarse grained and resistant rocks such as granites, dolerites and massive sandstones, the advanced stages of cryoplanation may result in the formation of distinct frost-riven tors or bedrock outcrops which rise above the surrounding cryoplanation surface. In the final stages of cryoplanation, downwearing as opposed to backwearing becomes the dominant mode of evolution, helped by wind action. Movement by solifluction is restricted by the low angles of slope and instead, frost action in the surface sediments leads to local sorting and patterned ground development.

Lithology plays an important role in the development of the various forms of cryoplanation. They are usually best developed in areas underlain by relatively

Fig. 7.11 *Stages in the development of cryoplanation terraces in resistant rock, according to J. Demek (1969a)*

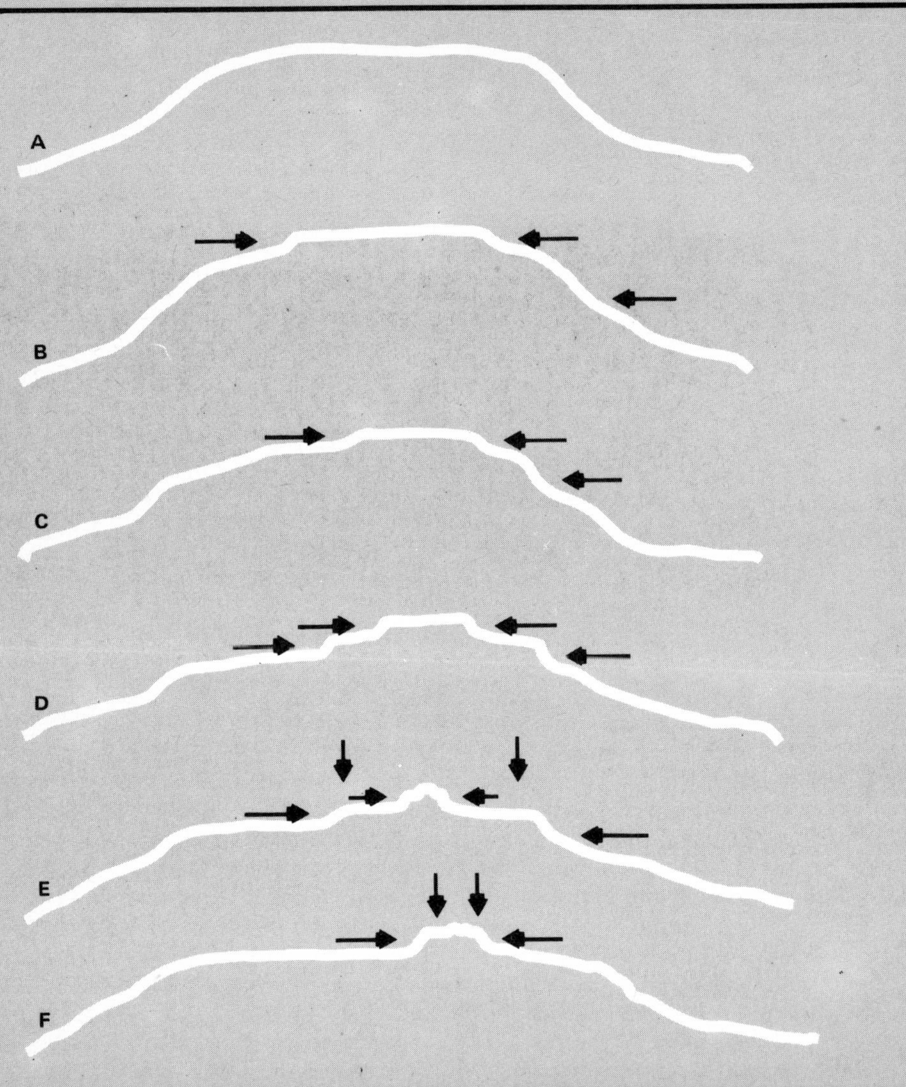

A — Original topographic surface
B — Stage of nivation hollows
C — Stage of initial cryoplanation terrace
D — Stage of mature cryoplanation terrace
E — Initial stage of cryoplanation surface
F — Cryoplanation summit surfaces
➡ Direction of surface modification

resistant rock, which, at the same time, allows the penetration of meltwater into the rock and so permits effective frost shattering. For example, on Ellef Ringnes Island in the Canadian Arctic, St-Onge (1969) has described the different forms which develop in gabbro, sandstone and shale. In gabbro, the evolution of the terraces is slow in comparison to their formation in other rocks since gabbro is a hard, compact rock with a low porosity, and frost shattering takes place only under the most favourable of circumstances. The terrace flats are usually between 10 and 15 m in width and the risers, with slopes of between 25 and 35°, are between 2 and 5 m in height. The terrace flats form giant, near horizontal steps which reflect the predominantly coarse boulders and few fine particles which result from the physical disintegration of the gabbro. Since the finer particles are quickly removed by water percolating through the boulders, the terrace is bounded by an apron of coarse, angular debris. These boulders act as a local base level for the terrace since they are immune to slopewash and solifluction. On adjacent sandstones however, the terraces are more subdued in form. The sandstones weather to silt, sand and sandstone aggregates, all of which are more easily moved by gelifluction, sheetwash and rillwash than the gabbro-derived material. The terrace flat becomes an inclined surface of 6–8° on which sheetwash and laminar gelifluction are important. Finally, in shales and silt-stones, a variety of features develop ranging from large amphitheatre-like hemi-circles to smaller hollows and ledges. These features reflect the ease with which frost shattering reduces soft shale to fine sand and silt, and the effectiveness of sheetwash in removing such material.

The significance of well developed cryoplanation terraces and tors goes beyond that of favourable lithological conditions for cryoplanation activity. In terms of periglacial climatic conditions, cryoplanation features require conditions of intensive frost wedging and sufficient snowfall for the formation of snowbanks, yet not too much snow such that the bedrock is insulated from the temperature variations in the air. According to Demek (1969a), cryoplanation terraces best develop in continental periglacial areas of moderate aridity. Finally, in terms of their Pleistocene significance, it is worth emphasising that the presence of permafrost is not essential for the formation of cryoplanation terraces or summit tors.

Slope evolution

Problems of slope evolution usually centre around the manner and rapidity of profile change with time. However, in spite of the quite considerable literature upon slope evolution in general, there is little which deals directly with such changes and even less which is concerned with present day, as opposed to Pleistocene, environments. This distinction between present day and Pleistocene slope evolution is important since there is also a difference in approach to slope studies. Pleistocene slope evolution is essentially one of slope reconstruction whereas, in present periglacial environments, the problem is that of developing process-response models and linking specific slope forms to specific geomorphic processes.

Souchez (1966) has developed one of the few process-response models of slope evolution specifically related to periglacial environments. The model dealt with slopes developing in coherent rock under the action of slow mass movement of the surface mantle and emphasised the role of plastic deformation and shearing failure. Under conditions of plastic flow, downslope regolith movement was assumed to be at a rate proportional to the angle of slope, and the ground loss proportional to convex curvature. The resulting model is one in which the slope declines but remains predominantly convex at all stages. Although entirely convex slopes are known to exist in periglacial regions (e.g. Budel, 1960) several reasons suggest this model to be of only limited applicability. First, the model does not account for the presence of erosional concave slopes (e.g. cryopediments). Second, the model does not apply to slopes developed in unconsolidated sediments. Third, the assumption that solifluction follows the laws of plastic flow is not necessarily true. There are instances where movement is not laminar in flow, and of course, an important component of solifluction is frost creep which is unrelated to plastic flow.

There are two other considerations which, in general terms, complicate the development of process-response models. First, it is unlikely that periglacial slope evolution has ever managed to run its full course. In many regions, slope forms are best interpreted as having been inherited from the previous glacial period, and are, therefore, in varying stages of disequilibrium. Only in areas which either (a) have experienced a long and uninterrupted period of relatively unchanging periglacial conditions, or (b) are underlain by soft, unresistant rocks, can one assume with any confidence that the slope forms represent a reasonably close adjustment with process. In reality, this means the only areas in the northern hemisphere in which periglacial slope development may have proceeded to its equilibrium condition are the unglaciated lowlands of northern and eastern Siberia, some of the islands in the western Canadian Arctic, and certain parts of the interior Yukon Territory and central Alaska.

Second, periglacial slopes experience not only processes which operate in relation to the base of that slope (e.g. slopewash, rockfalls), but also processes which operate without any dependence upon the main erosional base level of that slope (e.g. nivation and cryoplanation). Thus, while certain sections of a slope may be experiencing denudation and decline for example, other sections of the same slope may be experiencing steepening and parallel retreat. Moreover, two unique periglacial processes must also be considered; both fluvio-thermal erosion and thermokarst subsidence are capable of producing distinct forms of localised slope evolution. Thus, any one slope may be composed of a number of separate units, each evolving in its own way, and often bearing little relationship to the slope unit above or below.

Cryoplanation Notwithstanding these problems, it is generally believed that the change in slope form under periglacial conditions involves a smoothing and flattening of the profile, with net erosion occurring on the upper parts and net deposition on the lower parts, leading to an overall reduction in relief. This process is generally termed 'cryoplanation' and involves the parallel retreat of frost-riven scarps and the extension of low-angled slopes of transportation below.

Using the terminology proposed by Bryan (1948), these ideas were summarised by Peltier (1950) in his periglacial 'cycle of erosion'. Peltier envisaged an existing landscape formed under non-periglacial conditions being subject to intense frost action ('congelifraction') and solifluction ('congeliturbation'). The initial stage (Fig. 7.12) saw the downslope movement of soil and rock rubble by

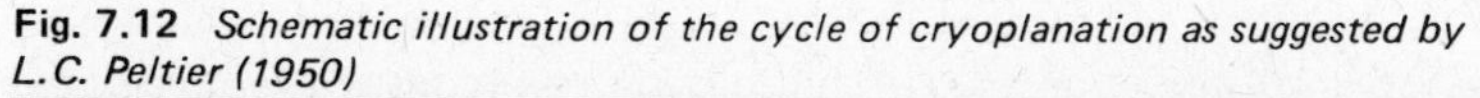

Fig. 7.12 *Schematic illustration of the cycle of cryoplanation as suggested by L.C. Peltier (1950)*

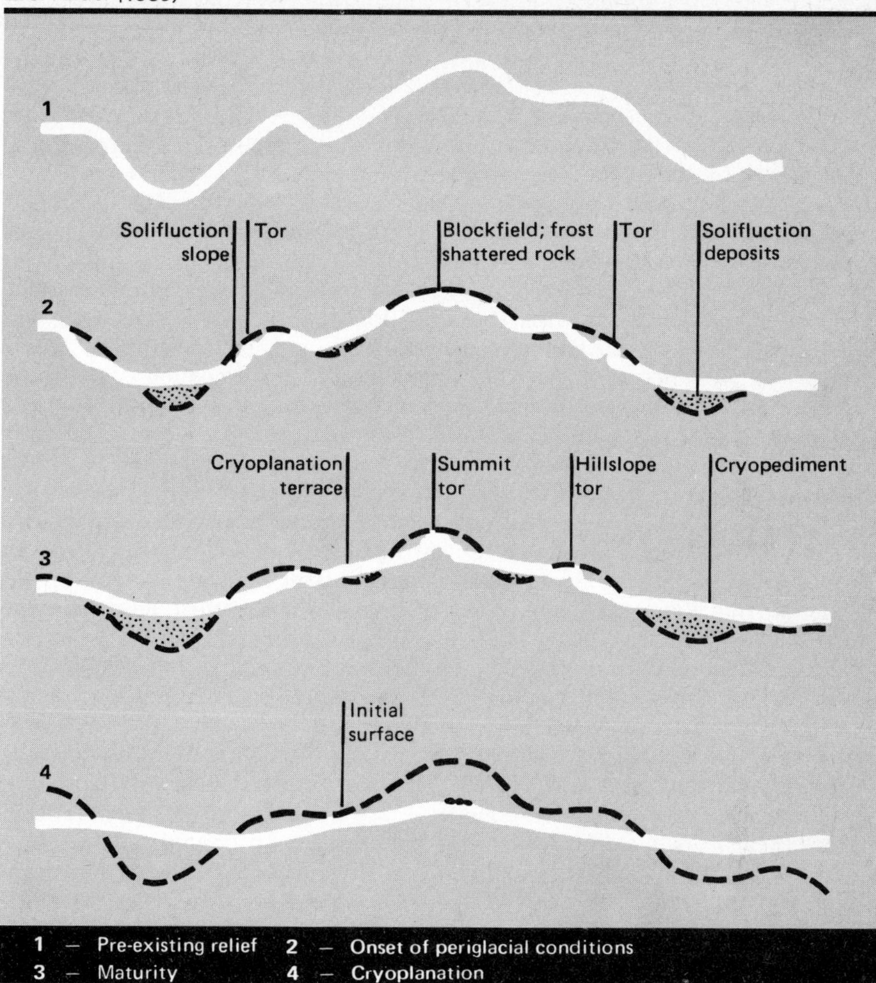

congeliturbation which exposed bedrock on the upper slopes. This bedrock was then exposed to frost shattering, rockfall and retreat with the accumulation of a talus or 'congeliturbate mantled' slope beneath. On adjacent upland surfaces, congelifraction produced extensive areas of angular frost shattered debris, forming blockfields or felsenmeer. With time, the continued retreat of the frost-riven scarp led to the complete disappearance of the cliff from the landscape, leaving only a residual tor-like feature at the summit. At this stage maturity was reached in which the pre-existing landscape had been consumed and the present landscape was covered with a continuous mantle of frost shattered and solifluncted material. In late maturity and old age, solifluction continued to degrade and flatten the summits while adjacent valleys and lowlands were progressively 'plugged' by the accumulation of the frost shattered and soliflucted debris. At this stage, the progressive comminution of debris through frost action enabled wind to assume a greater role in transportation and deposition in the landscape.

In terms of slope evolution, Peltier's idealised cycle represented an intuitive synthesis of field observations since it emphasised many of the typical slope forms described earlier in this chapter. It is too general, however, to provide anything more than an overall framework within which to view slope evolution. For example, no quantitative parameters are given for the landscape changes which are thought to occur, and there is no realistic discussion of the manner in which frost shattering and mass movement influence slope form. Moreover, the lack of attention paid to other processes, particularly running water, is a fundamental weakness. On the other hand, the general concept of slope flattening and cryoplanation through the parallel retreat of a frost-riven scarp has found wide acceptance with the recognition of cryoplanation terraces, summit tors, nivation hollows, and cryopediments. The importance attached to slopewash and 'thermoplanation' by Soviet authorities also finds acceptance within the general cryoplanation concept. It must be concluded that cryoplanation is, in general terms, a most useful concept in understanding landscape changes in periglacial environments.

Slope replacement In tropical semi-arid and arid areas, where the free face and talus slopes are well developed, it can be envisaged that slopes evolve by replacement, with scree slopes replacing the cliffs and these in turn being replaced by low-angled, concave, transportational slopes. If weathering and removal of fine material from the scree occurs in addition to the cliff retreat, an equilibrium condition may be achieved in which the cliff and the scree retain the same slope proportions during retreat (Carson and Kirkby, 1972, pp. 353–5). The rate of fines removed from the foot of the scree slope would also be proportional to the height of that slope. Therefore, as the cliff and scree slopes are reduced in height, the gradient of the low-angled basal slope becomes lower to compensate for the reduction in debris carried. This sort of development is illustrated graphically in Fig. 7.13 for the conditions of a broad plateau and a fixed stream which is exporting the debris reaching it from the lower concavity.

This model may be applicable to some of the more arid periglacial environments in continental and polar locations, where the similarity of certain slope forms with those of tropical semi-arid regions has already been mentioned. The cryopediments and the angular, pediment-like junctions at the foot of steeper

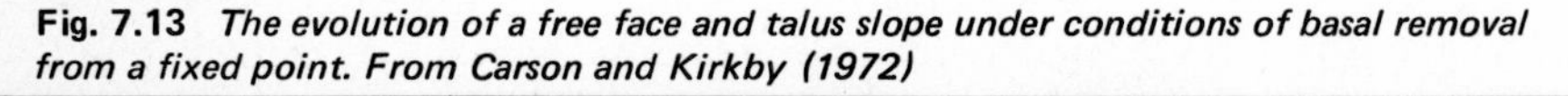

Fig. 7.13 ***The evolution of a free face and talus slope under conditions of basal removal from a fixed point. From Carson and Kirkby (1972)***

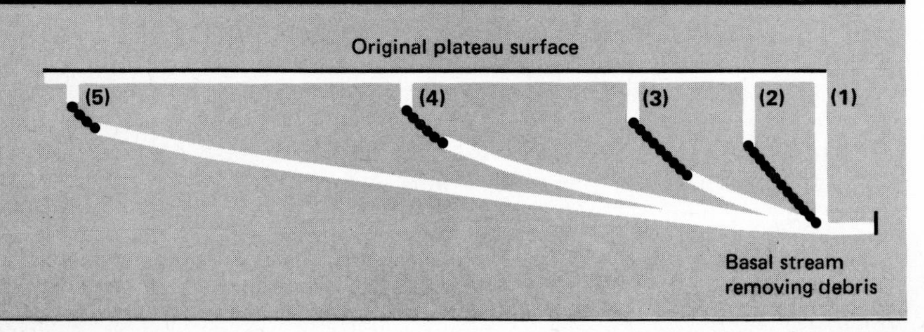

slopes may develop in this manner, with successive erosional regrading of the pediment surface. The difference between periglacial and tropical semi-arid slope forms may merely be one of degree. According to Carson and Kirkby (1972, p. 254) '. . . the much greater cliff retreat through frost action is the critical difference, leading to rounded forms with moderate relief under periglacial conditions, and to plains surrounding steep residuals in desert conditions'. While the assumption of greater cliff retreat under periglacial conditions is probably unjustified in view of our present state of knowledge (see Table 7.4), there is some evidence that slopes might evolve in the manner suggested in Fig. 7.13. For example, the variations in slope profile forms which occur beneath an ironstone scarp on Prince Patrick Island in the Canadian Arctic have been described by Pissart (1966b). Over a distance of 2 km, the ironstone scarp progressively departs from a small stream at its foot. The variations in the form of the scarp may be viewed, therefore, in terms of their evolution under conditions of progressive basal removal from a fixed point.

At the location where the small stream is continually removing material from the foot of the scarp, the stream has cut a small valley with slope angles of between 22 and 28° along the scarp. As the stream departs from the foot of the scarp, the slope changes its form. A basal slope of between 6 and 10° develops in the lower part of the slope while the upper part becomes constant in angle, at 25–27°. The transition between the two sections of the slope is marked by an abrupt junction. Further along the scarp, the slope develops an extensive concave section with angles of 5–7° while the scarp continues to maintain its angle of 25–27°. These observations suggest, therefore, that the lower concave section has undergone successive erosional regrading to lower its angle while the scarp face has continued to maintain its angle.

The rapidity of profile change It has been traditionally assumed that slopes evolve more rapidly under periglacial conditions than under non-periglacial conditions (e.g. Tricart, 1970, p. 112). This remains to be proven, however, since what little data that is available suggests that weathering and slope processes do not operate at a significantly faster rate in periglacial, as opposed to non-periglacial, climates. For example, the rate of rockwall retreat under periglacial conditions appears to be less than in rainforest, humid temperate and tropical semi-arid conditions (Table 7.4), and limestone solutional activity is no greater in Arctic regions than in other arid or semi-arid areas (Table 7.3). The assumption of rapid slope evolution probably results from the widespread assumption of the importance of frost shattering in periglacial climates, which, as has been demonstrated earlier (p. 40), is not necessarily true.

When considering slope evolution in any one area, it is necessary to consider (a) the regional and micro-climates currently being experienced, (b) the lithological conditions of the underlying rocks, and (c) the nature and type of the dominant weathering process. Cliff retreat is clearly faster on rocks which enable water penetration into and along joints and bedding planes than on coarse grained and massive rocks with few joints or bedding planes. Equally, as Souchez (1967) has stressed, frost wedging is less effective and slope retreat less rapid in areas of extreme aridity than in areas of greater humidity. Even within areas of uniform climate and lithology, micro-climatic differences on slopes of different orientations may induce the more rapid evolution of some slopes than of others.

The extreme example of such a situation is the formation of asymmetrical valleys (see pp. 178–83). For reasons such as these, it is probably unwise to assume that slope evolution is faster under periglacial conditions than non-periglacial conditions. More quantitative information upon both weathering processes and slope forms is required.

General conclusions on hillslope forms and processes

It is clear from the preceding discussion that our understanding of slopes in periglacial environments is limited. Field measurements of slope forms, together with detailed investigations of the link between form and process, are urgently required. Bearing this in mind, a number of points summarise the preceding discussion and serve as an illustration of our present lack of knowledge of hillslope forms and processes.

First, slopes evolve in periglacial environments primarily through the combined action of mass movement and running water, and not through 'unique' periglacial processes. Nivation, thermo-erosional wash and thermokarst activity are probably the closest to being regarded as uniquely periglacial processes.

Second, a variety of slope forms exist in periglacial regions and probably none are limited to periglacial regions. Slope form is primarily influenced by the lithological characteristics of the underlying rock, and in particular, its jointing and structural attitudes, and its porosity. The presence or absence of moisture, in determining the type and rate of weathering and weathering removal, is of prime importance.

Third, many slopes are best interpreted as having been inherited from a previous glacial period, and are, therefore, in disequilibrium with the present climatic environment.

Fourth, periglacial slope forms exhibit many similarities with slope forms found in tropical arid and semi-arid regions of the world. The difference may only be one of degree. Nivation and cryoplanation processes result in pediment-like slope forms of low angle which often border upon, or surround, abrupt and steep sided residual rock masses.

Fifth, in unconsolidated sediments, and in more humid periglacial environments, solifluction and slopewash are probably the main agents in promoting a landscape of exceptionally low relief with comparatively low angles of slope.

Sixth, slopes in periglacial environments probably undergo a progressive and sequential reduction of relief with the passage of time. Limited evidence suggests that this takes place by either cryoplanation or slope replacement from below.

Seventh, no evidence is yet available to verify the assumption that slopes evolve more rapidly under present-day periglacial conditions than under non-periglacial conditions.

The large-scale organisation of periglacial terrain is not unlike that of other regions. A reasonably well developed drainage network exists, even in those areas which have recently emerged from beneath Wisconsinan or late-Pleistocene ice. This fluvial network is most striking within the high Arctic landscape where, because of the lack of vegetation, the intricacies of the network are clearly visible.

The fluvial network is composed of essentially two types of rivers. First, a number of very large rivers and their tributaries, such as the Mackenzie River in North America, and the Ob, Yenesei and Lena rivers in Siberia, originate from non-periglacial regions or from deep springs in the discontinuous permafrost zone. These are some of the largest rivers in the world, flowing for several thousands of kilometres from their headwaters to the Arctic oceans. To a large extent, their discharge and sediment characteristics are independent of the terrain and climatic regions through which they flow. Second, there are the innumerable smaller rivers of varying sizes which constitute the overwhelming majority of the fluvial network. The drainage of these streams originates totally within periglacial regions and their discharge and sediment characteristics more truly reflect the environment in which they flow.

Since the geomorphic activity of the very large rivers reflect non-periglacial as well as periglacial conditions, the following comments are directed primarily towards those fluvial networks which originate within periglacial regions. Emphasis is further directed towards Arctic North America, and the Canadian Arctic in particular.

Fluvial processes

Our understanding and appreciation of the role of fluvial processes in periglacial environments has only developed in the last few years. There is still little information available as to either the hydrologic regimes of rivers or the rates of operation of fluvial processes. A number of isolated studies in parts of Alaska (e.g. Arnborg ***et al.***, 1967) and the Canadian Arctic (e.g. Cook, 1967; Church, 1972; McCann ***et al.***, 1972) constitute much of the quantitative data which is readily available at present. This overall lack of data is unfortunate since that which is available leaves one in no doubt as to the importance of fluvial processes in periglacial environments.

Arctic rivers flow only during the few summer months when temperatures rise above freezing, and for most of the year, there is little surface or subsurface water movement. Because of this, it had been thought that running water was of minor significance in fashioning the landscape of periglacial regions (e.g. Peltier, 1950). This is not the case. Recent studies suggest that flowing water in periglacial regions is capable of great denudational and transporting activity when compared with other geomorphological agents. The spring snowmelt is very rapid indeed, usually occurring over a period of 2–3 weeks, and gives rise to a prominent freshet, or flood. Thus, although the precipitation in many regions is slight and is spread throughout the year, between 25 and 75 per cent of total runoff is concentrated in a few days. Furthermore, in certain of the extremely arid regions, the majority of precipitation falls as summer rain and not as snow, during a few periods of prolonged rain. Because the ground is relatively bare of vegetation, and because permafrost inhibits the penetration of such rain into the ground, a high proportion of summer rain runs off directly on the surface, causing considerable rill and sheetwash erosion. In other areas, the presence of permanent snowfields or glaciers provide a near constant source of summer runoff, which varies with the local weather conditions.

Hydrology and summer weather A number of different types of river regimes occur in periglacial regions of high latitudes (Church, 1974). To varying extents, these river runoff regimes are dominated by the rapid melting of snow and ice in the short winter–summer transition period. In the Canadian Arctic, this occurs in late June or early July. During the rest of the summer, the runoff steadily decreases as less and less snow remains to be melted. This progressive decrease in runoff is periodically interrupted by subsidiary runoff peaks related to summer storms and direct surface runoff. Such a runoff regime is termed 'nival', and may be subarctic or Arctic in nature depending upon whether flow is maintained throughout the winter or not. A second type of river regime is a 'proglacial' one. In those watersheds where permanent snow or icefields occur, renewed melting occurs throughout the summer whenever warm and overcast conditions develop. Peak runoff under these conditions is often delayed until late July or early August, and the nival freshet is not so important a component of the runoff regime. A third type of regime is termed a 'muskeg' type. Because of the water retaining capacity of muskeg or lowland grassy tundra, and the high resistance to runoff presented by it, flood flows are attenuated in such drainage basins. Examples of the various periglacial river regimes are illustrated in Fig. 8.1.

The Arctic runoff season can be divided into four seasons; breakup, the snowmelt period (the 'nival flood'), late summer, and freeze-back. In detail, however, each season is determined by the pattern of local summer weather for each year. Usually, breakup begins in early June, when air temperatures are still below 0°C, through local snowmelt brought about by direct solar radiation. The meltwater percolates to the base of the snowpack and into snow choked stream courses, where it refreezes. It is not until two or three weeks later that sufficient melt has occurred for flowage of saturated snow to occur. The river channels turn to slush and runoff begins over the snow and ice in the stream bed. Usually after two or three days of intensive runoff, the winter ice on the stream bottom has been melted and runoff continues on the stream bed proper. There is usually a trigger which initiates the runoff; this is commonly a period of warm weather or a heavy storm which flushes an appreciable length of the stream channel.

In nival watersheds the majority of snowmelt occurs in the spring period. Runoff subsequently decreases from the nival freshet and comes to be dominated by storm runoff, either after heavy cyclonic activity or following periods of prolonged overcast and drizzly conditions. The hydrologic response is rapid, illustrating the importance of the presence of permafrost and the absence of vegetation. Superimposed upon these short term storm-controlled fluctuations are diurnal fluctuations, which are well developed in midsummer. Discharges usually increase in late afternoons and early evenings, or following periods of

Fig. 8.1 *Types of runoff regimes of periglacial rivers. From Church (1974)*

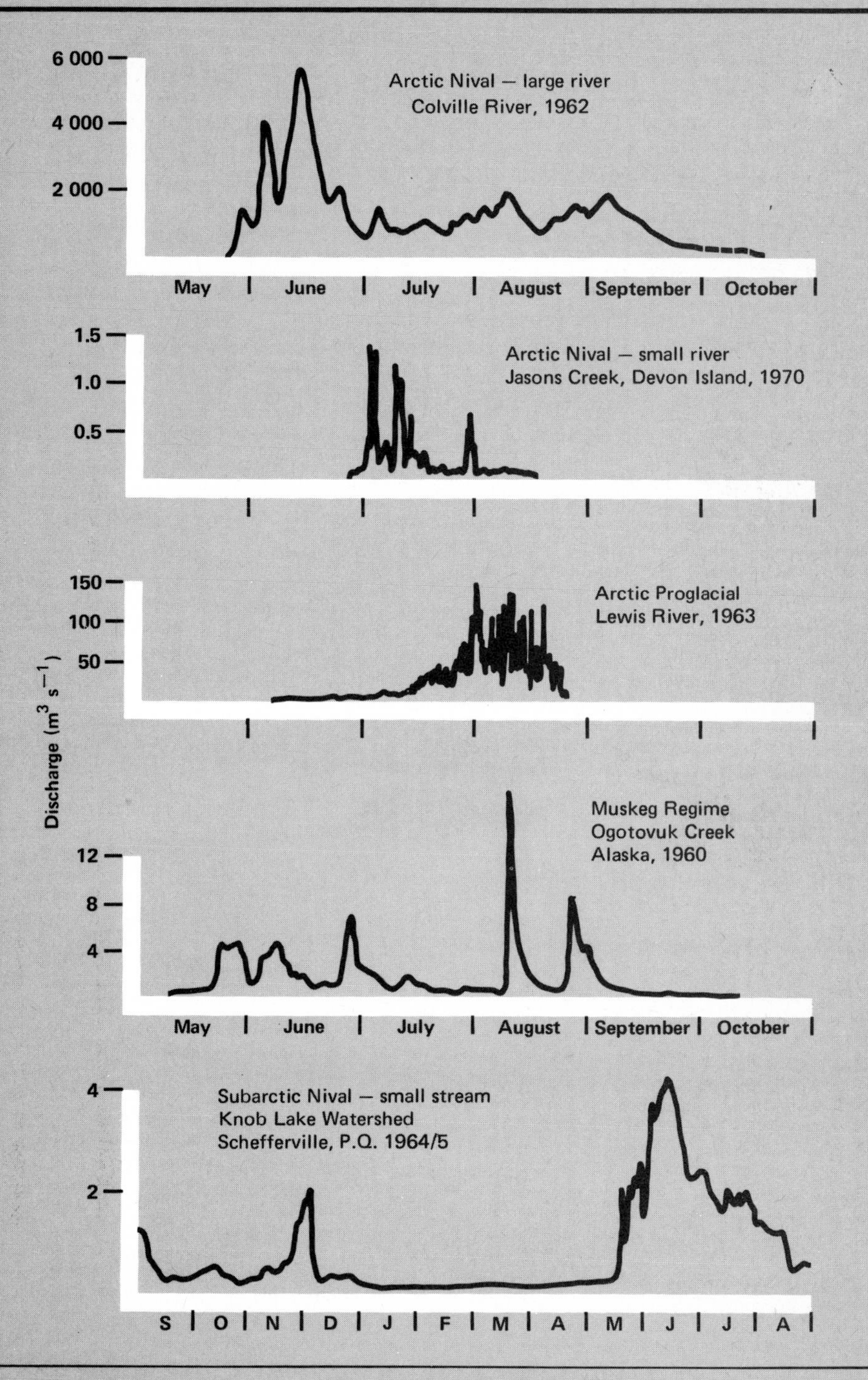

 uninterrupted solar radiation. 'Jason's Creek' on Devon Island (McCann ***et al.***, 1972) is fairly typical of these conditions (Fig. 8.1).

During late summer, many of the smaller streams are virtually without flow, and by the freeze-back period at the end of the summer there is very little melt for runoff. That streamflow which does exist is maintained almost entirely by recession flow from the groundwater discharge. Since the active layer is thin, flow is low. Figures 8.2 and 8.3 illustrate the difference in runoff conditions in a small stream on eastern Banks Island in late June and mid-August of the same year.

In proglacial watersheds, the importance of the nival freshet is much reduced and peak flow is usually delayed until late July or early August. This was the case for the Lewis River in 1963 (Fig. 8.1). According to Church (1972, p. 32), this indicated that (1) the early season heat deficit near the glacier surface had been overcome, (2) the albedo of the glacier surface had been lowered by melt of seasonal snow and exposure of darkened glacier ice, and (3) a drainage network had extended over the entire glacier watershed. During the melt, the daily runoff closely reflected daily heat inputs. The most extreme daily ranges of flow occurred during fine weather conditions when there was a strong diurnal control on glacier melt. By contrast, damp, stormy weather reduced the diurnal range of discharge.

The magnitude and frequency of flow events in both nival and proglacial

Fig. 8.2 ***A small stream in eastern Banks Island in late June, at time of spring snowmelt runoff.***

Fig. 8.3 *The same stream as in Fig. 8.2 approximately 2 km downstream and in mid-August.*

streams is difficult to determine, given the present lack of long-term hydrologic data. In proglacial streams, an interesting extreme event is the occurrence of jokulhlaupes. These are localised flood surges caused by the bursting of ice dams along the margins of the ice. In nival watersheds, true jokulhlaupes do not occur. However, Church (1972, p. 38) has described a similar, but small-scale event, which occurred in the South River at Ekalugad Valley in 1965. The discharge curve possessed all the characteristics of a jokulhlaupe, but was caused, instead, by a combination of very heavy rain and rapid melting occurring together. He stressed that this was not an abnormal event, but merely appears so in the short-term observations. Observations by St-Onge (1965, Fig. 13) on a small stream on Ellef Ringnes Island would support this view.

Sediment movement Sediment movement occurs through a combination of solution, suspension and bedload transport. In periglacial regions there are several sources of sediment for such transport. Material is moved off slopes to the stream channel by rockfall, solifluction and slopewash processes. Many areas, furthermore, possess an abundance of unconsolidated surficial sediments inherited from the previous glacial period which are easily picked up by streams through the undercutting and erosion of banks. Finally, the action of frost weathering provides rock particles suitable in size for transport.

The only comprehensive study of sediment movement in streams in the Canadian Arctic has been undertaken by Church (1972). Values of total

Table 8.1 *Sediment movement in some Baffin Island streams (from Church, 1972)*

A – *Total sediment movement*

River	Total sediment transport (metric tons)	Sediment yield (metric tons/km^2 of watershed)	Volume removed (m^3)	Equivalent surface lowering of watershed (mm/1 000 years)
Lewis River 1963–65 (mean)	151 021	735	56 989	277
Middle River, Ekalugad Valley, 1967	64 461	609	24 325	230
Upper South River, Ekalugad Valley, 1967	104 851	1 168	39 566	441

B – *Proportional distribution of sediment load*

	Solution	Suspension	Bedload
Lewis River, 1963	0.004	0.181	0.816
Lewis River, 1964	0.008	0.222	0.770
Lewis River, 1965	0.005	0.207	0.788
Middle Ekalugad, 1967	0.004	0.060	0.936
Upper South Ekalugad, 1967	0.004	0.042	0.954

sediment discharge, by solution, suspension and bedload transport for the Lewis and Ekalugad rivers on Baffin Island are summarised in Table 8.1. The solution and suspended sediment values were derived by traditional methods of field sampling. The bedload values however, were computed as values of 'potential sediment transport' at full supply and may be higher than the actual bedload transport achieved. Bearing this consideration in mind, the data nevertheless shows the dominance of bedload sediment transport. At Ekalugad River in 1967, bedload accounted for over 90 per cent of the load, while at Lewis River, over a three-year period, bedload accounted for 79 per cent of the total load. By contrast, the amount of movement in solution was insignificant, being less than 1 per cent each year, and suspended sediment made up the balance, varying between 5 and 25 per cent. Some insight into the significance of fluvial processes in fashioning the periglacial landscape is also indicated. Computations involving the total amount of material transported from the various watersheds studied indicate average ground surface lowering of the order of 200–450 mm/1 000 years.

Until more comparative data, of equivalent quality, is available from other periglacial areas, it is unwise to accept these values as being typical of all fluvial networks in all periglacial areas. Church points out that much of the present sediment yield is derived from the removal of unconsolidated glacial sediments and hence, sediment yield bears no relationship to present rates of sediment production in these watersheds (Church, 1972, p. 63). Baffin Island, as with most other recently deglaciated parts of arctic Canada, is currently going through a 'relaxation' process or disequilibrium phase in which streams are actively redistributing a mass of detrital materials laid down at the various ice marginal positions. It is probable that normal sediment production, and therefore, average ground lowering is at least one order of magnitude lower than the present yield rate. For example, the rate of 52 mm/1 000 years measured by Arnborg ***et al***. (1967) on the Colville River in northern Alaska is probably more typical of unglaciated periglacial areas or those which have completed their relaxation process.

The dominance of bedload transport appears to be a characteristic of periglacial rivers which is independent of recent glacierisation. In northern Alaska, Arnborg ***et al***. (1967) found that over 80 per cent of the sediment transfer in the Colville River was of either a bedload or suspended nature. This dominance of bedload movement in all of the Arctic streams so far studied is of interest, since it goes a long way towards explaining the distinctive flat-bottomed form of many periglacial stream valleys, which are discussed later in this chapter. In most rivers in other environments, bedload transport is thought to be relatively minor, and even mountain streams in alpine situations apparently move little bedload (see Church, 1972, p. 61).

From a geomorphic point of view, it is of interest to know when, and under what conditions, the majority of the sediment was transported. This usually bears some relationship to discharge. In the Canadian Arctic, Robitaille (1960) first drew attention to the importance of the nival freshet in transporting material on Cornwallis Island, and this was subsequently confirmed by stream flow measurements by Cook (1967) on the Mecham River. On the other hand, in the more arid areas of the western high Arctic, both Rudberg (1963) and Pissart (1967a) have stressed the fact that the nival flood does not necessarily give rise to extensive transport of material. Not only is the actual discharge associated with the nival freshet rather small, on account of the small amount of snow present in such regions, but the majority of the flood is completed by water flowing over the winter ice in the stream bed, effecting little or no streambed erosion.

Data from the Lewis and Ekalugad rivers illustrates the variability of sediment transport with discharge in both nival and proglacial environments. Table 8.2 lists the proportion of total seasonal runoff and sediment transport of each type listed cumulatively on the highest 1, 2, 3, 4 and 5 days of flow/transport for the years 1963–65 (Lewis River) and 1967 (Ekalugad River). The data clearly indicates the very high concentration of sediment movement events achieved by streams dominated by nival and storm events. For example, the South Ekalugad River in 1967 showed that over 75 per cent of total sediment movement occurred in less than 8 per cent of the runoff time of record. If typical, one must conclude that a major proportion of total sediment transport is accomplished in a very restricted period of time in periglacial environments.

Thermal erosion One of the more distinctive aspects of fluvial activity in periglacial regions is linked to the thermal effects of running water. In addition to the normal mechanical or abrasive effects, running water possesses the ability to thaw permafrost. This phenomena has already been discussed at some length in an earlier chapter (see Chapter 6, pp. 125–8), and needs little amplification. In summary, thermal erosion operates predominantly laterally, by undercutting river banks and sea cliffs, to produce thermo-erosional niches. This process is

Table 8.2 *Proportional distribution of discharge and sediment transport for the highest flow events in nival and proglacial streams, Baffin Island, Canada. Data from Church (1972)*

River	Days of record		Cumulative proportional discharge/transport during highest flow days				
			1	2	3	4	5
A – *Nival type*							
South Ekalugad, 1967	66	Q	0.103	0.155	0.200	0.245	0.288
		Q_c	0.059	0.099	0.135	0.171	0.207
		Q_s	0.407	0.471	0.530	0.575	0.612
		Q_b	0.477	0.579	0.653	0.724	0.793
		Q_t	0.466	0.562	0.635	0.702	0.767
B – *Proglacial type*							
Lewis River, 1963–65	206	Q	0.032	0.059	0.081	0.102	0.122
		Q_c	0.015	0.029	0.042	0.054	0.066
		Q_s	0.074	0.121	0.161	0.200	0.232
		Q_b	0.080	0.139	0.184	0.226	0.262
		Q_t	0.078	0.135	0.179	0.220	0.255

Q Discharge; Q_c – Solution; Q_s – Suspended; Q_b – Bedload; Q_t – Total movement.

important for at least three reasons: (a) the subsequent collapse of river banks provides material for bedload transport and deposition downvalley, and is also instrumental in the rapid coastal retreat which occurs in certain areas (see pp. 211–12), (b) it is one factor aiding the development of the flat and braided valley bottoms which are characteristic of many permafrost areas (see pp. 176–8), and (c) it helps explain the relative efficiency of lateral stream corrosion and subsequent stream migration in permafrost areas (see pp. 178–80).

Channel forms River channels may be of a single or multiple channel way. In periglacial regions, both types of channel patterns exist. However, the most distinctive and common is the multiple or braided one. Examples can be seen in Figs 8.4–8.6. This predominance of braided channel ways necessitates a brief consideration of the controls over braiding. We may identify several factors as being significant. First, the majority of braided channels occur in high energy streams flowing in noncohesive sediments and carrying heavy sediment loads. Bank erodibility is clearly a factor since excessive lateral erosion in weak sediments will not only lead to very wide channels in which shoaling occurs on central bars and thus, to the development of multiple channels, but also to the entrainment of large quantities of debris. Braiding is also related to rapid and large variations in runoff. In terms of hydraulics, an appreciable bedload transport appears to be the essential factor for braiding (Fahnestock, 1963). Once such a situation is present, any decrease in the competence of the stream will result in bar development and multiple channels. This decrease in competence may be caused by a variety of factors such as slope variations, discharge fluctuations, and variations in channel width and depth.

All these factors which determine braiding are present, to varying degrees, in periglacial regions. First, thermal erosion is a particularly effective agent of lateral bank erosion. The large-scale collapse of river banks adds sediment to the river as well as broadening the channel. Second, in recently deglacierised areas, the current redistribution of glacial sediments means a further source of material for transport, and in unglaciated regions where unconsolidated sediments outcrop widely, such as the western Canadian Arctic, a similar source of abundant sediment is present. Finally, the discharge of both nival and proglacial streams is subject to rapid and extreme fluctuation, as has been outlined in an earlier section.

As a result, the floor of many stream valleys in the Canadian Arctic is essentially flat, dissected by shallow braiding channels which are cut in alluvial sediments ranging from coarse gravel to medium sand. The banks of the channels are abrupt, giving a shallow box-like profile to the valley bottoms (Fig. 8.4). At times of flood, the whole of the valley floor becomes covered with a layer of turbulent water (Fig. 8.2), in which bedload transport dominates and the whole river becomes a moving mass of debris. Undercutting of the bank also occurs at this stage. As discharge decreases and as competence drops, the coarser material is redeposited, and the river once again assumes a new braided pattern.

Fig. 8.4 *Oblique air view from 150 m elevation of small box-shaped valley with stream incised within fluviatile terrace, eastern Prince Patrick Island.*

 Sometimes, the major channel may be relocated on the far side of the valley floor opposite to its position the previous year. Thus, given time and repeated adjustments to the braided channel pattern, all parts of the valley floor experience both flood conditions and braided stream activity. The channel floor is constantly being reworked.

The importance of abundant bedload sediment in producing braided stream channels is illustrated by the absence of well developed braided stream channels in those areas where bank erodibility is limited and where debris suitable for transport is limited. For example, on Banks Island, large rivers originating in the morainic areas flow west and northwest across the lowlands of the interior and western parts of the island. The present drainage commonly dissects redeposited glacial materials and/or late Tertiary sands and gravels. As such, there is no lack of coarse bedload sediment for transport and all of the major rivers possess well developed braided channels. The fluvial regime is distinctly nival and the fluctuations of discharge and variations in competence best explain the braided channel patterns. In certain places, however, braided streams change to become meandering for a certain length. This appears to be related to the exposure of the underlying fine sands of the Eureka Sound Formation in the river banks. The relative absence of coarse sediments at this point results in the stream becoming overcompetent and, to compensate and equalise energy, the streams adapt a meandering course.

Single and well defined channel ways are best developed in the larger rivers. Despite the fact that the surface of these rivers freezes over during the winter months, flow is maintained throughout the year and discharge fluctuations are not so extreme as in the smaller streams. The thermal influence of the water often promotes sub-river taliks or unfrozen zones, and an abrupt shelving of the permafrost table. Both of these factors favour the development of steeply inclined channel sides and a deep well defined channel. Braided patterns only develop on a large scale where these rivers exit into the sea or to large inland water bodies. In such deltaic situations, the energy available to the river rapidly diminishes and aggradation and multiple channels develop. The Mackenzie Delta is the classic example of such conditions (Mackay, 1963a) and similar conditions undoubtedly exist elsewhere along parts of the Arctic coasts of Alaska and Siberia.

Periglacial valley sandar

A sandur (plural: 'sandar') is an Icelandic term used to refer to alluvial surfaces formed by rivers carrying meltwater away from the fronts of glaciers (Krigstrom, 1962). Two types of sandur are recognised; valley sandur ('dalsandur') and plain sandur ('Slattlands-sandur') which identify their topographic locations within large valleys, or as outwash plains in proglacial positions. Sandar are characteristic of rapid aggradation and are crossed by braided streams that are continually shifting their pattern, as has been described in the previous section. Sandar surfaces developed extensively along the margins of the Pleistocene ice sheets in both North America and Eurasia. Most of the valley sandar developing today are in recently glacierised valleys, such as in Baffin Island (e.g. Church, 1972), the western Cordillera (e.g. Fahnestock, 1963), and Iceland (e.g. Price, 1969).

Fig. 8.5 The Ballast Brook, northwest Banks Island; an example of a large braided channel with periglacial valley sandur.

In certain periglacial environments, far removed from present-day glacial activity, extensive but shallow valley fill deposits exist of a coarse clastic nature. In places, they cover many square kilometres and form some of the most prominent physiographic features of the landscape (Fig. 8.5). To all intents and purposes, these depositional surfaces are plain sandar. To avoid confusion, however, the descriptive term 'periglacial sandur' is used in this text. It refers not only to the braided stream but also to the materials and the channel morphology. Periglacial sandar can be distinguished from classical alluvial flood plains, since the latter include layers of finer sediments which are deposited by a river which, in the short term, is relatively stabilised in a well defined channel. Furthermore, braided streams characterise active sandar surfaces, and channel scars on older deposits indicate a similar channel pattern in the past. By contrast, alluvial flood plains develop innumerable thaw ponds and lakes which progressively migrate across the floodplain surface to leave shallow, enclosed depressions (see Chapter 6, pp. 122–5).

Periglacial sandar are particularly well developed in the broad valleys which drain towards the Beaufort Sea in the western Arctic, and occur extensively in the coastal lowlands of Ellef Ringnes, Prince Patrick and Banks islands. According to Church (1972), three environmental factors seem to be necessary for the deposition of coarse gravelly valley fill. First, an abundance of coarse detrital material is required. Second, the stream gradient should be sufficient to move the material in the water courses. Third, the hydraulic regime should be characterised by relatively frequent and high floods, enabling large quantities of material to be moved. These characteristics are present in possibly four environments; (1) semi-arid, (2) periglacial, (3) high mountain, and (4) proglacial. All are characterised by frequent flooding which, in the periglacial (i.e. nival) context, is the result of the spring melt and/or summer rainstorms. The unique feature of the western Arctic coastal lowlands for sandur development is undoubtedly the abundance of coarse sediments available for transport, resulting from the underlying Tertiary formations and in particular, the Beaufort Formation.

Valley forms and slope asymmetry

In view of the relative absence of quantitative data on the form of slopes and the nature of fluvial processes, it is not surprising that there is also a lack of accurate information relating to the features of stream valleys in periglacial regions. In fact, the few studies which are available and which relate to valley forms have been made within the context of slope asymmetry. Thus, to judge by the literature on valley forms in periglacial regions, one would conclude that asymmetrical valleys are the only type present. This, of course, is not the case, since the lack of asymmetry in many regions does not apparently attract similar attention.

The presence of asymmetrical valleys in areas currently underlain by permafrost was first noticed by Shostakovitch (1927) in Siberia, where the steeper slopes commonly face north. To judge from the literature (Table 8.3), this appears to be the 'normal' for higher latitudes. Steeper north-facing slopes have been reported from Disko, Greenland, from northwest and central Alaska, from Southampton Island in the central Canadian Arctic, from the Mackenzie District of the N.W.T., and from northern Siberia. However, in extreme high Arctic localities, such as Spitsbergen (latitude 78°N) and northwest Banks Island (latitude 74°N) no such regularity exists, and instead, steeper slopes have been reported to face either southwest, west, south or east.

The most probable explanation for the 'normal' asymmetry with steeper north-facing slopes involves greater solifluction activity on south-facing slopes and asymmetric lateral stream corrasion (e.g. Bronhofer, 1957; Currey, 1964; Gravis, 1969). None of the reported examples of this type of asymmetry lies appreciably north of latitude 70°N. Thus, there is a distinct change in the inclination of the sun throughout the polar day in these regions. Greatest insolation is received on slopes oriented south and southwest, and these slopes often possess a deeper active layer than others. Where valleys are aligned approximately east—west, the unequal debris arrival at the base of slopes forces the stream to migrate to the foot of the north-facing slope, which is undercut and steepened.

Two rather different interpretations of periglacial valley asymmetry have been described by the writer from Banks Island (French, 1971a), and by Kennedy and Melton (1972) from the Caribou Hills, near Inuvik in the Mackenzie District. They illustrate the inadequacy of any single origin for the development of asymmetrical valleys under periglacial conditions.

On the Beaufort Plain of northwest Banks Island a network of strikingly

Table 8.3 *Some characteristics of valley asmmetry in northern regions*

Source	Locality	Valley alignment	Orientation of steeper slope
East Greenland			
Poser[1]	Wollaston–Vorland	E–W	N
Malaurie (1952)	Disko	E–W	N
West Spitsbergen			
Dege[1]	Andreeland	E–W	S
Dege[1]	Conwayland	N–S	W
Klimaszewski[1]	Kaffioya–Ebene	E–W	S
Klimaszewski[1]	Brogger–Halbinsel	N–S	E
Siberia			
Schostakovitch (1927)	Yakutia	E–W	N
Presniakow (1955)	Yakutia	E–W	N
Gravis (1969)	Yakutia	E–W	N
Northern Canada			
Bronhofer (1957)[2]	Southampton Island	E–W	N
French (1971a)	Banks Island	NW–SE	SW
Kennedy and Melton (1972)	Caribou Hills, NWT	E–W	N, S
Alaska			
Hopkins and Taber (1962)	Central Alaska	E–W	N
Currey (1964)	Northwest Alaska	E–W	N

(1) Quoted in Karrasch (1970), p. 205.
(2) Quoted in Bird (1967), p. 250.

asymmetrical valleys exist which bear no relation to structure (Fig. 8.6). The steeper slopes are oriented towards the west and southwest, and the asymmetry is clearly anomalous in terms of the 'normal' asymmetry of high latitudes. The generalised relationship between slopes, soils, surficial materials and micro-relief features in the asymmetrical valleys is summarised in Fig. 8.7 and Table 8.4. The origin of this distinctive terrain was investigated by a number of micro-climatic measurements which suggested that strong differences in temperature and moisture exist between northeast- and southwest-facing slopes. Contrary to theory, the south- and west-facing slopes are cooler than other slopes of equal inclination, and possess shallower active layers. Several reasons are suggested for this. First, these valleys are located at nearly 74°N. As such, the changing inclination of the sun throughout the polar day is less than in areas further south, and the preferential insolation received by slopes of southerly exposure is minimised. Second, the dominant winds in this part of the Arctic are from the west. During winter, snow is blown off the upland areas and deposited in gullies

Fig. 8.6 *Oblique air view of asymmetrical valleys incised within the Beaufort Plain, northwest Banks Island, Canada. The steeper slopes face towards the southwest and possess intense gully dissection. Northeast-facing slopes possess snowbanks beneath which extend gentle solifluction slope (mid-foreground).*

and lee slope positions, i.e. on east-facing slopes, while west-facing slopes are kept clear of snow. In the summer, the continued action of strong westerly winds promotes evaporation and latent heat loss from the exposed westerly slopes. West-facing slopes are therefore, both drier and cooler than east-facing slopes. Furthermore, the melting of the snowpatches on the east-facing slopes provides an additional source of moisture for that slope throughout the summer, which also has a deeper active layer because it is warmer and not subject to the same degree of evaporation and heat loss as the opposite slope. Thus, solifluction and nivation processes are particularly active upon the east-facing slope. Consequently, the stream moves laterally to the slope producing the least colluvium, and the southwest-facing slope is then undercut and steepened in angle through the operation of fluvio-thermal erosion processes. An asymmetry of the valleys is subsquently produced in which the two slopes are constantly adjusting to each other and to the basal stream channel. The asymmetry is regarded as an equilibrium form which is closely related to the climatic and geomorphic environment of the area.

In the Caribou Hills, Kennedy and Melton (1972) investigated slope forms in a number of different geomorphic environments. The study area is at a latitude 68°N and there is a progression from Arctic to subarctic conditions as one moves from the plateau top to the base of the hills. Valley side slopes were

Fig. 8.7 *Generalised relationship between slopes, soils and materials in asymmetrical valleys of the Beaufort Plain, Banks Island. From French (1971a)*

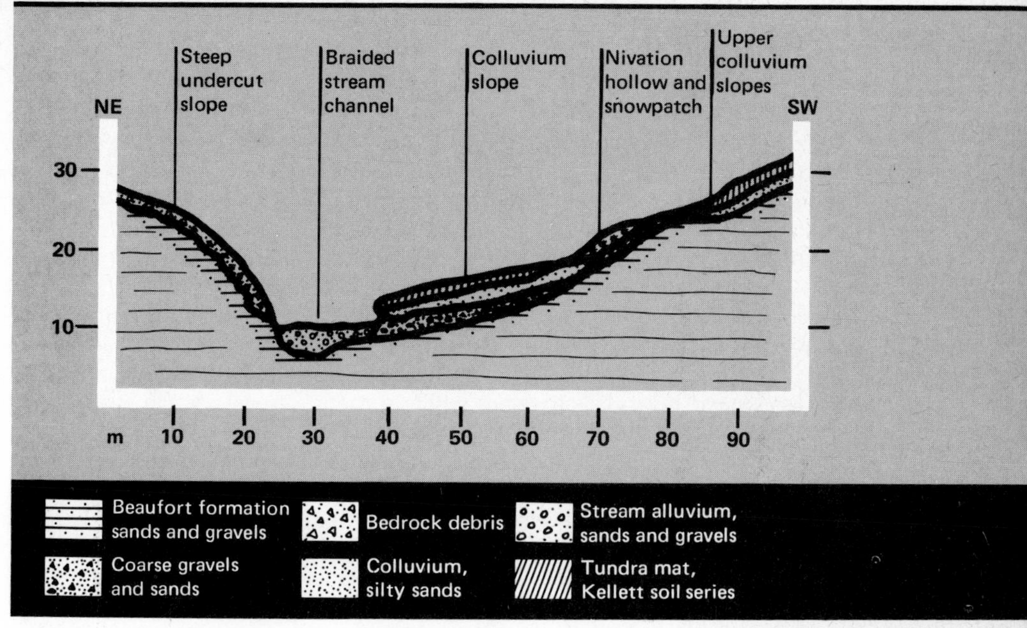

Table 8.4 *Generalised relationship of slopes, soils, surficial materials and micro-relief features in asymmetrical valleys, northwest Banks Island, Canada. From French (1971a)*

	Southwest-facing slope			**Northeast-facing slope**		
Soil series	**Polar desert (Storkerson)**	**Regosol**	**River sands and gravels**	**Meadow tundra (Kellett)**	**Upland tundra (Bernard)**	**Polar desert (Storkerson)**
Angle (°)	0–3	10–30	0	3–5	5–11	0–3
Drainage	Well-drained	Well-drained	–	Imperfect	Imperfect	Well-drained
Parent material	Beaufort Formation	Beaufort Formation	Fluviatile sediments	Colluvium	Colluvium, wind-blown materials	Beaufort Formation
Surface relief	Ice-wedge polygons, desert pavement	Gully dissection	Braided channels	Non-sorted polygons, frost boils	Soil hummocks	Ice-wedge polygons, desert pavement
Vegetation	–	–	–	Carex, salix, mosses	Dryas	–
Depth to permafrost	0.7–1.0 m	1.0 m+	?	0.3–0.5 m	0.5–0.7 m	0.7–1.0 m

measured in three altitudinal zones reflecting this transition. In each zone there was a significant difference between mean maximum angles of north- and south-facing slopes. However, there was a marked reversal in the aspect of the steeper slopes between the two upper zones and the lower zone. In the areas of more severe climate and less available relief, the asymmetry is 'normal' with steeper slopes facing north. In the lower valley zone of greater relative relief and less severe climate, south-facing slopes are steeper. Apparently, this reversal of asymmetry between the upper and lower valleys is brought about by a decline in angle of north-facing slopes and a steepening of south-facing slopes. According to Kennedy and Melton, three factors are responsible for this change. First, on the upper slopes and plateau top, snowpatches are common on north-facing slopes where they occupy much of the profile of shallow slopes. They lead, through nivation, to a general steepening of that slope. In the lower valleys, snowpatches are less important since they occur at the extreme upper part of the profile. Second, basal stream undercutting in the upper valleys is particularly effective against north-facing slopes while in the lower valleys, the streams are braided and neither slope experiences significant basal stream corrasion. Third, in the lower valleys, south-facing slopes experience high relative insolation which favours rillwash and gullying and the maintenance of steep angles. They conclude that asymmetry develops through snowpatches forming at the base of low north-facing slopes which are then steepened through nivation. As the valleys deepen, basal stream undercutting further steepens the north-facing slope, but at the same time, as slope length increases, the importance of nivation decreases. Further downvalley, as stream activity decreases, north-facing slopes decline through creep while south-facing slopes are maintained by rillwash and gullying.

The lack of detailed micro-climatic or process measurements makes this interpretation rather theoretical. For example, no reason is given as to why the streams should undercut north-facing slopes in the upper valleys, and the decline of north-facing slopes in the lower valleys is not proven. Nevertheless, this study highlights the necessity of considering the simultaneous variations through space of both micro-climate and relief as they influence geomorphic processes. It further illustrates that the relationship between aspect, micro-climate and basal stream activity is critical for the development of asymmetrical valleys.

These various examples enable certain comments to be made which are relevant to the discussion of Pleistocene valley asymmetry in Part III (pp. 253–6). First, stream action and in particular, lateral stream migration consequent upon unequal debris arrival at the basal channel, is an important factor in the development of slope asymmetry in present-day periglacial regions. At times of the nival freshet or storm flood runoff, the combined effects of mechanical and thermal erosion cause undercutting and thermo-erosional niches to develop which subsequently lead to collapse and slope steepening. The development of the 'normal' asymmetry of high latitudes is essentially a process in which both slopes evolve in relation to each other and to the basal channel.

Second, the effect of aspect upon slope evolution and geomorphic processes in high latitudes is delicate. It must be emphasised that high latitudes do not experience the strong diurnal rhythm of insolation which was characteristic of middle latitude regions during the Pleistocene. Notwithstanding this comment, it would appear that there is still a sufficiently large daily change in the inclination of the sun, in subarctic locations especially, to favour greater solar radiation reception by south- and southwest-facing slopes. However, these effects are relatively weak and in extreme high Arctic localities may be overridden by more local conditions. For example, the major conclusion to be drawn from the Banks Island study is that, in that area, the wind is more important in determining local micro-climatic conditions than is insolation, and that the theoretical temperature variations with orientation, as described in many texts, do not necessarily occur.

Micro-relief features 9

In this chapter the nature of the soils and micro-relief of periglacial environments is briefly outlined. Frost action gives rise to distinctive morphological characteristics of soils not found in other regions. At the same time the surface micro-relief is often composed of a complex variety of patterns and small-scale relief configurations which are commonly termed 'patterned ground'. However, it must be stressed that patterned ground phenomena are not restricted to periglacial regions. Furthermore, although the soils of the tundra and high latitude regions are obviously different to those of other regions, they reflect basically the same sorts of relationships between lithology, climate and topography that are found elsewhere.

Patterned ground

The detailed study of patterned ground is more appropriate to pedology than to geomorphology since patterned ground does not result in important landforms. Instead, patterned ground is merely a surface 'decoration' of the soil regolith. For various reasons however, patterned ground has attracted much attention in the periglacial literature. The early explorers and scientists of the tundra and Arctic regions, who travelled on foot or by dog team, had ample opportunity to obtain first hand acquaintance with such phenomena. The lack of vegetation made the various patternings of the ground particularly distinctive, and the botanical explorers quickly identified the intimate relationships existing between patterned ground and vegetation habitats. This early emphasis upon rather descriptive studies of patterned ground was unfortunate in that interest was directed away from the more fundamental aspects of the landscape such as slopes, and rivers. A second reason for this early emphasis upon patterned ground studies was the recognition of active miniature forms in alpine locations, and the desire to make comparisons.

Today, in spite of the voluminous literature, there are few field studies of patterned ground phenomena which are more than purely descriptive accounts. The processes responsible for the formation of patterned ground still remain largely unproven, although there is certainly no lack of hypotheses. In his classic review paper, Washburn (1956) listed no less than 19 hypotheses for patterned ground development. His conclusions are no less true today than at that time. These were '(1) the origin of most forms of patterned ground is uncertain; (2) patterned ground is polygenetic; (3) some forms may be combination products in a continuous system having different processes as end members; (4) climatic and terrain interpretation of patterned ground, both active and "fossil' is limited by lack of reliable data about formative processes' (Washburn, 1956, p. 823).

Description The terminology proposed by Washburn to describe patterned ground is now widely accepted. It is both simple and unambiguous. Patterned ground is classified on the basis of (a) its geometric form, and (b) the presence or absence of sorting. The main geometric forms recognised are circles, polygons, stripes, nets, and

Fig. 9.1 *Diagram illustrating the change from circles to stripes as slope angle increases, according to J. Budel (1960)*

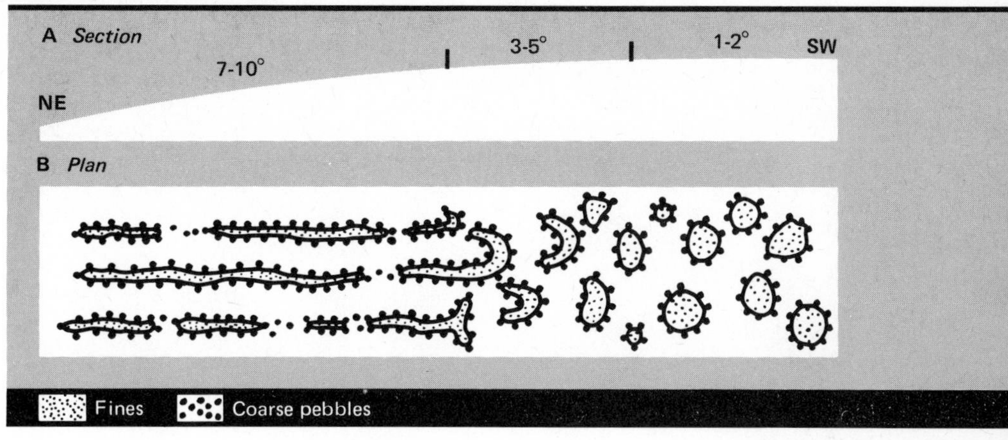

steps, all of which may be sorted and unsorted, thus giving ten principal categories of patterned ground. Circles, polygons and stripes are self explanatory. A net refers to a mesh which is intermediate between a polygon and a circle in plan. A step is a bench-like feature with a downslope border of vegetation or stones embanking an area of relatively bare ground upslope. The relation between these various forms and angle of slope is illustrated in Fig. 9.1. Circles, polygons and nets usually occur on flat or nearly flat surfaces. As slope angle increases to 2–3° however, these forms become elongated due to mass wasting and, depending upon local conditions, may change to stripes further down the slope. Steps are the transitional form in this sequence.

Illustration of the variety of patterned ground forms that exist is given in Figs 9.2–9.6. All the photographs were taken on either Banks or Prince Patrick Islands in the Canadian Arctic. The patterns typify the diversity of forms which may be encountered in different areas on different lithologies. In detail, the main features of circles, polygons and stripes can be readily identified. Sorted circles are defined as features . . . 'whose mash is dominantly circular and which have a sorted appearance commonly due to a border of stones surrounding fine materials' (Washburn, 1956, p. 827). They may occur singly or in groups, and commonly vary from 0.5–3.0 m in diameter. All varieties of sorted circles exist. For example, on steep, rubble covered slopes concentrations of fine material may appear amid blocks and boulders. These are called 'debris islands'. Non-sorted circles are essentially similar to sorted circles except that they lack a border of stones. They are usually bounded by vegetation and give the appearance that the fine material of the central area has been intruded from below. Other names for such features include 'mud circles' or 'mud boils' (Mackay, 1953; Bird, 1967). They are commonly dome-shaped and unvegetated at their centres. A particularly striking non-sorted circle is illustrated in Fig. 9.2. Here, the unvegetated central area of fines is completely surrounded by a closed rim of tussock grass. The raised tussock grass rim probably reflects the wetness of the area and the segregation of ice beneath the tussock vegetation.

Fig. 9.2 *Non-sorted circle surrounded by tussock rim, eastern Banks Island. The immediate site is a poorly drained lowland of meadow tundra vegetation.*

Fig. 9.3 *Earth hummocks, Prince Patrick Island.*

Polygonal forms can be distinguished from circular ones. Non-sorted polygons lack a border of stones. They can be categorised into small forms (diameter less than 1.0 m) and large forms (diameter greater than 1.0–2.0 m). The large polygons, which often possess ice wedges along their borders, form the ice-wedge polygons (tundra or Tamyr polygons) which have been discussed at some length in an earlier chapter (see pp. 84–93). The smaller non-sorted polygons commonly form meshes or nets over wide areas (Fig. 9.3). They are delineated by furrows and cracks to give a hummocky micro-relief of between 0.1 and 0.3 m. North of the treeline, many of the polygons may be as small as 5 cm but more often are 20–50 cm in diameter. Often, Dryas clumps develop on the drier and more elevated polygon centres, while mosses and lichens occupy the damper depressions. Terms such as 'tundra hummocks', 'earth hummocks', and 'thurfurs' have been used to describe these features (e.g. Beschel, 1966; Raup, 1966; Corte, 1971), although they are probably not all of the same origin (see pp. 33–7; 42). South of the treeline, in the northern forest zone (see below), non-sorted polygons are slightly larger in dimensions, often being 1.0–3.0 m in diameter and 0.4–0.5 m in height, and form extensive areas of what is termed 'hummocky terrain' (e.g. Zoltai and Pettapiece, 1973). This has been discussed in Chapter 3. Sorted polygons possess a border of coarser materials. The polygons are usually no larger than 10 m in diameter and more often between 1.0 and 4.0 m, although some are small, being less than

Fig. 9.4 *Non-sorted stripes, eastern Banks Island.*

20–30 cm in diameter. The bordering stones tend to increase in size with the size of the polygons, but decrease in size with depth whatever the polygon dimensions. In contrast to non-sorted polygons which may occur on slopes of considerable gradient, sorted polygons are usually restricted to flat surfaces.

Stripes occur only on sloping terrain. Non-sorted stripes are defined as patterned ground possessing a . . . 'striped pattern and a non-sorted appearance due to parallel lines of vegetation covered ground and intervening stripes of relatively bare ground oriented down the steepest slope available' (Washburn, 1956, p. 837). Both large and small forms occur. Large non-sorted stripes are commonly 0.3–1.0 m wide (Fig. 9.4). The intervening stripes of vegetation occupy shallow hollows between the stripes of non-vegetated ground and may be of equal width as the stripe. In areas of fine grained sediments, many of the stripes show a well developed dentritic pattern which suggests that surface rill-wash is the cause of the patterning (Fig. 9.5). Sorted stripes show a well marked differentiation between lines of stones and intervening stripes of finer material oriented parallel to the steepest gradient (Fig. 9.6). They seldom occur on slopes of less than 3° and are most common on slopes of 5–15°. The stoney stripes are usually narrower than the fine stripes, and vary from 0.3 to 1.5 m in width. The

Fig. 9.5 *Air view from 200 m of stripes developed on hummocky morainic terrain of eastern Banks Island. Ground investigations reveal the stripes to be composed of mosses and lichens developed along lines of concentrated wash on slopes of 2–5°. Average spacing between stripes is 1–2 m.*

Fig. 9.6 *Sorted stripes, Prince Patrick Island.*

stones and boulders are commonly on edge with a long axis parallel to the line of movement of the stripe.

Origins There are many suggested origins for patterned ground and in this section no attempt is made to cover all possible interpretations. The interested reader is referred to the publications of Washburn (1956; 1969) for more detailed summaries of the various hypotheses. Today, most investigators generally agree with Washburn that (a) much patterned ground is polygenetic and (b) that similar forms can be created by different processes. Recently, Washburn (1970) has attempted a comprehensive genetic classification of patterned ground (Table 9.1). This classification is an all-inclusive one, not being restricted to periglacial phenomena, in which form and genesis are linked in a matrix. The major feature of the classification is the twofold division of the processes responsible for patterned ground into (i) those in which cracking is essential, and (ii) those in which cracking is non-essential. Washburn recognises that this classification is merely an attempt to bring order to a phenomenon which is still not understood, and that the matrix approach tends to oversimplify the reality of any one patterned ground form. It is, therefore, an extremely tentative scheme. However, it does provide a framework within which deductions can be made as to the origin of patterned ground.

Washburn's genetic classification suggests cracking to be important in the

Table 9.1 *Genetic classification of patterned ground, according to A. L. Washburn*

Geometric patterns		*Processes*					
		Cracking essential					
				Thermal cracking			
					Frost cracking		
		Desiccation cracking	**Dilation cracking**	**Salt cracking**	**Seasonal frost cracking**	**Permafrost cracking**	**Frost action al bedrock j**
Circles	*Nonsorted*						
	Sorted						Joint-crack circles (at crack inter sections)
Polygons	*Nonsorted*	Desiccation polygons	Dilation polygons	Salt-crack polygons	Seasonal frost-crack polygons	Permafrost-crack polygons, incl. Ice-wedge polygons. Sand-wedge polygons	Joint-crack polygons?
	Sorted	Desiccation polygons	Dilation polygons	Salt-crack polygons	Seasonal frost-crack polygons	Permafrost-crack polygons	Joint-crack polygons
Nets	*Nonsorted*	Desiccation nets incl.? Earth hummocks	Dilation nets		Seasonal frost-crack nets incl.? Earth hummocks	Permafrost-crack nets, incl. Ice-wedge nets and? Sand-wedge nets	
	Sorted	Desiccation nets	Dilation nets		Seasonal frost-crack nets	Permafrost-crack nets	
Steps	*Nonsorted*						
	Sorted						
Stripes	*Nonsorted*	Desiccation stripes	Dilation stripes?		Seasonal frost-crack stripes?	Permafrost-crack stripes?	Joint-crack stripes?
	Sorted	Desiccation stripes	Dilation stripes		Seasonal frost-crack stripes?	Permafrost-crack stripes?	Joint-crack stripes

Table 9.1 – *continued*

Processes

Cracking non-essential						
Primary frost sorting	**Mass displacement**	**Differential frost heaving**	**Salt heaving**	**Differential thawing and eluviation**	**Differential mass-wasting**	**Rillwork**
	Mass displacement circles	Frost-heave circles	Salt-heave circles			
Primary frost-sorted circles, incl.? Debris islands	Mass displacement circles, incl. Debris islands	Frost-heave circles	Salt-heave circles			
	Mass displacement polygons?	Frost-heave polygons?	Salt-heave polygons?			
Primary frost-sorted polygons?	Mass displacement polygons?	Frost-heave polygons?	Salt-heave polygons?	Thaw polygons?		
	Mass displacement nets, incl.? Earth hummocks	Frost-heave nets, incl. Earth hummocks	Salt-heave nets			
Primary frost-sorted nets	Mass displacement nets	Frost-heave nets	Salt-heave nets	Thaw nets		
	Mass displacement steps	Frost-heave steps?	Salt-heave steps?		Mass-wasting steps	
Primary frost-sorted steps?	Mass displacement steps	Frost-heave steps	Salt-heave steps	Thaw steps?	Mass-wasting steps	
	Mass displacement stripes	Frost-heave stripes	Salt-heave stripes		Mass-wasting stripes?	Rillwork stripes?
Primary frost-sorted stripes?	Mass displacement stripes	Frost-heave stripes	Salt-heave stripes	Thaw stripes	Mass-wasting stripes	Rillwork stripes

development of polygonal forms, and relatively unimportant in the development of circular forms. In the periglacial context, three types of cracking processes occur; desiccation cracking, thermal contraction cracking, and seasonal frost cracking. The two latter types of cracking have already been discussed in Chapter 3 (pp. 21–7) and need no further comment. Desiccation cracking however, is unrelated to either frost action or the presence of permafrost. In theory, desiccation can occur for a number of reasons; wind action may promote evaporation from exposed surfaces, the terrain may be subjected to a change in drainage conditions, and desiccation may occur in association with ice segregation, as demonstrated experimentally by Taber and Pissart (see p. 30). In reality, most investigators conclude that wind desiccation is the most likely cause of the extensively developed and regular non-sorted polygons of small sizes which characterise many upland tundra sites (e.g. Tricart, 1970, p. 91; Washburn, 1973, p. 140). Where they develop considerable micro-relief and become distinct hummocks, it is likely however, that additional processes such as differential frost heave, cryostatic pressures or mass displacement are operative (see pp. 33–5; 40–2).

Circular forms of patterned ground are most likely the result of a combination of non-cracking processes such as primary sorting, differential heaving, and mass displacement either through cryostatic pressures or changes in intergranular pressure. For example, the origin of non-sorted circles ('mud circles') appears to follow the injection of subsurface material into the surface layer. This may be the result of simple hydrostatic build up of pressure during the freeze-up period when unfrozen ground exists between the surface and the permafrost table (i.e. a 'cryostatic' hypothesis). The wettest and finest sediments are the last to freeze, and pressure is relieved if these sediments can move upwards towards the surface (Fig. 9.7). Cook (1956) has argued that horizontal pressures may develop in heterogeneous sediments causing areas of fines to migrate upwards. He also envisaged the central area of fines as a 'wick' or plug transferring moisture from the permafrost table to the surface. Sorted circles may also be formed by a combination of frost related processes. Detailed field measurements by Chambers (1967) led him to conclude that sorted circles resulted from a two-stage process (Fig. 9.8). In the first stage, upfreezing of coarse material due to frost action resulted in the accumulation of coarse, relatively dense material at the surface, overlying finer materials at depth. In the second stage, the denser material at the surface is responsible for load deformation that is expressed in the upwelling of plugs of fines. Where the plugs break through to the surface, sorted circles are formed. One of the points in favour of this hypothesis is that it requires no assumptions as to the concentration of fines and stones in the active layer before the pattern forming processes began. It may also be that the surface load which brings about the displacement of the underlying fines may be due to differences in moisture content, reducing intergranular pressure at depth. Thus, the use of the term 'convection' to describe the processes involved in sorted and non-sorted circles is rather erroneous. Movement is not a circulation but instead, the upwelling of fines in plugs followed by a far more gentle and less obvious settling of the surface material. Once an equilibrium has been reached, movement will cease. According to Chambers (1967) this equilibrium stage may be represented by 'anchored' polygons, where the coarse material extends to the base of the active layer.

Fig. 9.7 *Diagram illustrating the development of non-sorted circles or 'mud-boils' according to a cryostatic pressure hypothesis*

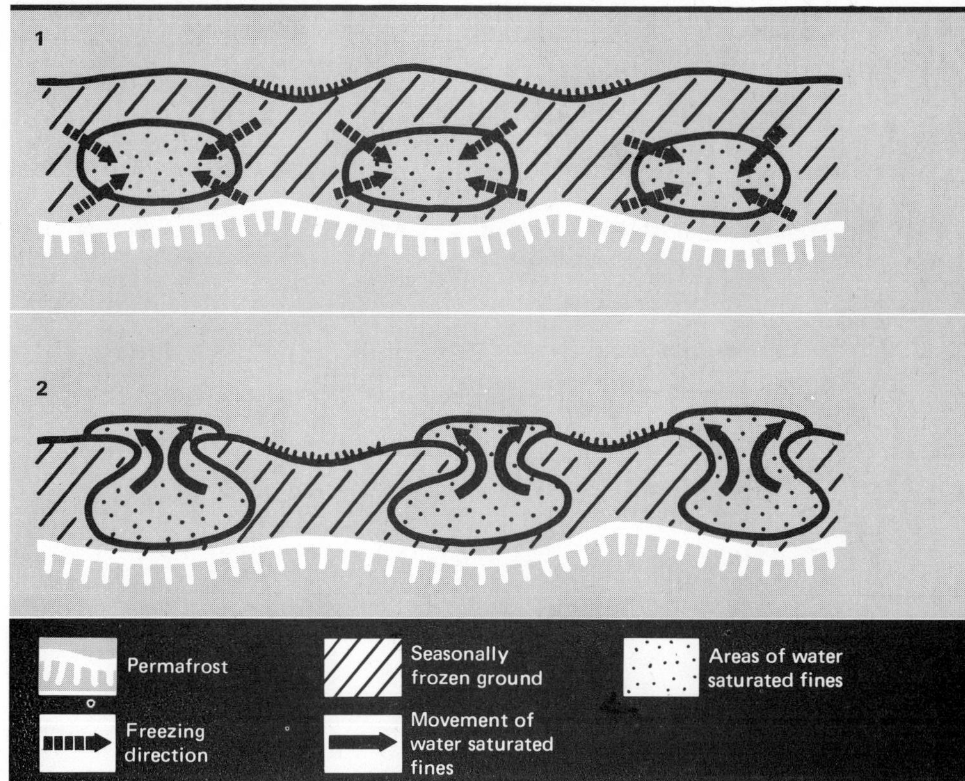

Stripes reflect the modifying influence of mass wasting processes upon circular and polygonal patterned ground forms. In both sorted and non-sorted stripes movement is greatest at the centre of the stripe, and decreases towards the sides and with depth (see Solifluction, pp. 135–41). Absolute rates of movement vary, depending upon gradient, moisture, and lithology. In non-sorted stripes, movement is confined to the bare, unvegetated areas and there is virtually no movement at all of the vegetated stripes. On Banks Island on slopes of 3°, French (1974a) recorded average rates of 1.5–2.0 cm year^{-1} for non-sorted stripes (see Table 7.2, p. 138), and rates on slopes of higher angle are probably greater. In sorted patterns, both the coarse and the fine stripes move, with the latter being appreciably faster than the former. On Signy Island for example, Chambers (1967, p. 30) reported that fines were moving about 15 cm year^{-1} relative to stones in stripes developed in wet areas on gradients of between 6 and 15°. These values for sorted and non-sorted stripes illustrate the range of movement that occurs within and between stripes. Sections excavated across sorted stripes indicate that the stone stripes commonly have a V-shape in section. This is because coarser material is continually being swept to either side of the more rapidly moving finer stripe. Since movement within stripes also

Fig. 9.8 *Diagram illustrating the stages in the development of sorted circles, according to Chambers (1967)*

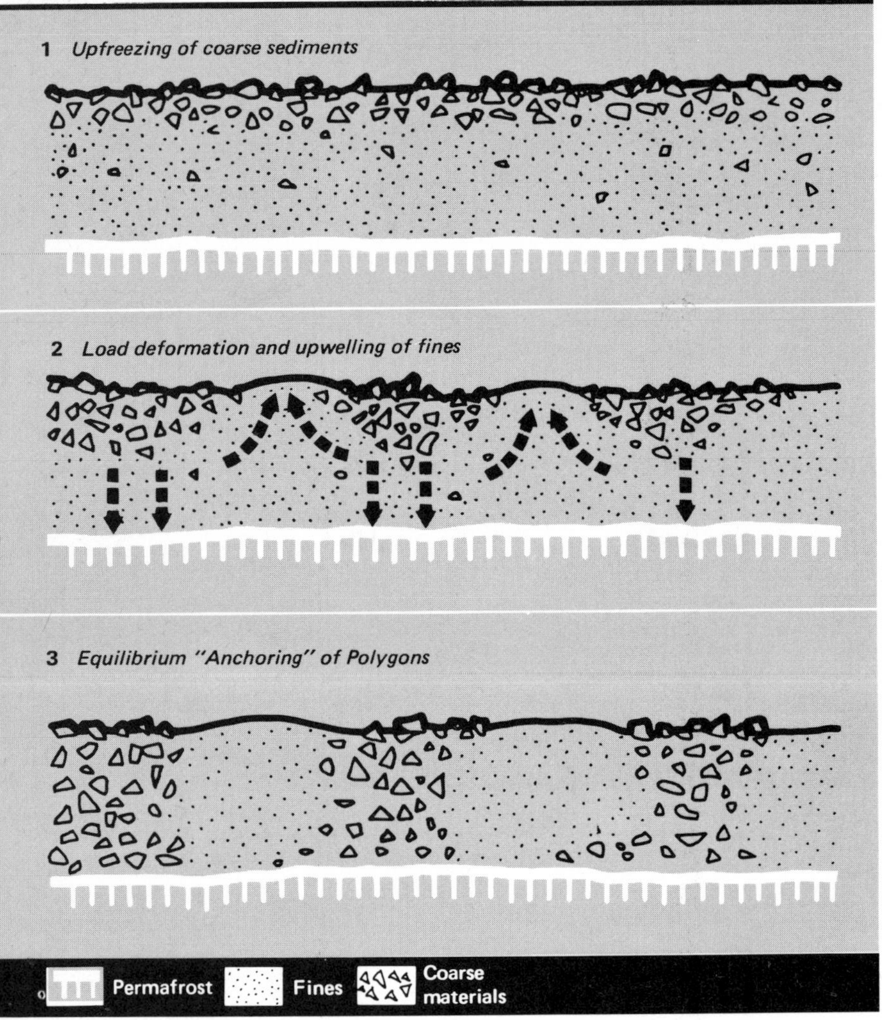

decreases with depth, the supply of stones to either side will also be less with increasing depth.

The environmental significance of patterned ground phenomena is not clear in view of the variety of origins that are possible, as revealed in Washburn's classification. The only really valid indicator of periglacial conditions is the presence of thermal contraction cracks since they indicate both (a) permafrost and (b) extreme cold. Their significance has been discussed elsewhere (pp. 21–5; 91–3). Most other patterned ground processes are not unique to periglacial regions. For example, desiccation cracking obviously occurs in other arid regions of the world, and seasonal frost cracking is known to occur in middle latitudes and is not necessarily associated with permafrost. Equally, mass displacements, if produced by changes in intergranular pressure, are controlled primarily by moisture differences and neither freezing temperatures nor permafrost need be involved.

On empirical grounds, the southern limit of active patterned ground of a periglacial nature has been suggested by Williams (1961) to correspond to a mean annual air temperature of +3°C. In terms of the relatively large scale and conspicuous patterned ground forms discussed above however, this limit is too far south, and Bird (1967, p. 197) suggests that in northern Canada for example, the southern limit of such phenomena is more closely linked with the −4°C isotherm. On the other hand, Williams's proposed boundary relates well to the distribution of the various miniature patterns that have long been recognised as occurring in less severe climatic zones where the larger forms are absent such as the Alps, the Pyrénées, and the uplands of northwest Europe. However, the occurrence of these miniature patterns is not necessarily a periglacial phenomena, although conditions for their formation are particularly favoured in such environments. The various processes such as rillwash, upfreezing, frost creep, needle ice formation, and gelifluction that have been invoked probably play only a secondary role. The primary miniature patterning is more likely connected to a network of contraction cracks which occur along the same lines of weakness year after year probably by desiccation. At Chambeyron in the French Alps for example, Pissart (1964a) has described the radial movement of stones across the surface of a large sorted circle over a 16-year period. The stones crossed several miniature patterns and moved from the fine centres to the coarse borders and on again. The network of miniature patterns is therefore independent of the movement of the surface particles.

Some insight into the origin of various patterned ground forms can be gained from experimental or laboratory studies. One such illustration is provided by the recent work of Corte (1971). Using trays of layered sediments which were subject to repeated freezing and thawing, he produced domes and upwellings which were clearly analogous to sorted and non-sorted circles. He termed these deformations 'freeze thaw extrusion structures' (p. 175). In his experiments, three types of sediment were used; two, a silty clay and a silty sand, were frost susceptible while the third, a crushed quartz of sandy silt size, was less so. The samples were placed in horizontal layers but the vertical arrangement of each type was different in each experiment. Where the crushed quartz was the middle layer, a hummocky topography progressively developed after 28 freeze–thaw cycles and by cycle 57 sand particles were extruded to the surface of some of the mounds from the lower layers (Fig. 9.9). In another experiment where the quartz was the surface layer, a single dome developed after 57 cycles by extrusion from the finer layers below. The coarser surface sediments moved to the sides, thus forming a simple sorted circle or 'debris island'. In section, cryoturbation-like structures (see Fig. 3.10) had formed. They are interesting in that they are further proof of the point already made (see pp. 43–4) that such structures do not necessarily need permafrost for their development. In this case, Corte attributes such structures to the differential frost heaving characteristics of the various layers involved.

Further careful laboratory studies of this nature are clearly required before one will be in a position to fully evaluate the importance of the various processes which contribute to patterned ground. At the same time, there is a need for

Fig. 9.9 *Sections through frost mounds and extrusion features developed after repeated freezing and thawing in a tray of layered sediments.* **A** — *Crushed quartz was the middle layer.* **B** — *Silt was the middle layer. From Corte (1971)*

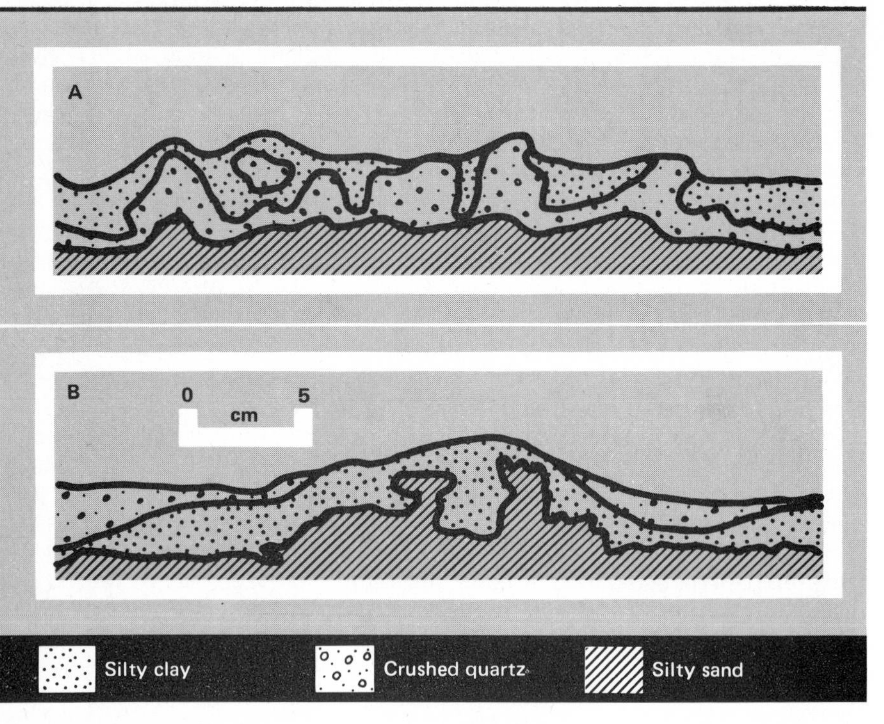

more long-term and accurate measurements of the processes operating within sorted and non-sorted forms. The rate of growth of patterned ground, and why certain patterned ground becomes inactive are additional problems which can best be treated by an approach combining both laboratory work and field measurement.

Soils

Although the tundra has traditionally been recognised as one of the five natural soil zones of the world, pedogenetic studies in Arctic North America and Antarctica have only been undertaken relatively recently, mainly by J. C. F. Tedrow and his associates (e.g. Tedrow, 1973; 1974). In spite of significant advances, our understanding of northern soils is still incomplete. Generally speaking, the soils of such regions reflect the harsh climate in which they occur; for the most part, they are shallow and poorly developed since the rate of weathering and soil formation is slow. They do not possess well defined soil profiles but instead, are characterised by an instability of the mineral soil, the result of frost action; finally, because of the presence of permafrost, the soils are predominantly poorly drained and partly water-logged with glei-ing a common occurrence.

From a geomorphological point of view, the various genetic soils of the northern periglacial regions can be grouped into sequences representing drainage catenas. Furthermore, at least three latitudinal zones can be recognised in terms of the basic soil and vegetation sequences. These are (1) the Northern Forest Zone, (2) the Main Tundra Zone, and (3) the Polar Desert Zone. While the differences between these three zones reflects essentially climatic and soil-forming-potential differences, the variations within each zone are predominantly controlled by local relief and drainage.

The Northern Forest Zone In the boreal and taiga forest zones, two types of soil dominate. First, the brown wooded and arctic brown soils represent the most highly developed soils of periglacial regions. They are the northward extensions of the northern forest podzolic soils. They occur in predominantly well drained environments such as escarpments, ridges, terrace edges and anywhere the active layer is reasonably thick, between 1.0 and 2.0 m. Second, in the poorer drained lowlands, often underlain by permafrost at shallow depths, glei soils are widespread.

In North America, the most extensive development of brown wooded and Arctic brown soils is at higher elevations in the southern Arctic foothills of Alaska, the Yukon Territory and the mainland of the Northwest Territories. Further north, in the Main Tundra Zone and the Polar Desert Zone, the brown wooded ceases to exist and the arctic brown becomes rare, occurring only in isolated topographic positions. The glei soils of the forest zone are shallow, poorly drained and are covered with a thick organic matt of sphagnum moss and, on drier sites, Cladonia lichens. In terms of aerial extent, the glei soils are by far the most widespread. Because of their poorly drained nature, they are subject to intense frost heaving and cryoturbation which leads to the dislocation of horizons and the formation of hummocky terrain. The nature of this terrain, and the 'drunken' appearance to many of its trees has been discussed earlier (pp. 34–5; 156–7). In particularly poorly drained areas, even stunted scattered spruce are unable to survive and bog soils or treeless organic terrain ('muskeg') develop. The latter is important in terms of its insulating and thermal properties and has been discussed in the context of permafrost (pp. 60; 63–4).

The pedological importance of the arctic brown lies not in its spatial extent, which is small, but in its zonal or podzolic nature. In profile, arctic brown soils average 0.4–0.6 m in depth. The upper mineral horizon is a dark brown colour but with depth, colour grades through yellow-brown and grey. Usually, the podzolic nature of the soil can only be determined by chemical and mineralogical analyses since visible horizon differentiation is weak. It would appear that the strong brown colour of the solum is the result of the release of small quantities of various iron compounds which, translocated from the *A* horizon because of a low surface pH, combine with organic matter in the *B* horizon.

The Main Tundra Zone Frost-action processes and patterned ground phenomena are most clearly visible north of the treeline, in the Main Tundra and Polar Desert Zones. In many ways, the Main Tundra Zone is transitional between the two other zones. In the Main Tundra Zone, the most widespread soil is the tundra soil. It tends to mantle undulating and sloping landscapes from ridge top to valley bottom, and also flat coastal plains. In the more arid Polar Desert Zone to the north, the tundra soils are restricted to the lower parts of valley side slopes, especially below snowpatches and other localised areas of abundant

Fig. 9.10 *Genetic soils of the Main Tundra Zone arranged in a drainage catena sequence. From Tedrow and Cantlon (1958)*

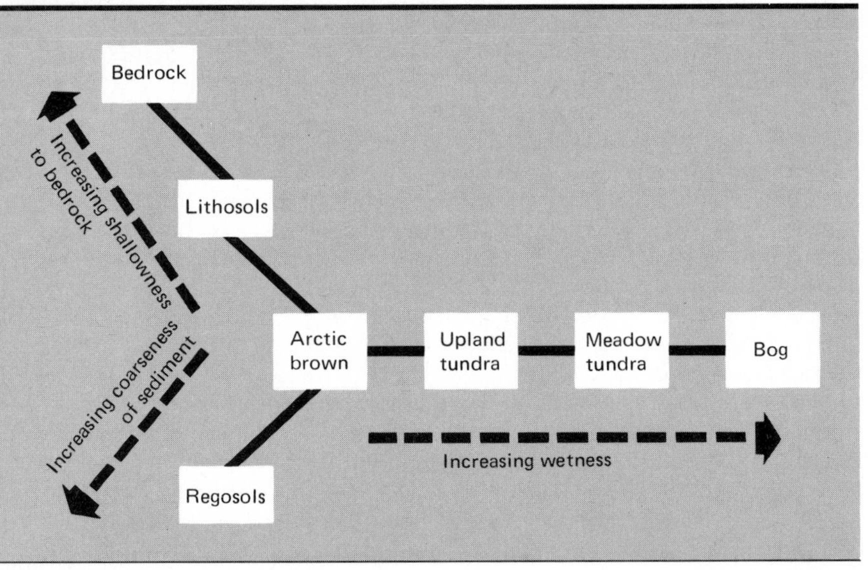

moisture or poor drainage. In the Northern Forest Zone, the tundra soil grades into the glei soils of hummocky terrain.

Drainage is an important factor in soil distribution in the Main Tundra Zone. Figure 9.10 summarises the various soil catenas which occur in Arctic Alaska, an area typical of the Main Tundra Zone. Commencing with the arctic brown or lithosol, the basic sequence involves a transition to upland tundra, to lowland tundra, and then to bog soils. Moving progressively north or south of the Main Tundra Zone, the soil-forming potential either decreases or increases, and modifications to this tundra sequence occur. North of the arctic brown climatic optimum, the soil-forming processes weaken and the solum becomes shallower. Eventually, pedotranslocation processes reach a minimal level of operation in the arid polar deserts of the high Arctic. South of the arctic brown climatic optimum, there is an increase in podzolisation processes and subarctic brown wooded soils change to minimal podzols in the northern forests as temperature and precipitation increase.

The dominant characteristics of the tundra soils are that (a) they are poorly drained and (b) subject to frost action and solifluction. In profile, tundra soils show considerable variation and the application of conventional horizon designations has little meaning. Three basic divisions can be recognised; (a) the active layer, (b) the upper part of the permafrost zone, and (c) the parent rock, i.e. permafrost. Because of their poorly drained nature, the active layer is thin and the soil is shallow, often less than 0.4–0.6 m. In the active layer the upper 2–15 cm is composed of an acid, fibrous and wet organic mat, below which brown and grey silts extend to the permafrost table. These silts are commonly waterlogged and gleisation is the main soil-forming process. The presence of abundant moisture makes tundra soils highly susceptible to frost action. Ice wedges and veins occur in the underlying permafrost and cause deformations in the active layer adjacent to the fissures. Differential heave and localised ice segregation occur beneath grassy hummocks. Non-sorted circles appear as 'islands' of unvegetated terrain in flat areas while on slopes, non-sorted stripes commonly develop. In places, small organic matter particles, such as roots and stems, become incorporated in the active layer. Thus, the active layer of the tundra soil is characterised by cryoturbations and mobility of the mineral soil.

Tundra soils have been differentiated into upland tundra and lowland tundra, a distinction based primarily upon drainage differences. Upland tundra, is the drier variety, and is common on the upper parts of slopes. Small non-sorted polygons and earth hummocks are the characteristic surface micro-relief. Lowland or meadow tundra occurs on the more poorly drained localities towards the bottom of slopes and in valleys and possesses a more luxurious grassy matt. It is often underlain by ice wedges.

Bog soils represent the most poorly drained soils within the drainage sequence previously outlined. They occur in valley bottoms and other localities wherever the terrain is saturated with water for much of the summer. The water-logged conditions prevent organic matter decomposition and material, commonly sedges, sphagnum and grasses, accumulates to thicknesses of 1.0–2.0 m. In areas where deposition has continued without interruption throughout the majority of middle and late Wisconsinan time, such as central and western Banks Island, organic deposits in excess of 10 m in thickness have been observed. In profile, bog soils are characterised by a dark fibrous mat, with small ice lenses, which extends to permafrost. Because of the thermal properties of such material, summer thaw rarely penetrates more than 0.2–0.4 m. Large ice wedges develop in such terrain because of the abundance of moisture. However, fluvio-thermal action along the ice wedges often results in polygonal mound development in the bottom of many valleys, with the result that many of the organic deposits are now in comparatively dry sites. The importance of these bog deposits in terms of geomorphic and climatic reconstruction is great since they are capable of ^{14}C dating and many are suitable for palynological analyses.

The Polar Desert Zone In the high Arctic islands and in Antarctica, not only are there fewer areas with abundant moisture and vegetation suitable for tundra and bog soil development but also the soil-forming potential decreases. Soils developed in such environments are termed polar desert soils (Tedrow, 1966b; 1974). They are defined as those mature well drained soils of the ice free polar regions having a sparse covering of higher plants (Tedrow, 1966b, p. 382). As such, and in contrast to all other genetic soils of the periglacial zone, they experience neither podzolic nor glei influences. Furthermore, the almost complete absence of vascular plants and the low temperatures and precipitation mean that the organic content scarcely enters the soil system.

The major genetic soils within the polar desert landscape can be shown within a simplified drainage catena sequence (Fig. 9.11). Tundra and bog soils occupy only a small percentage of the landscape. The uplands are free of snow for much of the winter since the snow is redistributed by winds into adjacent gullies and depressions leaving the higher surfaces bare. These uplands probably have as little as 3–5 cm of water percolating into the soil during the course of a year. Exposed positions are also subject to losses by evaporation. The soil distribution tends, therefore, to follow the snow accumulation patterns. The

Fig. 9.11 *Major genetic soils of the Polar Desert Zone arranged in a drainage sequence with the associated slope micro-morphology. Modified from French (1971a) and Tedrow (1974)*

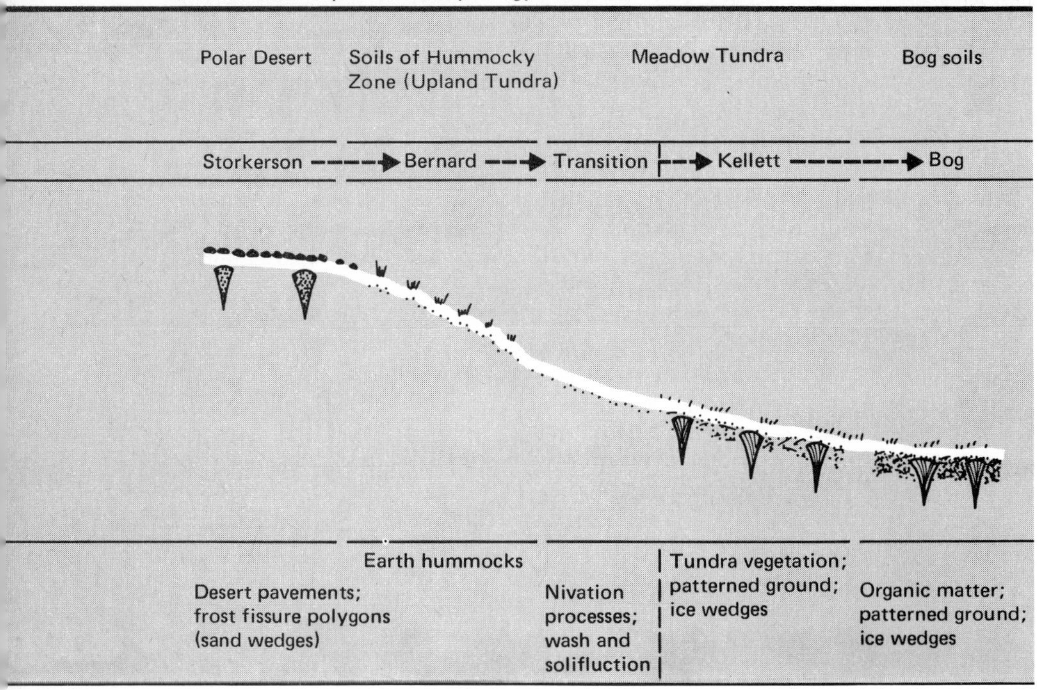

exposed snow-free areas are occupied by polar desert soils while the depressions and valleys are mantled with a web like pattern of upland and meadow tundra soils and, occasionally, bog soils. Closely associated with the polar desert soils are 'soils of the hummocky ground' (Tedrow, 1974), which represent a transition from polar desert to tundra conditions. Earth hummocks (see Fig. 9.3) occur widely in the high Arctic, nearly always on sloping terrain where there is a supply of water from snow or general seepage. The mounds result from a combination of frost-action processes (see pp. 33–44; 195) and surface soil erosion by concentrated wash. Sometimes, the transition from polar desert to tundra conditions is via a zone without hummocks but which is exceedingly wet in the early summer and then becomes very dry later in the summer. A favoured position for this to occur is at the site of a high level snowpatch hollow.

A common morphological feature of terrain underlain by polar desert soils is the formation of a veneer of coarse particles resting on the surface. This is termed a 'desert pavement' and is a common feature of many arid regions of the world. In the high Arctic, it probably reflects two factors. First, since the polar desert soils are developed on higher ground which is bare of vegetation, the ground surface is exposed to wind action which removes the finer particles (see pp. 204–5). Second, vertical sorting and upthrusting of the larger stones by frost action also concentrates the coarser fragments at the surface.

A second characteristic of many polar desert soils is their predominantly saline or alkaline reactions (pH 7.0–8.0). Salt crusts and efflorescences commonly develop at the surface, after several days of rainless and windy conditions which promote, through evaporation from the surface, the rise of capillary moisture. Salt crusts are particularly noticeable on materials with high silt and clay fractions.

Frost fissures and sand wedges are characteristic of polar desert soils. They form extensive polygon networks, 10–15 m in diameter, giving a shallow micro-relief to the ground surface.

The growth of sand wedges and their relationship to polar desert soil evolution has been examined in Antarctica by Ugolini ***et al***. (1973). There the polygon network is less clearly developed on older terrain than on younger terrain. It would appear that the higher the degree of pedogenesis the more poorly developed are the sand-wedge polygons. In the areas of older terrain, the soils of the polygon centres contain more silts, clays, salts and free iron than the soils in the younger terrain. Thus, the more immature soil condition and the lack of structure and consistency in the younger terrain allows sand to move into the troughs and infiltrate the thermal contraction cracks. As a consequence wedge growth is unrestricted by sand supply. On the older terrain where the polygons appear less sharply developed, the better structure and consistency of the soil inhibits the unrestrained growth of the sand wedges. The higher proportions of silt and sand, and the greater salt and sesquioxide contents of the more mature soils give a compound structure which is unable to flow and fill the cracks. Thus, as pedogenesis progresses, the availability of free moving sand is progressively reduced. Consequently, the rate of growth of the sand wedges is reduced and they become less clearly defined. It follows that estimates of the age of land surfaces based upon the extrapolation of the annual growth rate of sand wedges is not straightforward.

10 Wind action and coastal processes

The majority of processes operating within the periglacial environments are not unique to such environments. On the other hand, the activity of certain processes is either enhanced or inhibited by the climatic conditions of these areas. This is the case of wind action and of coastal processes and justifies their treatment in a separate chapter. It is not the intention however, to present a systematic treatment of these topics. Instead, the approach is to highlight those aspects of wind action and coastal processes which either are peculiar to, or achieve their greatest importance in, periglacial environments.

The role of wind

It is widely believed that wind action played a dominant role in both erosion and transportation in temperate regions which experienced severe periglacial conditions during the Pleistocene, such as central and eastern Europe and the north-central United States (e.g. Tricart, 1970, pp. 143–50; Embleton and King, 1968, pp. 564–85). Evidence includes the presence of wind modified pebbles and blocks ('ventifacts'), stone pavements, frosted sand grains, and extensive loess and aeolian deposits. The factors thought to have favoured intense wind action included the absence of vegetation, extensive deposits of fine grained sediments laid down at the margins of the ice sheets, and the large amounts of comminuted debris produced by the intense frost action.

While this general conclusion may very well be correct for Pleistocene environments (see pp. 249–53), it should not be assumed that a similar situation is necessarily true for the present-day periglacial environments of polar and continental locations. In these regions, it seems best to regard the direct action of the wind as of only minor importance, giving rise to relatively small-scale and localised weathering effects, niveo-aeolian deposits, and certain patterned ground phenomena (Pissart, 1966a; Bird, 1967, pp. 237–41; Washburn, 1969). On the other hand, the indirect effects of wind action are probably more important than has been thought previously. For example, the wind plays a primary role in the redistribution of snow from exposed locations into gullies and lee slopes during the winter months. It exerts, therefore, a fundamental control upon the operation of nivation processes, including solifluction, on the slopes below the snowbanks. In the summer months, wind action results in significant evaporation and latent heat loss from exposed slopes, which in turn, influences the depth of the active layer and the dampness of the slope. This may lead to slope and valley asymmetry (see pp. 178–80). A further example of the indirect effect of the wind is to be found in the oriented nature of many thaw lakes in alluvial tundra lowlands (see pp. 122–5). Finally, the wind is important in influencing the movement of sea ice and in wave generation. It is clear therefore, that the wind operates in a number of indirect ways to influence landforms. Since these indirect effects of wind action are dealt with in more detail in the appropriate sections, the following discussion centres around the less important but direct role of wind in periglacial environments.

Wind erosion If wind is to carry out significant erosion, there must be a source of abrasive material suitable for transportation by the wind. Silt and fine sand particles, picked up from the ground surface during the summer months, are one source of abrasive material. However, in the forest and tundra zones, there is a general lack of exposed surfaces from which it can be picked up, and only the vegetation-free polar desert zone is well suited for the pick up of abrasive material. Furthermore, the time duration for such erosion is short, being limited to 1–2 months in middle and late summer, and the ease of detachment of surface particles is often limited by the presence of a salt crust or hard pan at the ground surface. For these reasons, wind erosion in the summer months is limited.

Probably the majority of wind erosion in high latitudes is carried out during the 6–10 months of winter by wind driven snow particles. It is well known that the hardness of ice increases as the temperature drops, such that at −50°C ice has a Moh hardness of about 6 (Bird, 1967, p. 237). It is also during the winter months that winds are strongest and most constant in direction since well developed high pressure systems become established over the polar landmasses.

In the polar deserts of the northern hemisphere, ventifacts and wind erosional features have been reported from ice-free Peary Land in northeast Greenland (e.g. Fristrop, 1952) and from northern Siberia (e.g. Sverdrup, 1938). In the Canadian Arctic however, ventifacts are relatively rare (Pissart, 1966a) and in general, there is an absence of wind erosional features in northern Canada (Bird, 1967). Significant wind eroded landforms in the form of badland topography is limited to soft sandstones or silts (e.g. St-Onge, 1965).

The greatest variety and frequency of occurrence of wind erosional features occurs in the ice-free areas of Antarctica (e.g. Nichols, 1966, pp. 35–6; Sekyra, 1969, p. 282). According to Sekyra, conspicuous aeolian corrasion in the form of wind troughs and mushroom-like forms occurs in the inland oases and on old Pleistocene morainic accumulations in mountain areas. In some instances, chemical weathering processes associated with wind action combine with mechanical disintegration through the growth and expansion of salt crystals to produce a number of weathering-induced honeycomb structures. These are called 'taffoni' and are best developed in coarse grained crystalline bedrock and morainic boulders (e.g. Selby, 1971). Case hardening for example, is a particular process by which the exterior of a boulder or rock is made more resistant to weathering by the evaporation of mineral bearing solution from the rock surface leaving a thin cementation layer. When combined with cavernous weathering beneath, case hardening may produce very striking shell-like structures (Fig. 10.1). Instances of case hardening and cavernous weathering have also been reported from Greenland in the northern hemisphere (e.g. Washburn, 1969, p. 32). The origin of taffoni structures is complex and the role played by wind not clearly understood. Many areas of taffoni show a preferred orientation which may, or may not, relate to present-day wind patterns. At McMurdo Sound for example, Cailleux and Calkin (1963) have noted that the preferred orientation of taffoni in that region is not consistent with present-day wind direction and may be unrelated to wind action.

Many questions are still unanswered with respect to wind erosion in present-day periglacial regions. The apparently greater frequency of wind erosional features in Antarctica than in Arctic North America and Eurasia is not

Fig. 10.1 *Types of cavernous weathering, 'taffoni' structures, and wind erosional features. Modified from Selby (1971)*

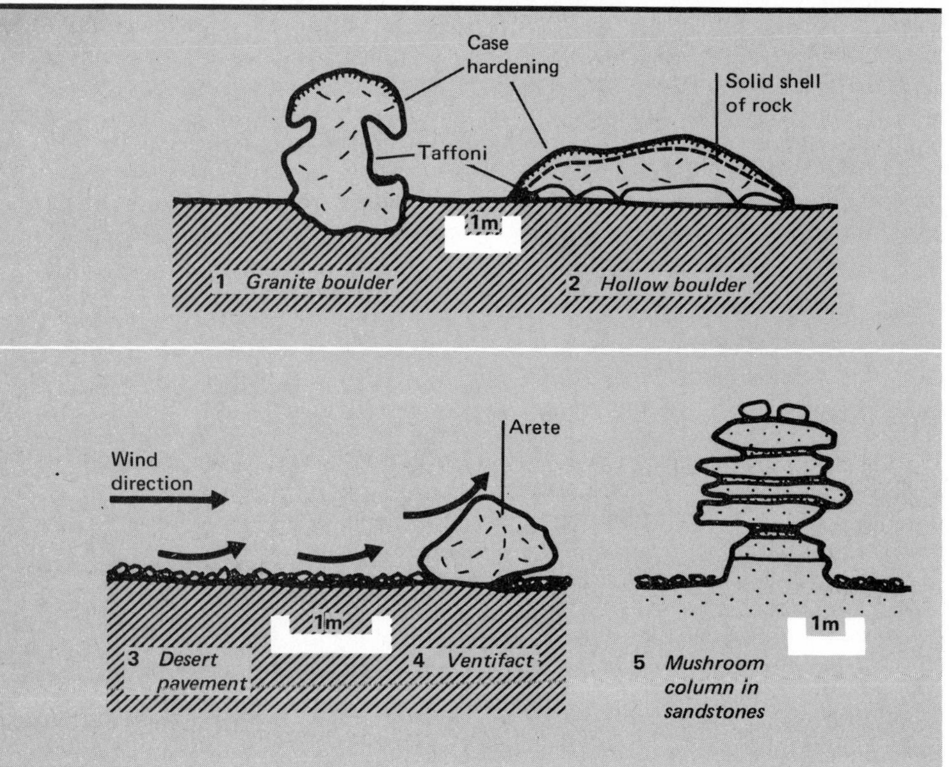

readily explainable. It may be that the winds are stronger and more constant in direction in the Antarctic. A second possibility is that the ice-free areas of Antarctica have experienced a longer period of stable and uninterrupted climatic conditions.

Wind deflation Deflation, the second aspect of wind activity, is the winnowing out of fine particles and their transportation by the wind. Since plant cover is one of the major controlling factors, deflation probably reaches its greatest intensity in the unvegetated and ice-free areas of the polar deserts, and not in the tundra and forest zones. In the tundra zones, deflation is limited to sparsely vegetated areas such as valley sandar, recent deltas and terrace deposits.

The most obvious indicator of deflation in the polar deserts is the widespread presence of a lag gravel or desert pavement at the surface (see p. 200). Deflation may also produce shallow vegetation-free depressions or blow-outs, usually only a few centimetres deep and 1–3 m wide. As with direct wind erosion, it is probable that the majority of deflation activity occurs during the winter months on exposed, snow free surfaces. Evidence for this is to be found in the nature of the snowbanks which accumulate in the gullies and depressions during the winter. They are often composed of layers of snow

Fig. 10.2 *Niveo-aeolian deposits associated with snowbank melting in a small gully, Thomsen River, central Banks Island.*

separated by fine dirt bands. These deposits are termed 'niveo-aeolian' and may result in a significant concentration of water-lain silt-size particles in snowbank locations. In mid and late summer, as the snowbanks melt, it is not uncommon for the surface of the snowbank to assume a grey, almost black colour, and a pitted, small-scale thermokarst-like relief similar to glacier and other snow surfaces (Fig. 10.2). Intense deflation may also occur in late summer with localised dust storms brought on by surface heating and instability. Dust clouds may rise several hundreds of metres into the air, and blanket the surrounding terrain with a thin cover of fine sand particles. Areas of sandy sediments exposed on valley sandar and along the flood plains of the major rivers are particularly suited to this aeolian activity. In the Pleistocene literature, the term 'cover sand' (e.g. Maarleveld, 1960) is used to describe similar fine sands of aeolian origin (see p. 251).

In areas where there are extensive bare alluvial sediments or till plains, long continued deflation activity may result in the deposition of a considerable thickness of windblown silts over the surrounding terrain. These deposits of well sorted, homogeneous and unstratified sediments are termed loess. Loosely coherent grains of between 0.01 and 0.05 mm in diameter form the dominant grain size fraction or 'loess fraction'; this often exceeds 50–60 per cent of the deposit. Loess may also contain a significant percentage, between 5 and 30, of clay size particles (less than 0.005 mm in diameter) and also 5–10 per cent of

 Fig. 10.3 *Typical grain size distribution for loess and cover sands*

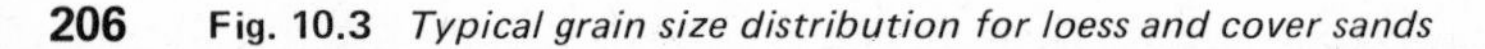

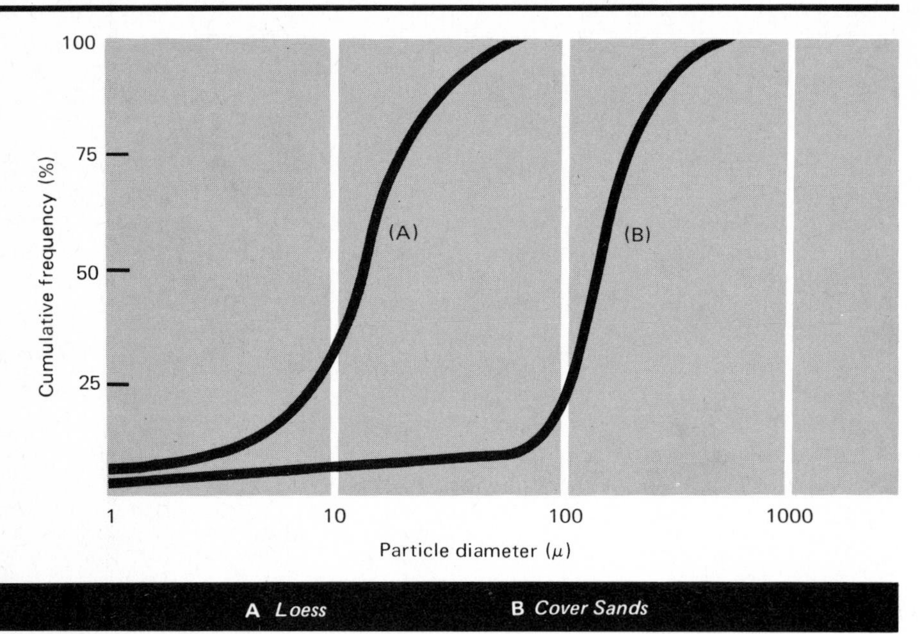

A *Loess* **B** *Cover Sands*

sand particles (greater than 0.25 mm in diameter). The typical grain size distributions of wind blown loessic sediments and cover sands are illustrated in Fig. 10.3.

Areas of current loess deposition are to be found around the margins of the major deserts of the world, particularly the polar desert and that of central Asia. In North America, thick deposits of Holocene loess occur in central Alaska and parts of the Yukon Territory of Canada. In the Fairbanks region of Alaska, there is evidence of deposition from the Middle Quaternary onwards through to present day. The silts mantle the upland surfaces and originated from the outwash plains to the south of Fairbanks and the flood plain of the Tanana River (Péwé, 1955). Subsequently, they have been retransported to the valley bottoms by solifluction and incorporated with organic debris and now constitute between 1 and 60 m of perennially frozen ground.

Pleistocene loess deposits mantle over extensive areas of temperate latitudes in Eurasia and North America and may exceed 50–100 m in thickness. They resulted from the deflation of large vegetation free outwash areas and till plains left by the retreating ice sheets. They are discussed more fully in Chapter 12.

Coastal processes

In the northern hemisphere, the Arctic Ocean is almost completely landlocked. Extensive stretches of coastline occur in Alaska, Canada, Greenland, and along the whole of the northern margin of the Eurasian landmass from the Kola Peninsula in the west to the Chuckchi Sea in the east. In addition, there are the various island groups such as the Canadian Arctic archipelago, Spitsbergen, Franz Josef Land, Novaya Zembyla, the North Land, the New Siberian Islands and

Wrangel Island. In Antarctica, the majority of the coastline is composed of ice shelves. However, there are significant stretches of coastline bounding ice-free terrain in the Grahamland Peninsula, and the South Shetland and South Orkney Islands. The extent of coastlines currently experiencing periglacial conditions is, considerable therefore, and it is not surprising that there are significant differences in environmental conditions as regards ice cover, wave climate and tidal range, not to mention structural and lithological variations. Nevertheless, there are several characteristics of all of these coastal areas which, in varying degrees, reflect the periglacial environment in which they occur. These characteristics may be summarised under three broad headings: first, the presence of sea ice and its effect upon wave generation, second, the effects of ice upon the beach regime, and third, the influence of permafrost and ground ice upon coastal development.

Sea ice and wave generation The major characteristic of the majority of the above-mentioned coastlines is the long season when they are locked in by ice, and the very short season during which coastal processes operate. In certain sheltered areas, such as the various islands located in the middle of the Canadian Archipelago, wave action can be restricted to as little as 8–10 weeks a year (Owens and McCann, 1970). In other areas, open water conditions may exist for several months, such as in northern Alaska from Point Barrow southwards to Cape Thompson (Hume and Schalk, 1964; 1967). At the other extreme of the spectrum, the warm North Atlantic Drift enables certain parts of the west coasts of Greenland and Spitsbergen to be virtually ice free for most of the year. A second factor to be considered is that, even when there is open water, the nearby presence of the permanent pack ice drastically reduces the distance of fetch and thus, the magnitude of wave action. Moreover, the movement of sea ice during the open water season may further reduce the distance of maximum fetch in any one situation. A third restraint upon the effectiveness of wave action is the presence of a narrow strip of ice which is frozen to the shore and unaffected by tidal movements. This land-fast ice is called an 'ice foot' and is a common feature on many Arctic beaches (e.g. McCann and Carlisle, 1972). It is important from a geomorphic viewpoint in that it further protects the beach from the limited wave action that does occur, and decreases the effective duration of the open water season. As a result of all these factors, Arctic beaches are essentially low energy environments in which normal wave action and coastal processes are limited.

In this context, the magnitude and frequency of periods of significant wave action assume great importance. It would appear that the majority of beach reworking and sediment transport occurs during the few major events which have a frequency of occurrence of once or twice a year. For example, Owens and McCann (1970, pp. 399–403) concluded, using wind data and calculations of fetch distances from published ice survey maps at Resolute Bay in the central Canadian Arctic, that in the years 1961, 1962, and 1963, there were only 3, 5 and 2 periods respectively when waves of over 0.5 m in height could possibly have been generated. This conclusion was substantiated by field observations over three field seasons between 1968 and 1970 when only one major storm occurred which produced waves capable of really significant action on the beach (McCann, 1972). During this one storm, the whole of the beach was combed down and the profile lowered by 0.3–0.6 m in the mid to high tide zone. There was also erosion of the bluff which backs the modern beach and large amounts of material were transported alongshore. Similar observations emphasising the role of major but low frequency storms in shoreline change have been made by Hume and Schalk (1967; 1972) for the Point Barrow area of Alaska. There, the period of open water is greater than at Resolute Bay. Variations in coastal cliff retreat between 1948 and 1968 appear to be related to the frequency of west wind storms per year which have changed from approximately two per year in the years 1948–62 to one per year in the years 1962–68. It would appear that a 50 per cent decrease in west wind storms during the open water season resulted in a decrease in cliff retreat from approximately 4 m year^{-1} to 1 m year^{-1}. However, there are many problems inherent in isolating and defining the geomorphic effectiveness of single, major events. At Point Barrow for example, unusually high cliff retreat rates for 1968–69 are attributed to beach borrow associated with the construction of a new airport and the removal of coarse materials. Under these conditions, the effectiveness of a storm of a given magnitude would increase. Likewise, on Devon Island, the effectiveness of the single storm observed by McCann was accentuated by a preceding period in which large waves had broken up the remaining beach fast ice (see below) which normally protects the beach.

As a direct result of the relatively limited wave action on many Arctic beaches, beach processes and the transport and redistribution of material operate at a slow rate. In general, beaches are poorly developed and narrow, often composed primarily of coarse sands and cobbles. Beach sediments are poorly sorted and possess low roundness values as compared to other beach environments (Table 10.1). However, there is abundant evidence that considerable longshore transport of material does take place on Arctic beaches. This is particularly true of the western Arctic and other areas with relatively long open water periods. Along the northern coast of Alaska, Hume and Schalk (1967) report approximately 10 000 m^3 year^{-1} of net sediment transport as being typical, and complex depositional features such as spits and offshore bars can develop (Fig. 10.4). Even in the very enclosed beach environments of the central Canadian Arctic, where sediment transport is less and depositional features not so well developed, systematic variations in sediment size and shape along and across beaches have been attributed to the breakdown and abrasion of beach material by wave action and its selective transportation along the beach (McCann and Owens, 1969). Thus, despite the limited period of open water, wave action still plays an important role in coastal evolution.

The effects of ice on the beach The ice foot plays an important role in limiting beach changes. It develops in the upper part of the intertidal zone and often extends well below high water mark. It forms in the autumn freeze-up period when swash or spray from breaking waves freezes on contact with the beach. Once a layer of ice has formed on the surface of the beach, wave induced movement of beach material will cease. Likewise, in the spring, considerable amounts of ice may remain frozen to the beach after the sea ice breaks up and may prevent the direct action of waves for several weeks. The presence of ice on the beach before the development of the winter sea ice and after the spring break-up means that conventional freeze-up and break-up dates do not necessarily apply

Table 10.1 *Some roundness values for beach materials in various environments in periglacial and non-periglacial regions. Data taken from McCann and Owens (1969)*

Locality	Beach environment	Rock type	Cailleux roundness values
Periglacial			
Devon Island, NWT, Canada	Sheltered	Limestone	25–267
Hall Beach, NWT, Canada	Sheltered	Limestone	216
Jacobshaven, west Greenland	Sheltered	Quartz	90–105
Kuggsa Dessa, west Greenland	Exposed	Quartz	135–160
Godthaab, west Greenland	Exposed	Quartz	270–388
Non-periglacial			
Western Mediterranean (various sites)	Enclosed sea	Limestone	355
Lake Ontario, Canada	Enclosed sea	Gneiss	388
Finnestere, northwest France	Exposed	Quartz	250–270 400–460

to the Arctic beach zone (Owens and McCann, 1970). The size and extent of the ice foot varies from year to year depending primarily upon the sea conditions the previous autumn; moderate to strong wave action is an optimum for ice foot accretion. The tidal range and beach slope are other factors which influence ice foot widths and thicknesses. The greater the tidal range, the greater is the zone of spray and swash accretion, while the steeper the beach profile, the smaller is the width and thickness of the ice foot (McCann and Carlisle, 1972). The ice foot usually ablates and erodes away during the early part of the open water season. Initially, a shore lead develops some 5–15 m off-shore to isolate the sea ice proper from the ice foot. Then the ice foot is breached by water draining seawards from the melting snow in the backshore zone.

The more direct effects of ice on the beach occur when iceflows or the sea ice pack are forced, by strong onshore winds, to impinge upon the shoreline. If a rigid ice foot is present, buckling may occur at the seaward edge of the ice foot. If the ice foot is not rigid or is absent, there will be a general piling up of sea ice and beach ice on the beach zone to form mounds of contorted ice (Owens and McCann, 1970). In the following summer, these ice mounds begin to ablate and develop a veneer of beach gravels and coarse sands which enable the ice to persist into midsummer. More commonly, the impinging ice merely pushes and scours beach material to form irregular lobate ridges and scour depressions in and above the high tide zone (Hume and Schalk, 1964; Owens and McCann, 1970). These ridges may be as much as 1.0–2.0 m in height but are usually much less.

Similar ice push features occur around the shores of the larger inland water bodies. However, because of the absence of a tidal range, there is no protection

Fig. 10.4 *Coastal features of part of south west Banks Island showing extent of spit and offshore bar development*

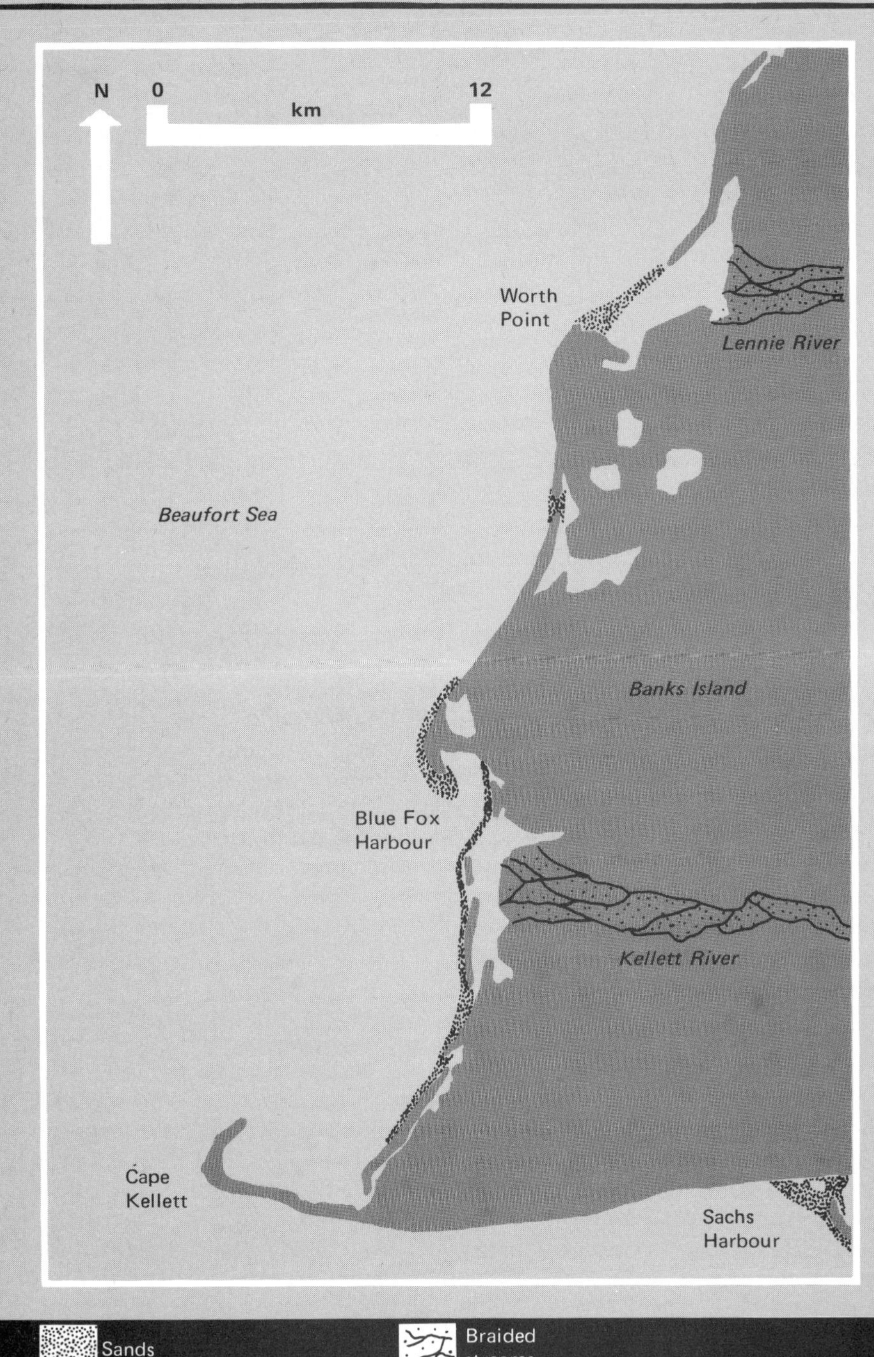

by a well developed ice foot, and the ice push ridges are formed at the same elevation each year. As a result, they are often remarkably regular in plan and form. In general, beach ridges are not long term features of periglacial shorelines since they are systematically reworked every few years.

The influence of permafrost and ground ice Where coasts are developed in permafrost terrain composed of unconsolidated and ice-rich sediments, rapid and dramatic coastal retreat may be associated with the melting and release of the excess ice. This is accomplished in at least two ways. First, in those areas where the frozen sediments are sands and gravels, beaches can form which protect the bluff from direct wave action. In these situations, retreat of the bluff is relatively slow, occurring through the subaerial melting of the permafrost and the transport of the thawed material to the beach by solifluction and rainwash. The finer particles are then removed and transported in suspension along the coast, leaving the coarser pebbles and cobbles as a beach gravel. Second, in areas where the frozen sediments are silts and clays, beaches do not form and the bluff is exposed to direct wave action. Since fine grained sediments are usually associated with high ice contents (see pp. 27–8), storm waves result in either the exposure of the ice bodies or the undercutting and formation of thermo-erosional niches at high tide level (see pp. 126–7). The former induces rapid melting and slumping of the ice face while the latter may lead to the later collapse of the bluff, often in blocks delineated by ice wedges. In both cases, the sediment within the permafrost is released and carried away in suspension. A more detailed discussion of these processes is given in Chapter 6 (pp. 125–8).

The shorelines developed around the Beaufort Sea in the western Arctic are probably the best example of rapid coastal retreat. For example, in the vicinity of Barrow, Alaska, average retreat rates of between 2 and 4 m per annum appear typical (Hume ***et al.,*** 1972) for sections with no protecting beach gravels but where beach gravels are present, annual retreat is much less. Along the Yukon coast and in the Mackenzie District of Canada, a similar pattern of rapid coastal retreat can be observed (Mackay, 1963b; McDonald and Lewis, 1973). Measurements at the Yukon–Alaska border indicate that the coast at that point has retreated 43 m since 1912, while the southeast side of Herschel Island has probably undergone 1–2 km of retreat in postglacial times. In the vicinity of Tuktoyaktuk, coastal retreat of over 150 m has occurred since 1935 (Mackay, 1972a). This has caused the draining of a lake and the growth of a number of pingos in the drained lake bottom (see p. 99).

As a result of the abundance of fine grained sediments released by this rapid cliff recession, and aided by the effect of the wide shallow platform offshore, much of the Beaufort Sea coastline exhibits classical depositional features. Due to the rapidity of headland retreat and the shallowness of the near shore zone, offshore bar development is particularly common. Simple spits and bay bars are the other major depositional features although in certain localities, such as Point Barrow, Alaska, and Cape Kellett, Banks Island (Fig. 10.4), complex depositional features have developed in response to local coastal conditions. An additional source of sediments is provided by the discharge of the many large rivers, notably the Mackenzie, which enter the Beaufort Sea and form aggradational coastlines in their own right (e.g. Mackay, 1963a, pp. 55–7).

The rates of coastal retreat observed around the Beaufort Sea of the western

Arctic are clearly exceptional, being the result more of thermal erosion than true coastal erosion. For the most part, wave action and coastal processes play the subsidiary role of transportation of thawed sediments away from 'source' areas towards 'sinks' or zones of deposition. Similar conditions with respect to permafrost, ground ice and open water conditions exist in Siberia along the coastline bordering the Laptev Sea, where retreat rates of between 4 and 6 m per year are not uncommon (Grigorev, 1966; Are, 1972).

Coastal retreat rates in ice-rich permafrost terrain can be compared to retreat rates recorded in unconsolidated sediments in non-permafrost environments and where wave action is year-round. For example, in England, cliffs developed in boulder clay at Holderness are estimated to have receded 2 miles since Roman times (i.e. nearly 2 m per year), while those developed in Tertiary sands at Barton-on-Sea, Hampshire, are retreating at a rate of approximately 1 m per year (Small, 1970, p. 442).

Finally, it must be stressed that ground ice and unconsolidated sediments, and the associated rapidity of coastal change, are not typical of all coastlines in periglacial environments. Large extents of coastline in the central and eastern Arctic, Greenland, Spitsbergen and Antarctica are developed in relatively resistant and coherent rock in which retreat rates are slow. The absence of fine grained sediments limits depositional features. Abrupt coastal cliffs with scree slopes beneath are probably the most typical form. Within this context, the coastal conditions of the ice-rich and unconsolidated sediments of the Siberian and western Arctic lowlands are rather unique.

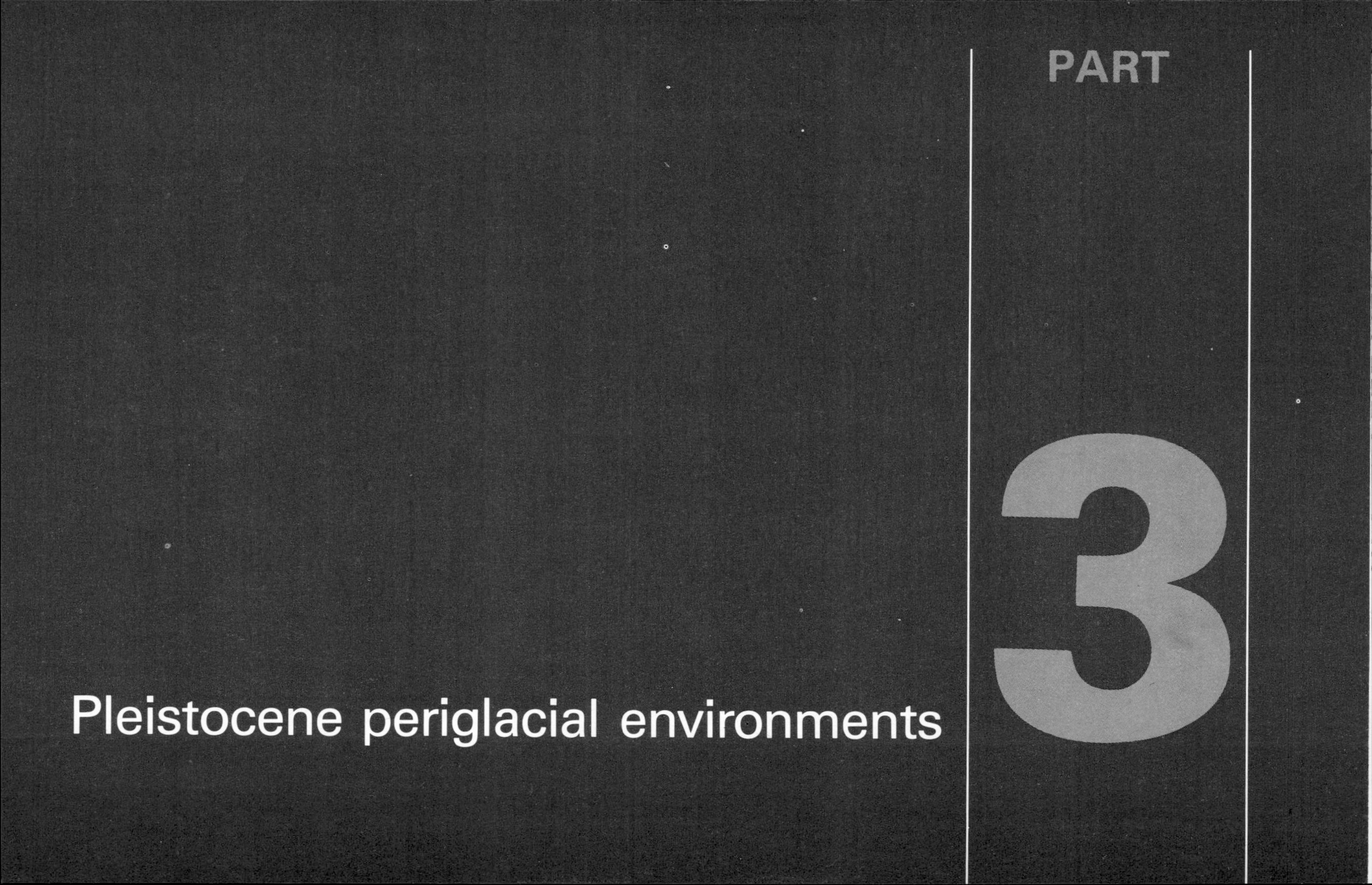

PART 3

Pleistocene periglacial environments

11 Pleistocene periglacial conditions

Introduction

In Part 2, the geomorphic processes and landforms of lowland periglacial environments, notably the western North American Arctic and central Siberia, were emphasised. It is believed that these environments represent a close analogy to the periglacial environments which developed in mid-latitude lowland areas of Europe and North America during the Pleistocene. In this chapter, and the two subsequent ones, an attempt is made to briefly describe these mid-latitude conditions, to compare them with present-day high latitudes, and to evaluate the role played by periglacial processes in the development of mid-latitude landscapes.

By necessity, emphasis is upon Late Pleistocene periglacial conditions in Europe and North America, since it is these which are best understood and for which there is most evidence. Since mid-latitudes are now temperate and periglacial conditions no longer prevail, the approach must be through the identification of relic periglacial features, which can be described in terms of their morphology and distribution. The interpretation of such features, however, necessitates climatic reconstructions and a certain understanding of the nature of periglacial processes. Both identification and interpretation can be difficult and ambiguous. In particular, attempts at Quaternary climatic reconstructions are complex and rely heavily upon detailed stratigraphic and biologic investigations together with considerations of Quaternary vegetation and fauna successions. Since these are the fields of Quaternary geologists and biologists, involvement in such detail is kept to a minimum, although it cannot be denied that such information is extremely relevant to the correct interpretation of periglacial phenomena. Instead, the approach adopted concentrates upon morphology and attempts to interpret through comparison with known forms and processes in high latitudes.

The time scale and climatic fluctuations

Some comments on terminology and the time scale involved need to preface any discussion of Late Pleistocene conditions. In broad outline, the various glacial and interglacial stages are known for both North America and Europe. The generalised temperature curve, time scale, and most common terminology for the Middle and Late Pleistocene is illustrated in Fig. 11.1. The temperatures reflect surface water conditions as determined by oxygen-isotope methods and indicate, in a relative way, the probable air temperature fluctuations. It is believed that periglacial climatic conditions roughly coincide in time with the various advances and retreats of the ice sheets, as indicated by the temperature curve fluctuations. For example, the stratigraphic positions of wedge casts, involutions and various solifluction deposits in south and east England certainly suggests a number of periods during which either frost action or permafrost, or both, were present (Table 11.1).

From a practical point of view, we need only be concerned with the cold conditions associated with the last two glacial stages. Surface

Fig. 11.1

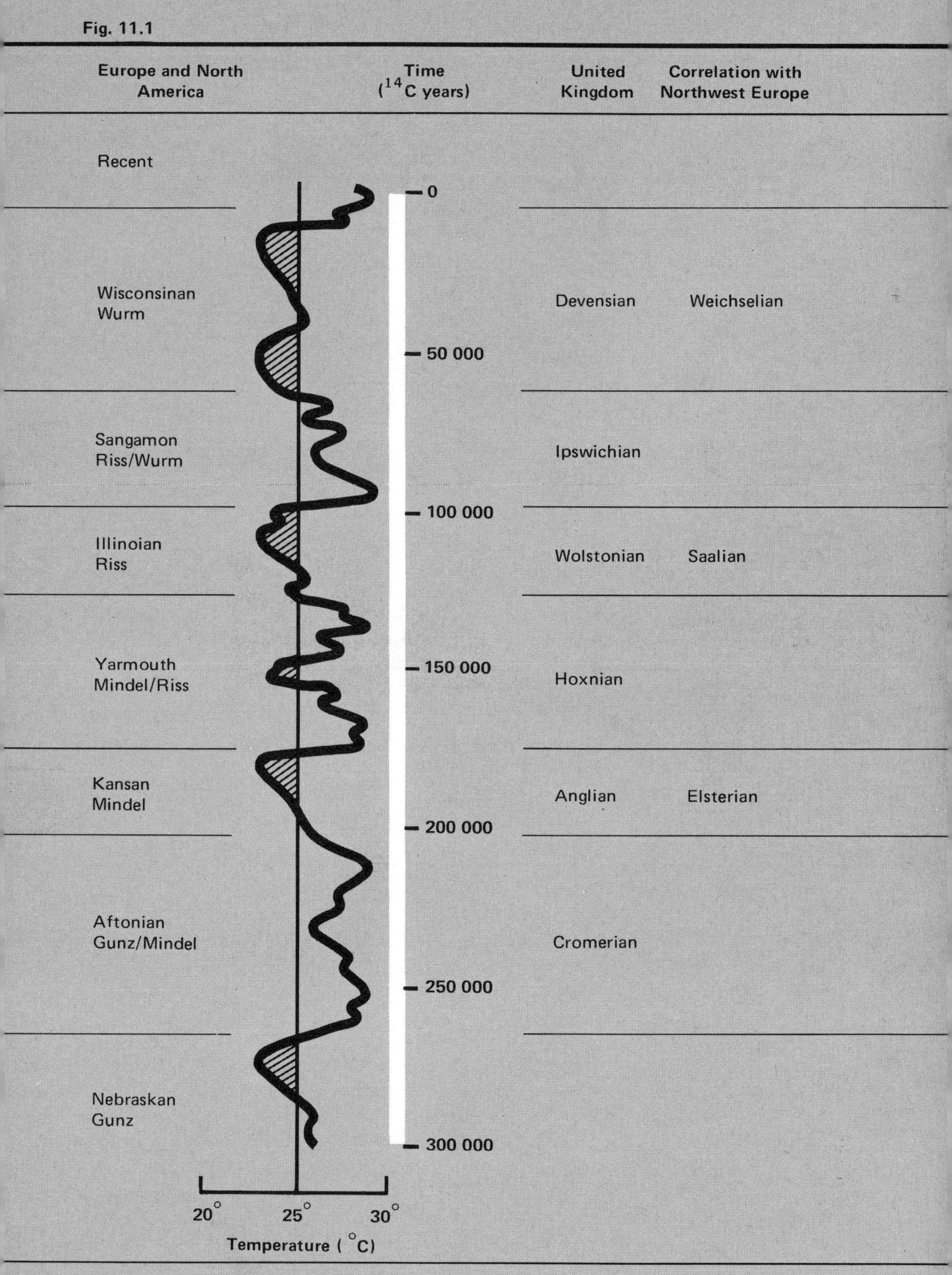

Europe and North America
Time (^{14}C years)
United Kingdom
Correlation with Northwest Europe
Recent
0
Wisconsinan Wurm
Devensian
Weichselian
50 000
Sangamon Riss/Wurm
Ipswichian
100 000
Illinoian Riss
Wolstonian
Saalian
Yarmouth Mindel/Riss
150 000
Hoxnian
Kansan Mindel
Anglian
Elsterian
200 000
Aftonian Gunz/Mindel
Cromerian
250 000
Nebraskan Gunz
300 000
20°
25°
30°
Temperature (°C)

Table 11.1 *Stratigraphic position of various periglacial phenomena in south and east England. Data from West (1968), Table 12.8*

Locality	Features	Periglacial conditions	Age
Kent, various localities	Snow meltwater deposits; chalky rubbles, coombe cutting	Freeze–thaw	Late-glacial Zone III
	Chalky rubbles; meltwater deposits	Freeze–thaw	Late-glacial Zone I
	Loess	Eolian activity	Middle Weichselian
	Coombe rock, involutions, meltwater deposits	Freeze–thaw	Middle Weichselian
Wretton, Norfolk	Ice-wedge casts penetrate involutions	Permafrost; freeze–thaw	Middle Weichselian
Interglacial – Ipswichian			
Beetley, Norfolk	Ice-wedge casts underlie Ipswichian deposits	Permafrost	Wolstonian
Ebbsfleet, Kent	Coombe rock, solifluction	Freeze–thaw	Wolstonian
Interglacial – Hoxnian			
W. Runton, Norfolk	Ice-wedge casts and involutions underlie late Anglian till	Permafrost; freeze–thaw	Anglian
Interglacial – Cromerian			
W. Runton, Norfolk	Ice-wedge casts and involutions underlie Cromerian deposits	Permafrost; freeze–thaw	Beestonian
Interglacial – Pastonian			
Suffolk; Norfolk	Involutions and coombe rock below Pastonian deposits	Permafrost; freeze–thaw	Baventian

periglacial features formed in earlier cold periods are unlikely to have survived to the present day without having been modified beyond recognition by the later cold stages or by the intervening interglacial periods of more temperate conditions. In both Europe and North America, the penultimate cold stage occurred some time prior to 100 000 years B.P. The last interglacial period is generally believed to have lasted until 60–70 000 years B.P. after which the climate deteriorated. The final cold stage lasted until approximately 10 000 years ago. For simplicity in the following chapters, the more general terms Weichselian and Wisconsinan are used in preference to the more specific and local terms which are frequently used in Europe (including the British Isles) and North America.

During the last glacial stage, environmental conditions fluctuated considerably, both in time and space, with the onset of the climatic deterioration, and the growth, advance and retreat of the ice sheets. Various small-scale climatic oscillations gave periods of relatively more temperate, or interstadial, conditions.

In addition, there were the periods of climatic transition before and after the glacial stage which were also more temperate. These fluctuations meant that periglacial conditions varied in intensity and were by no means continuous throughout the whole of the last glacial stage. In northwest Europe for example, where our understanding of these events is furthest advanced, the various Weichselian climatic fluctuations resulted in Arctic and subarctic conditions for only about 25–30 000 years during the last 60–70 000 years. At least two major ice advances are believed to have occurred, one in early Weichselian times, the other in middle Weichselian times, separated by a definite period of relatively warmer conditions. There is evidence to believe that broadly similar fluctuations occurred in North America. For example, in the western Arctic, one can identify both early and late Wisconsinan ice advances. The latter attained its maximum extent approximately 25–16 000 years B.P., and corresponds to the second Weichselian period of cold conditions. It seems reasonable to assume that severe periglacial conditions only developed at the height of these cold stages, and that the interstadial periods saw repeated fluctuation between periglacial and non-periglacial conditions.

The late glacial transition to temperate conditions well illustrates the oscillating nature of cold climate conditions during the Pleistocene. In Europe, the late and post glacial stages of the last glaciation can be divided into pollen zones, indicating the dominance of different vegetation sequences (Table 11.2). For example, fluctuations in the arboreal and non-arboreal pollen ratio indicate that

Table 11.2 *Generalised sequence of Late Glacial and Post Glacial (Flandrian) vegetational successions in England and Wales. Modified from West (1968)*

Stage	Period	Date	Pollen zone	General characteristics of vegetation (dominants)
Post Glacial (Flandrian)	Sub-Atlantic		VIII-modern	Afforestation
		AD 1 000 AD/BC	VIII	Alder, oak, birch, beech, ash
		1 000 BC 2 000 3 000	VIIb	Alder, oak, lime —deforestation Elm decline
	Atlantic	4 000 5 000	VIIa	Alder, oak, elm, lime
	Boreal	6 000 7 000	VI	Hazel, pine (c) Oak, elm, lime (b) Oak, elm (a) Hazel, elm
			V	Hazel, pine, birch
	Pre-Boreal	8 000	IV	Birch, pine
Late Glacial (Weichselian)	Younger Dryas	9 000	III	Park tundra (Dwarf birch)
	Allerod	10 000	II	Birch with park tundra
	Older Dryas Bolling Early Dryas	11 000 12 000	Ic Ib Ia	Park tundra/tundra

during deglaciation, two periods of relatively warmer conditions occurred, the Bolling and Allerod interstadials, at which time the tundra vegetation was replaced temporarily by park tundra (dwarf birch) and birch forest. Then, in late glacial times, the climate rose to a peak of warmth and dryness between 5 000 and 3 000 years B.P. In all probability, therefore, strict periglacial conditions were only present during zones I and III, since by zone IV a forest cover was becoming well established.

In North America, the pollen stratigraphy is less advanced in terms of the interpretation of climatic and vegetational successions. The reason is probably related to the greater complexity of the North American tree flora (Flint, 1971, pp. 390–1). Thus, the vegetational sequences in late and post glacial times are more general. However, they do suggest a strong synchrony of climatic events with Europe. For example, the vegetational fluctuations in the 'cooler' portion of the sequences have apparent correlations with the Bolling (zone IB) and Allerod (zone II) fluctuations in Europe. The pollen record suggests, however, that the tundra belt beyond the receding ice margin was much narrower and moister than in Europe, and was intermixed with forested areas. Since the climatic amelioration was accentuated by the development of large proglacial water bodies in many areas, the tundra belt was quickly replaced by spruce-fir and then by pine. Only upland areas retained an open tundra environment for any length of time. It seems reasonable to assume, therefore, that the late Wisconsinan periglacial environment of North America was neither so extensive nor so marked as in Europe.

A more detailed description of Late Pleistocene climatic and vegetational conditions is beyond the scope of this book. There is an abundant literature available, and the interested reader is referred to the works of Flint (1971), West (1968) and Sparks and West (1972). In summary, it is clear that (a) periglacial conditions occurred many times during the Late Pleistocene, each occurrence being of varying duration and intensity, (b) the more recent periods of periglacial conditions occurred during the middle and late Wisconsinan/Weichselian cold stage.

Geomorphic considerations

From a geomorphic point of view, there are a number of important differences between the Pleistocene periglacial environments of mid-latitudes and those of present-day high latitudes.

The major difference is undoubtedly a difference in solar conditions. While both middle and high latitudes experience both summer and winter, the contrast between the two seasons and the rapidity of the change from one to the other is not as marked in middle as in high latitudes. In middle latitudes there is no equivalent of the 'arctic night' or 'arctic day', each of which lasts for periods of several months without interruption. Instead, the daily rhythm of day-time and night-time is the dominant condition. During the Pleistocene, therefore, this diurnal solar radiation regime was far more important in mid-latitudes than it is today in high latitudes. It probably led to a considerably greater frequency of freeze–thaw cycles of shallow depth and of short duration. An average frequency of between 50 and 100 cycles per year might be a realistic figure, bearing in mind the number of freeze–thaw cycles currently occurring in alpine mid-latitude environments such as the Colorado Front Range of the United States (see Table 3.1). If correct, frost shattering and frost creep processes, including needle ice, were probably several times more intense than in high latitudes today, and contributed greatly to the mechanical breakdown of rock and the transport of frost shattered debris, as solifluction, downslope.

The effect of orientation, with respect to solar radiation inputs, was also more marked than in high latitudes today. The fact that the sun disappeared below the horizon at the end of each mid-latitude day meant that in the northern hemisphere the north- and northwest-facing slopes were exposed to very low angles of inclination of the sun. Other things being equal, this would favour lower air and ground temperatures on such slopes and deeper frost penetration. If permafrost were present, it would also imply a thinner active layer. The effect of orientation is not that simple, however, since the various climatic fluctuations would promote a space-time change in the micro-climatic conditions on any one slope. For example, with the deterioration of climate from an interglacial or interstadial period, the north- and east-facing slopes would have begun to experience freeze–thaw conditions first. But during the period of maximum severity of periglacial conditions, the south- and west-facing slopes would probably experience the greatest frequency of freeze–thaw cycles, while the colder slope would remain frozen for a longer period of time. Then, as the climate progressively ameliorated, frost-action processes would linger longest on the cooler slope. It is possible, therefore, that depending upon the severity of the temperature drop, frost-action processes would be favoured either on south- and west-facing slopes (being 'warmer' slopes with greater solar radiation enabling ground temperature to rise temporarily above 0°C) or on north- and east-facing slopes (being 'colder' slopes with temperatures falling temporarily below 0°C). In all probability, the frost shattering and creep occurring on south-facing slopes repeatedly rising above freezing during the summer months would be the more potent geomorphic situation. This is because the south-facing slope would be in a near constantly frozen state, with moisture being available at each occasion of thaw either from the melt of segregated ice lenses in the active layer or from the melting of the permafrost.

Related to solar radiation inputs is the question of permafrost. In high latitudes the long Arctic night enables intense cooling of the ground surface and the formation of permafrost. However, the relatively short duration of the mid-latitude winter and the fact that even in winter temperatures followed a diurnal pattern, probably meant that the extreme of winter temperatures, as experienced by polar and continental environments today, did not exist. This would have been particularly true of the milder, oceanic environments. On the other hand, in continental areas radiation cooling during the winter nights would have created temperature inversions which might have persisted through the daylight hours. It has been suggested (p. 222) that temperature depressions as great as 10–12°C might have existed in the central European lowlands. In areas under such climatic conditions, permafrost would have formed but in the more oceanic locations, it is unreasonable to think of the formation of thick permafrost layers. In western Europe in particular, permafrost was probably thin and discontinuous, probably less than 30 m in thickness at most times. The evidence concerning the nature and extent of permafrost is examined in more detail in the next chapter.

A second major difference between mid-latitudes during the Pleistocene and high latitudes today arises from the fact that global atmospheric circulation patterns were very different during glacial times when large ice sheets extended into mid-latitudes. This is in addition to the normal, latitudinal, differences that one would expect when comparing mid- and high-latitudes. In all probability, there was an increased intensity of the climatic gradients equatorwards away from the ice margins. The mid-latitude westerlies, while being displaced latitudinally, would also have been stronger. At the same time, intense anticyclonic conditions probably developed over the ice sheets and in continental locations free of ice. In Europe for example, the development of strong high pressure systems, especially in winter, may have led to the 'blocking' of travelling disturbances in the westerlies. These would have been diverted either northwards towards Iceland and Spitsbergen or southwards towards the Mediterranean Sea. If this were correct, central and eastern Europe would have experienced below average precipitation amounts while the western fringes of France, southwest England and southern Ireland would have experienced above average amounts for that time.

Several reasons suggest a greater importance of wind action in middle latitudes during the Pleistocene. First, as already mentioned, wind gradients were strong at the southern margins of the ice sheets due to the concentration in space of the various climatic zones. Second, anticyclonic conditions which developed over the ice sheets led to strong local winds blowing off the ice sheets. Third, the periglacial zone developed in an ice-marginal position, when extensive outwash and till plains were exposed as the ice sheets retreated. Abundant sources of fine grained and unconsolidated sediments were available, therefore, for transport and use as abrasive materials. Fourth, the effectiveness of wind action in middle latitudes was further enhanced by the general decrease in precipitation resulting from the colder conditions of glacial times and the anticyclonic blocking of travelling disturbances in the westerlies.

On a more specific level, the nature of fluvial processes in Pleistocene periglacial environments needs brief mention. In all probability, fluvial activity was greater than is usually imagined. Moreover, it was certainly different in nature to the fluvial activity currently experienced by high latitudes (pp. 167–76). In particular, the absence of a long and continuous Arctic night in which temperatures dropped well below 0°C, meant that running water was an important characteristic of the periglacial landscape for more than just two or three months, as is the case in high latitudes. In addition, there would not have been the same dramatic spring flood as occurs in high latitudes. Some of the winter snowfall would have melted and entered the drainage system during the winter day-time periods. If permafrost were absent, there would have been infiltration of water to the groundwater table, and there would have been less direct runoff from either summer storms or spring snowmelt. For these reasons, fluvial activity involving both slopewash and channel flow would have been of a more year-round importance, particularly so in the milder maritime environments.

In summary, we might speculate that the periglacial environments of middle latitudes during the Pleistocene were different to those of polar and continental high latitudes in the following ways. First, the solar radiation pattern was dominated by a daily, as opposed to a seasonal, rhythm. This led to a greater intensity of frost action processes, notably frost shattering and frost creep.

Second, the effects of orientation and local micro-climatic differences with respect to geomorphic processes were more obvious. Third, atmospheric conditions were different as a result of the large Pleistocene ice sheets. In particular, wind gradients were strong on account of (a) intense high pressure systems developing over the ice sheets, and (b) the concentration of the various climatic zones. Fourth, wind action, as both an erosive and transporting agent, assumed great importance on the extensive outwash and till plains adjacent to the ice margins. Fifth, permafrost may not have been such an important element of the mid-latitude periglacial landscape and, except in continental locations, was relatively thin and/or discontinuous. Sixth, the extremes of winter temperatures, as currently experienced in high latitudes, were probably never reached. Seventh, the fluvial regime was different in that streams may have flowed for much of the year, and there was not the same concentration of fluvial activity in a short period of time as there is in the high latitudes today. Finally, it seems reasonable to suppose that there were significant variations in periglacial climates, both in a spatial and temporal sense, in the same way that there are variations in the nature of present-day periglacial climates.

Problems of reconstruction

The spatial extent of the periglacial conditions which existed during the Pleistocene is difficult to assess for a number of reasons. These may be summarised as being either geomorphic or climatic.

From the geomorphic point of view, the major problem is related to the lack of reliable indicators of cold conditions. Most attention has been concentrated upon the identification of 'relic' frost action phenomena and/or the previous existence of permafrost. At first, a wide range of features were thought to be diagnostic of periglacial climates; Smith (1949a) for example, listed 13 types of features as being of climatic significance. Many of these features, however, such as landslides, superficial folds, earth mounds, stabilised taluses, dry valleys and asymmetric valleys can demonstrably be shown to occur under non-periglacial conditions as well, and their usefulness, therefore, is limited. Others, such as patterned ground phenomena, solifluction deposits, involutions, and blockfields, are more reliable indicators of frost action (Wright, 1961). But these features are still open to interpretation since some do not require permafrost for their formation, and others require different degrees of cold climates for their development. Probably, the only reliable indicator of past periglacial conditions is that of ice and sand wedge casts, since only they unambiguously demand permafrost for their formation and also require severe winter temperatures (see pp. 21–5). Thermokarst forms have also been suggested as valid indicators, but with the exception of pingos, their identification in the landscape of now temperate latitudes is extremely difficult indeed. Even when the past permafrost distribution is known, there is still no simple relationship between permafrost formation and air temperatures (see pp. 52–3). Equally, when the rather broader category of frost action phenomena (blockfields, involutions, etc.) are used as the diagnostic criteria, the relationship between the frost structures and air temperatures is also not clear, since local lithological, moisture, site, and vegetation conditions will influence the effectiveness of frost action at any one locality.

From the climatological point of view, a number of problems hinder attempts to calculate possible temperature depressions during the Pleistocene, and hence, to give quantitative parameters to the cold conditions of the time. Usually, temperature depressions have been calculated on the basis of modern lapse rates and the present snowline in alpine regions. For example, assuming a lapse rate of 0.5°C/100 m, a mean annual temperature at the snowline of 0°C, and an elevation difference between the present snowline and the inferred Pleistocene snowline of approximately 1 000 m, it has been calculated that there was a probable temperature depression of between 4 and 6°C in Europe during the last glacial period (Wright, 1961, pp. 966–70). This sort of analysis masks the considerable variability of modern lapse rates which occur in mountainous regions. Moreover, there is no guarantee that lapse rate conditions were similar during the Pleistocene to today and, furthermore, that the European alpine conditions are representative of the lowland periglacial zone lying between the Alps and the continental ice sheets. In fact, within the lowland periglacial zone, marked temperature inversions were probably very common up to 1 500 m in elevation on account of the very low ground surface temperatures brought on by anticyclonic conditions, terrestrial radiation and a negative heat balance. As a result, a mean annual air temperature 10–12°C lower than today is probably more realistic for the European lowlands during the last cold stage (e.g. Shotton, 1960). Estimates of the drop in elevation of the Pleistocene snowline are also difficult to make because (1) the Pleistocene snowline is usually identified upon morphological evidence, such as cirque heights or nivation hollows, which themselves may be of considerable amplitude or variability with respect to elevation, (2) the modern snowline is not necessarily at 0°C as is often assumed, but generally lower, and (3) in certain areas, such as southern Africa, a hypothetical snowline has to be assumed. Finally, there is little evidence from which to make deductions concerning precipitation amounts. In general, the colder conditions and the blocking of travelling disturbances in the westerlies would have led to an overall decrease in precipitation. Based upon the absence of cirques from southern Britain, Manley (1959) has suggested that the uplands of western Britain received only 80 per cent of today's amounts. If such areas were above average for Europe, then one might expect only 40 per cent of today's amount to have fallen in parts of central and eastern Europe. In Poland for example, where current precipitation amounts are approximately 400–500 mm per annum, this would mean less than 200 mm per annum, an amount comparable to many high Arctic regions today (see pp. 8–9).

Extent of Late Pleistocene periglacial conditions

Perhaps the most reliable criterion of the extent of periglacial climates during the Late Pleistocene is the distribution of plants, and the extent of the different eco-zones. In terms of periglacial climate identification, this means the analysis of the past migration and extent of the tundra, steppe tundra and forest tundra zones. As stated earlier, palynological studies have been particularly helpful in outlining late and post glacial vegetational successions in both middle and high latitudes. In addition, the study of Pleistocene steppe and tundra faunal remains, such as land mollusca and insects, has helped determine environmental conditions and absolute chronology. Unfortunately, the number of such studies

Fig. 11.2 *Climatic reconstruction and regionalisation of Europe during the Weichselian stage, according to J. Budel (1951)*

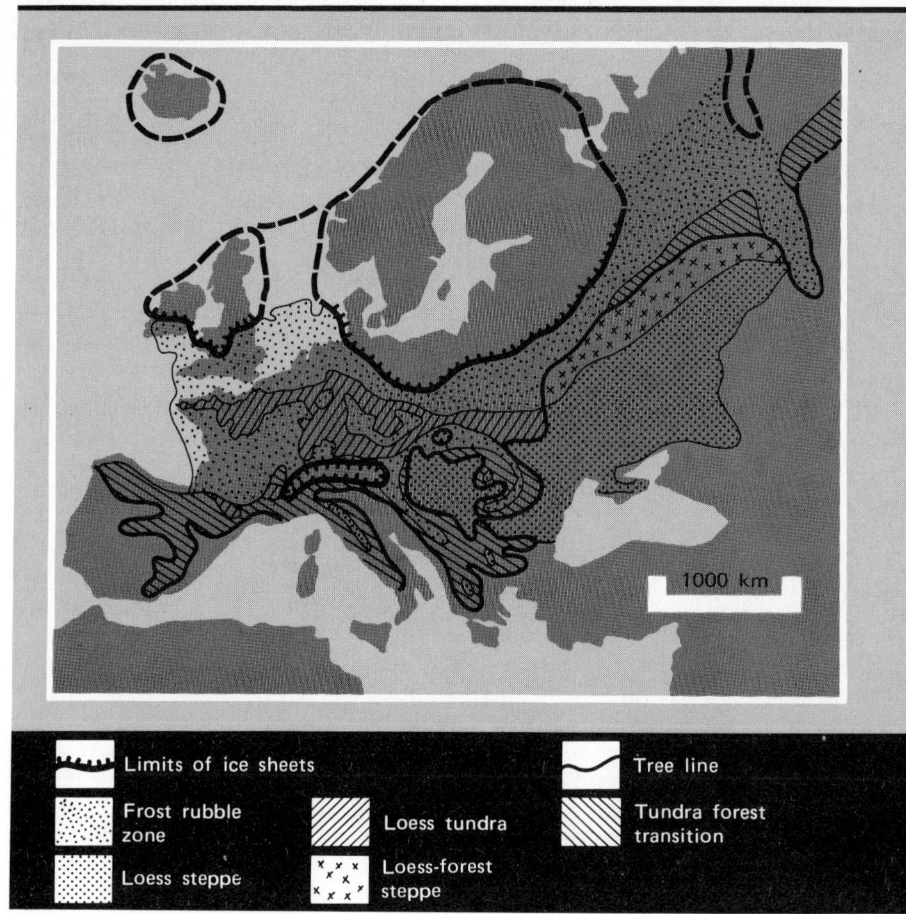

is limited, even in Europe, and the accurate delineation of Pleistocene periglacial climatic zones is far from complete. In many cases, heavy reliance is placed upon morphological evidence, some of which is open to differing interpretations.

For Europe, where most information upon Late Pleistocene climatic and vegetational changes is available, the best known attempt at morpho-climatic reconstruction is that by Budel (1951). This is illustrated in Fig. 11.2. It should be treated with extreme caution, however, since in spite of its claims, there is no evidence that palaeobotanical evidence was used in either the distribution or classification of the forested areas. It should be regarded as an intuitive synthesis which, in broad terms, is probably representative of conditions at the maximum advance of the Weichselian ice sheets in Europe. Budel identified three vegetation zones which, together, may well represent the extent of the periglacial domaine at that time.

First, a frost-rubble tundra zone ('frostschuttzone') lay to the south of the continental ice sheets. It was characterised by intense frost action and strong winds, and the spread of subarctic and Arctic plants and animals. This zone covered much of southern England, France and the north European plain extending eastwards through Poland into European Russia. A further characteristic, designated by Budel as a distinct eco-zone, was a loess tundra belt extending from western France eastwards into Poland. Frost action is recorded by ice wedge casts, involutions and solifluction/head deposits; wind action by the loess deposits, frosted sand grains and ventifacts. The absence of a forest cover, the presence of large unvegetated outwash plains, comminuted frost debris and exposed till sheets, all accentuated the action of the wind. Continuous or discontinuous permafrost underlay this zone.

Second, a forest tundra zone, generally more restricted in extent, lay to the south of the polar treeline and the frost-rubble tundra zone. Parts of southern France, the Iberian Peninsula, northern Italy and the areas between the Adriatic and Black Seas were included within this eco-zone. Because of the tree cover, wind action was reduced and permafrost development was discontinuous or absent. In upland areas, local frost-rubble tundra zones probably occurred beneath the Pleistocene snowline.

Third, a steppe zone of parkland vegetation lay to the east of the polar treeline, and extended over large areas of eastern Europe and European Russia as far as the Ural Mountains in the east and the northern shores of the Black Sea in the south. Drier, more continental conditions were probably the control over this particular eco-zone, and the sheltered Carpathian Basin represented the most westerly limit to this zone. Budel subdivided the zone into (a) loess steppe, representing the eastern extension of the loess tundra zone, and (b) loess forest steppe, which was a transition from forest tundra and loess steppe conditions.

One is led to conclude that much, if not most, of Europe to the south of the continental ice sheets was affected by periglacial conditions at the height of the last glacial advance. According to Budel, only the southern parts of the Iberian Peninsula and coastal areas bordering the Mediterranean escaped.

In North America, a similar attempt to map the various eco-zones which existed at the height of the Wisconsinan glaciation has been made by Brunschweiller (1962). However, the available data for North America is less than for Europe and insufficient to justify a definitive map. Figure 11.3 is a tentative reconstruction compiled from various sources, including Brunschweiller. Several considerations suggest that this map overemphasises the extent of the tundra and steppe eco-zones. First, the pollen record for many parts of the United States and southern Canada does not suggest a widespread frost-rubble tundra zone (see p. 218), and instead, indicates that forest zones intermingled with loess steppe zones. Second, the southern extent of the Wisconsinan ice sheets was further south than that of the Weichselian in Europe; as a consequence, the zone of most severe periglacial conditions was probably more restricted than in Europe due to the tighter alignment of the displaced climatic zones. Third, the retreat of the Wisconsinan ice sheet was accompanied by the development of extensive proglacial lakes along its southern margin, brought about by isostatic depression and the ponding of meltwaters or the influx of marine waters. These water bodies exerted modifying influences upon the regional climate, and restricted the amount of land surface exposed to sub-aerial conditions. Furthermore, upon their retreat several hundreds or thousands of years later, as isostatic

Fig. 11.3 *Tentative reconstruction of the mid-latitude periglacial zone in North America during the Wisconsinan glaciation. Modified from Brunnschweiler (1962). Glacial limits taken from Prest, et al., (1968)*

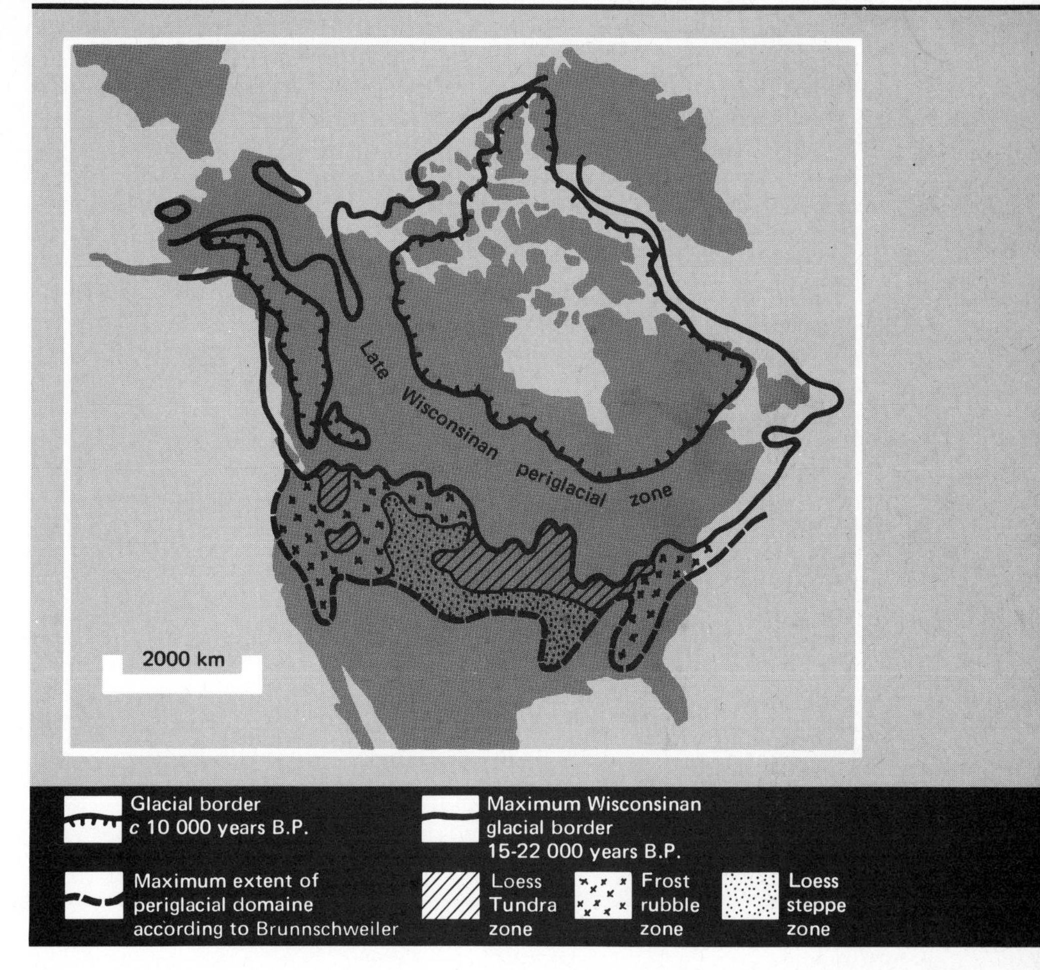

re-adjustment took place, the climate had ameliorated sufficiently to allow forest growth to re-establish itself.

In recent years, there has been recognition of Pleistocene periglacial conditions in other mid-latitude areas of the world besides Europe and North America. A large literature is now available to document this fact, much of which is summarised in the various issues of the ***Biuletyn Peryglacjalny***. Because of the wide ranging nature of much of the information, it is somewhat difficult to evaluate the importance of the various observations without personal acquaintance with the terrain. Of note however, are the claims for Pleistocene frost-action phenomena in the southern hemisphere, particularly Tasmania, New Zealand, and the High Drakensburg of southern Africa (e.g. Derbyshire, 1973; Butzer, 1973).

It should also be stressed that although Pleistocene frost-action conditions apparently existed in the southern hemisphere, there is not convincing evidence for the previous existence of widespread permafrost. For example, none of the periglacial phenomena described from Tasmania (e.g. talus, solifluction debris, blockstreams, nivation cirques, and bedded screes) necessarily require permafrost. Features which do, such as sand and ice wedge casts, pingo remnants, and thermokarst forms, have yet to be recognised. This suggests that the periglacial environments of the southern hemisphere did not experience the same magnitude of temperature depressions as did the periglacial environments of the northern hemisphere. For example, Derbyshire (1973) suggests a depression of only 6.5°C for the last cold period in Tasmania. Presumably, the much milder periglacial conditions of mid-latitudes in the southern hemisphere reflects their more oceanic surroundings.

Estimates as to the global extent of Late Pleistocene periglacial conditions is extremely difficult. The periglacial climates varied in time and space. Some were characterised by both intense frost action and permafrost; others, by intense frost action only. The two areas in which there is sufficient information of reliable quality to allow one to make estimates are Europe and, possibly, North America. If it is assumed that these areas, together with southern Siberia, were the major areas of Late Pleistocene periglacial conditions, a conservative guess would be that as much as 20 per cent of the earth's surface has experienced cold climate, non-glacial conditions in the past.

12 Relic periglacial phenomena

One of the problems of Pleistocene periglacial reconstruction is the recognition and interpretation of relic phenomena in the landscape of now temperate mid-latitudes. On the basis of the last chapter, one might expect Pleistocene periglacial environments to have been characterised by at least one, or all, of the following three conditions: (a) intense frost action, (b) the presence of perennially frozen ground, either continuous or discontinuous, and (c) strong wind action. In this chapter, the various morphological evidence thought diagnostic of periglacial conditions is discussed under these three headings.

Evidence for frost-action conditions

Since frost-action processes probably assumed a particularly significant role in the mid-latitude environments, it should be expected that frost-action phenomena are well developed. In spite of this, their recognition and interpretation may be difficult. To begin with, the effects of present-day frost-action processes must be assessed before a feature can be attributed to a past period of more intense frost action. Then, even when it is clear that present frost activity is not responsible for the feature concerned, its 'relic significance' will depend upon an evaluation of the lithological susceptibility of the rock in question to frost action, and the moisture, vegetation and local site conditions existing during that period of colder conditions. A second problem arises from the fact that certain features, such as tors, can be produced under both periglacial and non-periglacial conditions (see below, p. 231) and a frost-action genesis is difficult to prove. Finally, the effects of man in clearing and ploughing the land must be considered since frost heave structures may be modified or even obliterated in certain instances. Under certain conditions for example, it is easy to mistake a series of old plough furrows for a sequence of shallow involuted structures.

For these reasons, the recognition of relic frost-action phenomena requires caution. Few provide unequivocal proof of periglacial conditions and only where several different features occur together or in association is one justified in assuming a frost-action environment.

Frost-disturbed soils and structures Involutions and cryoturbation structures are some of the most widespread features thought indicative of Pleistocene frost action. They can often be seen in road cuttings, building excavations and other exposures, and were first described from central Europe where they were termed 'Wurgeboden' (strangle soils) or 'Brodelboden' (Troll, 1944). The variety of forms is considerable, ranging from relatively amorphous cryoturbation structures through to regularly spaced and uniform involuted structures (Fig. 12.1). They appear best developed on relatively flat surfaces where the absence of downslope movement by creep and solifluction prevented their destruction or modification.

The variety and frequency of occurrence of Pleistocene periglacial involutions in Europe is remarkable and contrasts with the relative

Fig. 12.1 *Flame-shaped involutions and frost shattered Chalk, Pays de Caux, northern France.*

absence of similar forms reported from present-day high latitudes.

The various types of involutions require brief description (Fig. 12.2). In plan, involutions usually display a constancy of spacing as well as of height and in this way are rather different to cryoturbation structures which are more localised and irregular in form. In distribution, involutions form small polygons, up to 1.5 m in diameter while the form of the involution appears closely related to the particle size of the sediments which are involved. Drip or 'plug' involutions are those where the upper sediments constitute the border of the polygons. 'Pocket' involutions, by contrast, are those where the upper layers reach down to form the centres of the polygons. Both plug and pocket involutions are typical of relatively homogeneous, fine grained sediments. Other types of involutions involving a wider range of particle sizes are more irregular. Flame-like and club-shaped structures involve the injection of finer sediments, usually silts and clays, into overlying sands and gravels. Stone pillars, blunt nosed cones, and 'festoons' are various types of involutions in which tongues of frost shattered and disintegrated rock rise upwards into the overlying soil regolith composed of finer particles. Viewed in cross section, festoons represent a form of sorted stripe.

It must be emphasised that the presence of periglacial involutions and related structures does not necessarily imply the former existence of permafrost. It is likely that only the very regular and simple structures are the result of straightforward cryostatic pressures, and that differential frost heaving and gravity controlled load adjustments are involved in the more complex structures. Where periglacial involutions are recognised, they are diagnostic indicators of a climatic regime in which seasonally frozen ground developed. Only if other indicators of permafrost (see below) are found in adjacent localities should one assume that periglacial involutions reflect past permafrost conditions.

Blockfields and frost weathered bedrock A phenomenon commonly interpreted as reflecting intense frost wedging of exposed rock surfaces is the presence of extensive accumulations of angular boulders and coarse debris. These rock and rubble accumulations are termed 'blockfields' or 'felsenmeer' and are clearly analogous to the debris mantled rock outcrops and upland surfaces of many Arctic regions. They are particularly common in areas of hard, relatively well bedded rocks such as shales and sandstones which are susceptible to frost wedging along major joints and bedding planes and which weather into coarse debris and angular boulders. More localised block accumulations which occur on slopes and in valley bottoms are variously termed boulder fields, or 'stone streams' (coulées pierreuses) and clearly demand some form of mass movement in addition. Scree slopes are a form of localised boulder field developed beneath vertical or near vertical rock faces. Pro-talus ramparts are sometimes identified also. These are ridge-like accumulations of coarse angular blocks which develop some distance from the base of the steep slope from which they are derived. They are thought to form at the base of a large semi-permanent or permanent snowbank which existed below the free face, with the blocks detaching themselves from the free face through frost wedging and then sliding down the surface of the snowbank.

Fig. 12.2 *Types of involuted structures*

A *Pocket and plug involutions*

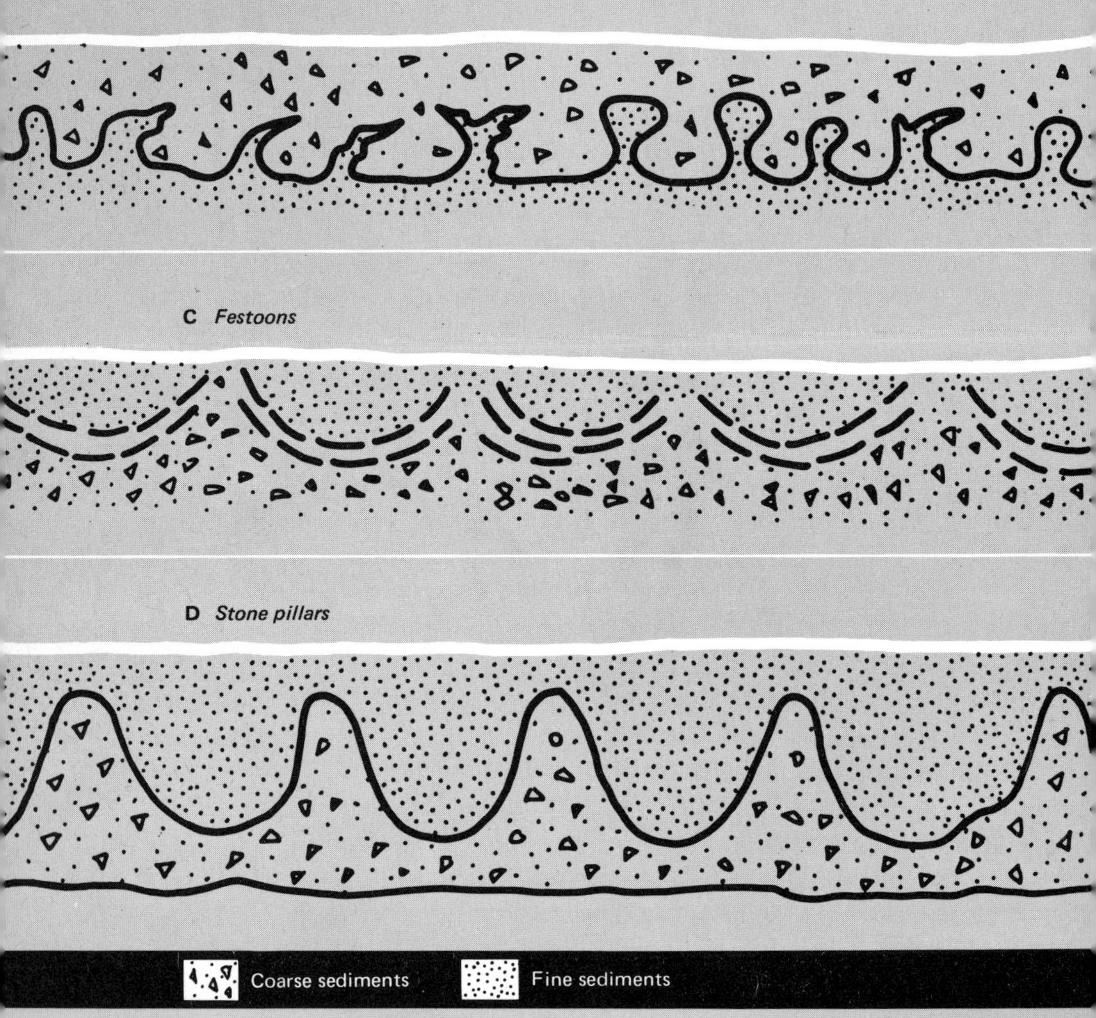

In Europe blockfields and related forms are known to occur in most of the upland massifs and low mountain ranges (e.g. Pissart, 1953; Dylik, 1956). In North America, numerous blockfields occur south of the glacial limits, especially in the central interior uplands and the Appalachian plateaus (e.g. Smith, 1962). In the southern hemisphere, blockfields, talus, and frost shattered debris are the most widespread of the periglacial phenomena described (e.g. Derbyshire, 1973).

The age and rate of formation of blockfields is still not clear and thus, blockfields and related forms are of limited use in Pleistocene periglacial reconstruction. To begin with, it has to be proven that these deposits are not being formed under present-day climatic conditions, or that if they are, the current rate is not sufficient to account for the magnitude of the deposits. For example, scree slopes in the English Lake District have been compared to similar forms in central Labrador–Ungava (Andrews, 1961). According to Andrews, the Lake District screes formed under periglacial conditions in relatively recent (i.e. post glacial) times. Further studies by Caine (1963) indicate, however, that movement on certain scree slopes in the Lake District is greater than 30 cm per annum, and that scree slopes are still very active. The same problem occurs in the case of blockfields; in the Appalachian Mountains of the United States for example, both actively forming and relic blockfields are thought to exist (e.g. Hack and Goodlett, 1960, pp. 31–2; Rapp, 1967). Thus, there is no proof that blockfields form only under periglacial conditions, and they cannot be regarded as diagnostic of such a climate.

A second problem concerning blockfields is the uncertainty as to whether they can form under glacial, as opposed to periglacial conditions. In the Narvik Mountains of Norway for example, it is not clear whether the felsenmeer formed since the last glaciation or whether it developed during a much longer period of time, either in a nunatak position or beneath an inactive and protective ice sheet (e.g. Dahl, 1966; Ives, 1966). Similarly, in parts of the English Lake District, some of the boulder spreads possess rounded blocks in addition to angular ones, and grooves, aligned in roughly the same direction, suggest that the boulders have been glacially fashioned. A final problem, stressed in the introduction to this chapter, and in Chapter 3, is that different lithologies and moisture conditions determine the effectiveness of a given frost action climate in promoting rock weathering and shattering (e.g. see pp. 37–9). Thus, the climatic significance attached to any blockfield can vary considerably.

Tors The existence of hillslope and summit tors in many mid-latitude regions (Fig. 12.3) has attracted considerable attention, but their climatic significance is also the subject of dispute. Well known tors and tor-like features have been described from the uplands of central Europe, notably the Sudetes Mountains (e.g. Jahn, 1962) and the Bohemian Massif (e.g. Czudek, 1964; Demek, 1964), and from western Europe, notably the Pennines of northern England (e.g. Palmer and Radley, 1961; Linton, 1964) and the Dartmoor Massif of southwest England (e.g. Linton, 1955; Te Punga, 1956; Palmer and Nielson, 1962). Tors are also known to exist in certain areas of the southern hemisphere (e.g. Caine, 1967).

One school of thought interprets mid-latitude tors as essentially angular landforms produced by frost shattering under Pleistocene periglacial conditions, with any rounding of the tor edges resulting from a subsequent change to more

Fig. 12.3 *The Devil's Cheesewing, a valleyside or hillslope tor, near Lynton, north Devon, England. The feature is 10 m high. Its form closely resembles hillslope tors on Prince Patrick Island, in the Canadian Arctic (see Fig. 7.8). Photograph courtesy of D. Mottershead, Portsmouth Polytechnic, England.*

temperate climatic conditions. They are interpreted, therefore, as being 'palaeo-arctic' tors (e.g. Palmer and Radley, 1961). Such an interpretation may be classified as a one-stage hypothesis involving weathering and simultaneous removal of the regolith. It assumes that the formation of tors is intimately linked to the evolution of slopes through the cryoplanation process. Under such a hypothesis, the tors reflect the remnants of a frost riven bedrock outcrop surrounded by a low angled altiplanation terrace across which the frost shattered debris is transported by solifluction and mass wasting processes (Fig. 12.4). A second school of thought argues that tors are prepared by deep weathering principally along bedrock joints to produce subsurface rounding of corestones ('woolsacks') under warm temperate or tropical conditions in Late Tertiary times. They have subsequently been exhumed by solifluction and mass wasting, and sharpened by limited frost action under periglacial conditions during the Pleistocene (e.g. Linton, 1955, 1964). In this context, tors are interpreted as being 'palaeo-tropical' forms analogous, but on a smaller scale, to the larger inselbergs and castle koppies of central Africa and other tropical regions.

It is very unlikely that a single hypothesis of tor formation is applicable to all tors. Instead, a variety of processes and environmental conditions may operate to produce essentially the same landform, and periglacial conditions merely represent one set. In parts of the Polish Sudetes, for example, there is convincing evidence that summit tors were formed by deep weathering of granite, chiefly during warm interglacials, and by exhumation under periglacial conditions during the glacial periods (Jahn, 1962). On the other hand, in the adjacent Hruby Jesenik Mountains of Czechoslovakia, Demek (1964) has described how tors and castle koppies of both a deep weathered (i.e. two cycle) origin and a frost shattering or periglacial (i.e. one cycle) origin exist in the same area. According to Demek, exhumation of the deep weathered tors had already begun at the end of the Tertiary, and continued during the Pleistocene, at which time the parallel retreat of frost riven cliffs caused the isolation of other tors and the formation of altiplanation terraces.

A similar situation may have existed in southern England, and may reconcile the apparently conflicting interpretations of the Dartmoor tors, as exemplified by Linton (1955) and Palmer and Nielson (1962). Decomposed granite, or 'growan', usually attributed to chemical rather than mechanical weathering (e.g. Eden and Green, 1971), is encountered in a number of localities in or near the main river valleys. In general, the majority of tors are also located in such positions and are peripheral, therefore, to the planation surfaces of supposed Tertiary age which occupy the higher interfluves. While these facts support Linton's (1955) two-cycle hypothesis of tor formation, they also suggest the localisation of the deep weathering process. Thus, the existence on the higher summits (e.g. Cox Tor) of one cycle tors rising above altiplanation terraces cut in bedrock and covered with a veneer of frost shattered rubble (Fig. 12.4) as described by Te Punga (1956), Palmer and Neilson (1961), and Waters (1962), is not contradictory. It is conceivable that both exhumation and modification of two-cycle tors and the formation of one-cycle tors could have occurred at the same time in different parts of Dartmoor depending upon the localisation of the deep weathering process.

Hillslope tors, occurring on the upper parts of slopes and in areas where

Fig. 12.4 *Altiplanation terrace at elevation of 480 m on Cox Tor, Dartmoor, England. The frost-riven scarp is 4–5 m high and developed in metadolerite; the terrace slopes towards the viewer. Photograph courtesy of D. Mottershead, Portsmouth Polytechnic, England.*

 there is little evidence for deep weathering, such as northern England (e.g. Palmer and Radley, 1961) and Bohemia (e.g. Czudek, 1964), are less ambiguous. They seem best explained by a one-stage development of slope retreat caused by frost action, with mass movement, nivation and wind action as subsidiary transportational and erosive processes.

In summary, it is clear that the majority of tor-like features in mid-latitudes cannot be solely attributed to periglacial conditions, although most have undoubtedly experienced some exhumation and modification by periglacial processes. Moreover, even if a one-cycle periglacial origin seems the more likely, it is unwise to immediately assume that tors reflect intense frost action. In current periglacial environments, frost shattering is not the only process capable of forming tors (see pp. 154–5). In Antarctica for example, both rounded and angular forms are developed side by side in dolerite (Derbyshire, 1972). The angular forms occur in depressions where snowpatches permit effective frost shattering. The rounded tors, on the other hand, occur in the more exposed locations where snowpatches are absent and where long continued and slow chemical weathering and exfoliation has been experienced. Thus, tors of different forms may exist adjacent to each other and develop under the same climatic conditions but by two different processes. Clearly, therefore, tors cannot be used as diagnostic indicators of Pleistocene frost action without careful field investigation of the morphology and weathering patterns of the associated terrain.

Stratified slope deposits and grèzes litées Intense frost action together with periglacial slopewash is generally believed to result in rhythmically stratified slope deposits (Dylik, 1960). These are termed 'grèzes litées' or 'éboulis ordonnés' in the French literature (e.g. Guillien, 1951; Malaurie and Guillien, 1953). The sediments involved are essentially frost shattered debris, the nature of which depends upon the bedrock. Such deposits are clearly distinguishable from those of solifluction since they possess coarse bedding and a certain amount of sorting. Occasionally, stratified deposits interdigitate with solifluction deposits. In the Charente region of France, grèzes litées occur on relatively steep slopes and are composed of small angular particles, up to 2–3 cm in diameter, together with finer fractions (Guillien, 1951). At Walewice in central Poland, rhythmically stratified slope deposits attain a thickness of over 10 m below an old terrace bluff and are composed of alternating sands and silts (Dylik, 1969b, pp. 371–2). The calcareous meltwater muds which mantle certain areas of the Chalk escarpments of southern England (e.g. Kerney, Brown and Chandler, 1964) are a form of stratified slope deposit. Bedded scree, composed of alternating stratified angular to sub-angular coarse talus, with stratification parallel to the slope, is another form of grèzes litées found in upland areas (e.g. Sanders, 1973; Derbyshire, 1973).

It must be admitted that the mechanism of formation of stratified slope deposits is not clearly understood. In part, this reflects a lack of recognition and study of similar deposits in present-day periglacial environments. The angular particles and the finer matrix within which they are incorporated, is generally thought to reflect the frequent, probably diurnal, oscillations of freezing and thawing which might have taken place upon climatically favoured steep slopes. The sorting and bedding, and the eluviation of the finer particles and their deposition downslope, is usually attributed to slopewash operating on vegetation free slopes below snowpatches, and to the melting of pore ice. At the moment, one is not sure whether stratified slope deposits and grèzes litées represent a diurnal or seasonal phenomenon.

In high latitudes, there is an apparent absence of stratified slope deposits, although they have been reported from west Greenland (Malaurie and Guillien, 1953) and observed by the author in the deep and sheltered Masik Valley of southern Banks Island. Probably, high latitudes do not experience the repeated freeze–thaw oscillations which favour their widespread development, and they only form under localised conditions. Thus, the occurrence of grèzes litées and stratified deposits in mid-latitudes may be significant in terms of Pleistocene frost action although, as stated above, their exact origin requires further investigation.

'Head' and solifluction deposits In present-day periglacial environments, frost heaving of the ground surface leads to the downslope movement of material by frost creep. This process, as described in Chapter 3, is one component of solifluction (see pp. 135–41). In areas where there is a relatively high frequency of freeze–thaw cycles, as in the Colorado Rockies, frost creep is thought to be a more important component of solifluction than gelifluction (e.g. Benedict, 1970). Given the probable importance of freeze–thaw activity in middle latitudes, one should expect to find evidence, in the form of solifluction deposits, of frost creep and gelifluction movement. Thus, many of the unstratified and heterogeneous 'drift' or superficial deposits which mantle the lower parts of valley side slopes are regarded as indicative of Pleistocene solifluction, and frost creep in particular.

In Europe, these deposits were first reported from southwest England where they were termed 'head' since they formed a capping to many coastal cliff sections (De la Beche, 1839). Later, Fisher (1866) gave the name 'warp' to the non-glacial drift deposits in eastern England, Reid (1887) described the Coombe Rock which infilled the valleys and mantled the foot of the steeper slopes of many of the chalk regions of southern England, and Prestwich (1892) documented the raised beach and overlying 'head' or 'rubble drift' deposits of southern England. Subsequently, the mapping of 'head' deposits in England became standard practice by both the Geological Survey and the Soil Survey (e.g. Dines ***et al.***, 1940; Avery, 1964). On the Continent too, the significance of 'pseudo-glacial' sediments was appreciated at an early date and the recognition of frost derived sediments in the lowlands of western and central Europe became commonplace (e.g. Troll, 1944; Budel, 1944).

The variety of solifluction and 'head' deposits is considerable. All gradations exist between 'head' and river deposits and it is sometimes difficult to draw the line between the two. However, two general characteristics need to be emphasised. First, 'head' deposits are composed of predominantly poorly sorted angular debris of local derivation. Stratification is poorly developed and often absent. Where present, the stratification at the top and the bottom of the deposit is not horizontal. Often pebbles are aligned in a downslope direction and tilted upwards. Second, some of the 'head' deposits contain faunal remains indicative of cold climate conditions at the time of deposition. At the present day, downwash and soil creep account for the various types of 'head' being formed in humid temperate mid-latitudes.

As general indicators of frost-action conditions, solifluction and 'head' deposits represent an important line of evidence. However, their direct interpretation in terms of frost action is not always easy. A fundamental problem lies in our inability, even in present-day periglacial environments, to differentiate between a solifluction deposit primarily the result of frost creep and retrograde movement, and one primarily the result of gelifluction. A second problem is that not all deposits contain readily identifiable cold climate faunal remains, and a third is that many deposits are often difficult to distinguish from tills. In the latter situation, the distribution and surface morphology (e.g. lobes, stripes, etc.) of the deposit is often a clue, but where landforms have been considerably modified since deposition, or where the deposit is derived from a glacial deposit and thus contains its erratics, it may be impossible to distinguish till from solifluction.

Evidence for permafrost conditions

The evidence in favour of the previous existence of permafrost in mid-latitudes is less ambiguous than that of frost action. Certain features demand permafrost for their development and, upon the thawing and disappearance of the permafrost, morphological and/or stratigraphic evidence is left as proof of their previous existence. This is the case of (1) thermal contraction cracks (i.e. ice- and sand-wedge casts), (2) pingos and other ice segregation features, and (3) thermokarst forms. In the case of these features, the problem is not that of their climatic significance; rather it is their initial recognition and their differentiation from similar features of different origins.

Ice- and sand-wedge casts The existence of wedge-shaped structures interpreted as casts or pseudomorphs of ice and sand wedges has frequently been reported from temperate latitudes. Their recognition is particularly important since not only do thermal contraction cracks require permafrost for their development but it is believed that the top surface of the permafrost must cool to −15 to −20°C for their widespread occurrence. In present-day periglacial environments, this coincides with a mean annual air temperature of −6 to −8°C (see p. 91). Differentiation between ice-wedge and sand-wedge casts is extremely difficult in the fossil form and, as yet, no satisfactory criteria are available. It is highly probable that some of the ice-wedge casts reported in the literature are, in fact, sand-wedge casts. The importance of this particular distinction lies in the fact that sand wedges require more specific climatic conditions, in particular colder and drier conditions, than do ice wedges (see pp. 23–5). The third type of thermal contraction crack, the ground wedge or seasonal frost crack, has not been widely reported or studied in mid-latitudes, and is not considered here.

The problems of interpretation of ice- and sand-wedge casts are accentuated by the fact that many workers in mid-latitudes, especially prior to the last 5–10 years, had little field familiarity with modern ice- and sand-wedge structures of high latitudes. Thus, their knowledge of what pseudomorphs of such features should look like was often erroneous. There is, furthermore, a need to distinguish thermal contraction cracks from other fissure structures such as (a) fissures originating during the upheaving of pingos, and (b) sand dikes generally formed through the injection of sand into the overlying deposits (e.g. Butrym ***et al.,*** 1964).

Fig. 12.5 *Oblique air photograph showing Weichselian ice-wedge polygon as revealed by differential crop markings. The pattern is a crude random orthogonal one, with some incompleteness to the pattern. Northwest of Boxted, Essex, England. Photo by permission of University of Cambridge.*

The outline of the frost fissure polygons can sometimes be seen by differential crop ripening on air photographs (Fig. 12.5). In Europe, areas from which such patterns have been reported include Sweden (e.g. Svensson, 1964), and southern England (e.g. Morgan, 1971). In North America, fossil polygonal patterns have been observed in the north and east central United States (e.g. Clayton and Bailey, 1970), and parts of southern Ontario and the St Lawrence Lowlands (e.g. Morgan, 1972). However, using this technique, one is only able to identify the most recent features. Older polygon nets are often concealed by overlying younger deposits. Also, present-day land use and ploughing activity may destroy the fissure pattern at the ground surface. Ultimately, therefore, the interpretation of thermal contraction cracks in fossil form requires the study of wedges and casts in cross section, bearing in mind the characteristics of their modern counterparts.

Our understanding of fossil frost cracks is furthest advanced in Poland and other parts of central Europe where both sand- and ice-wedge casts are commonly differentiated (e.g. Dylik, 1966; Gozdzik, 1973). Ice-wedge casts represent fissures

◀ **Fig. 12.6** *Ice-wedge cast developed in Riss fluvioglacial sands and gravels near Modlna, north of Lódź, Poland.*

or wedge structures of secondary infilling. They form when the ice wedge slowly melts, usually as the result of climatic warming (Fig. 3.6B). As this happens, there is a general collapse of sediments into the trough. The ground also expands as it warms and helps fill the void left by the ice. In general terms, therefore, ice-wedge casts are characterised by the penetration of material into the fissures from above and from the sides, the downward inflection of layers if the enclosing sediments are stratified, and various systems of miniature faults. An example of a simple ice-wedge cast developed in glacio-fluvial gravels is illustrated in Fig. 12.6. The majority of ice-wedge casts are much more complex than this, however. It must be stressed that the melting of the ice and the release of excess water from the surrounding sediments may result in considerable deformation of the structure, and a shape far removed from a simple wedge form. The irregular form of many ice-wedge casts (Fig. 12.7) is a major problem in their identification and interpretation. Often, they bear remarkable resemblances to load cast deformations.

Sand-wedge casts represent fissures or wedge structures of primary infilling. In contrast to ice-wedge casts, these wedges contain an infilling of sandy relatively well sorted sediments, which is markedly different from the enclosing sediment. The sands are often highly aeolianised. In many cases, sand-wedge casts possess a more truly wedge shape, being relatively wider in cross section than ice-wedge casts, with a clear junction between the wedge and the enclosing material (see Fig. 12.7). The lack of water, consequent upon the thawing of the permafrost and the non-icy infilling material, does not lead to the same amount of intermixing of the infilling material as occurs in ice-wedge casts. In sand-wedge casts, there is no necessity for a downturning of adjacent beds since the wedge does not need to be infilled after thawing. A further characteristic of sand-wedge casts, as observed in central Poland, is that the distance between casts is often much less than that between ice-wedge casts. Commonly, they are 3–5 m apart. This may mean that some sand wedges are initiated by desiccation cracking. Equally, it may reflect the more intense cold required for thermal contraction cracking in a well drained arid site (see pp. 23–5) which would lead to a higher order of fissure polygon development (see pp. 90–1).

The variety of frost fissure casts and the difficulty of distinguishing between sand- and ice-wedge casts is further illustrated by the recognition of composite wedges (e.g. Gozdzik, 1973). These are features which show evidence of both primary and secondary infilling (Figs. 12.7 and 12.8). Sometimes, two wedges are visible, one inside the other. The inner wedge is of primary infilling while the outer wedge has signs of secondary infilling. According to Gozdzik (1973, p. 112), the inner wedge would have developed in the active layer along with periglacial involutions, while an ice wedge would have developed within the permafrost beneath. In other composite wedges, the infilling sands may have almost the same structure and texture as the sands in typical fissures of primary infilling. However, there may also be the inclusion of material from the fissure walls and some inflection of the adjacent sediments bordering the wedge. In cases such as these, it seems that the fissures were alternately filled with sands and ice. Melting of the ice contained in the sands occurred at such a slow rate

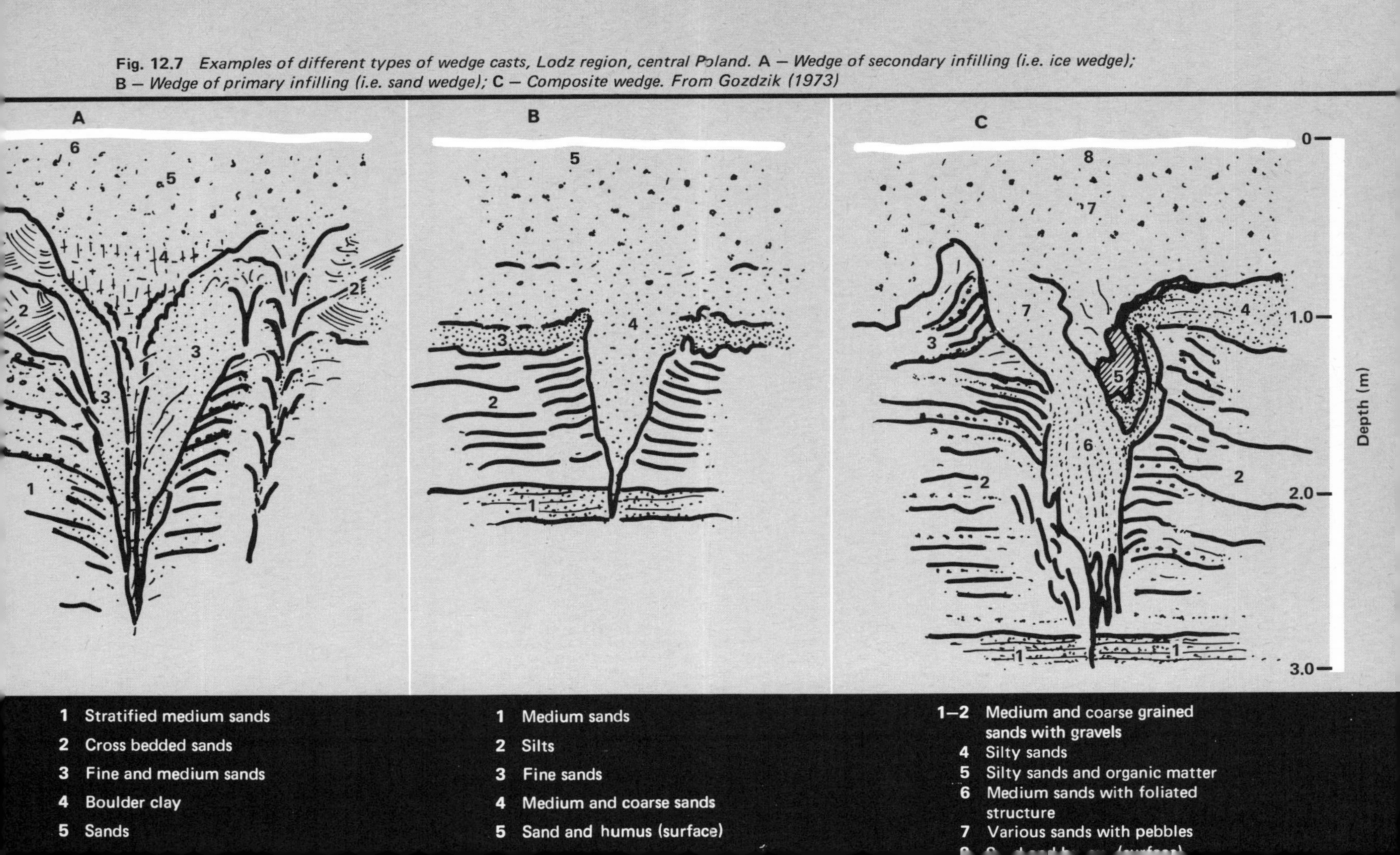

Fig. 12.7 *Examples of different types of wedge casts, Lodz region, central Poland. A – Wedge of secondary infilling (i.e. ice wedge); B – Wedge of primary infilling (i.e. sand wedge); C – Composite wedge. From Gozdzik (1973)*

that the foliated structure was not destroyed. The volumetric loss of water within the wedge, however, caused a downwards development of the adjacent sediments and a degree of infilling of the upper parts of the wedge.

The ice-wedge casts described by Morgan (1971) from the Wolverhampton area of England may be of this composite nature. Typically, they are developed in Devensian (i.e. Weichselian) till deposits overlain by ablation gravels (Fig. 12.9). The alignment of pebbles and the presence of gravels and coarser sediments immediately adjacent to the sides of the wedges indicate some degree of secondary infilling. On the other hand, their length–width ratio of 4.5:1, their close spacing (2.8–6.2 m), and the fact that the majority of the infilled material consists of medium grained sands are more typical of sand-wedge structures. Biological evidence, notably the beetle faunas which existed in the English Midlands during late Devensian times, also indicates a more continental, and hence drier, regime than at present (A. Morgan, personal communication, 1974). Since mean air temperatures may have been 10–12°C lower than present (Shotton, 1960), climatic conditions suitable for sand-wedge formation almost certainly existed in the Midlands during late Devensian (Weichselian) times.

In Europe, the distribution of ice- and sand-wedge casts has been mapped by several workers, notably H. Poser (1948), and J. Budel (1953). Poser observed that the depth and width of the various wedge casts which lay to the south of the limits of the last ice sheet tended to increase to a maximum in central Europe. While such characteristics may very well reflect the greater age of wedges in locations where the duration of periglacial conditions may have been longer, Poser chose to interpret these dimensions as indicating a severe periglacial climate in central Europe and a gradual amelioration of conditions westwards and southwards. Since wedge casts are also recognised from other areas south of the maximum limit of the last ice sheet in France, southern Germany, and southern England, it seems reasonable to assume that permafrost was quite widespread in Europe during the last glacial stage. Probably, it was best developed in central and eastern Europe where it may have been continuous, while in the milder environments of western Europe, it was probably discontinuous and restricted to the upland plateaus. Wedge casts are also known to occur in areas within the limits of Weichselian ice sheet, as in the English Midlands (see above) and in parts of Scandinavia (e.g. Johnsson, 1959). Their occurrence indicates that, as the ice sheet withdrew, a periglacial zone underlain by permafrost developed peripheral to the ice sheet.

The extent of ice- and sand-wedge cast distribution in temperate latitude North America has been recognised only recently, in the last 10–15 years. Their distribution suggests that a similar permafrost zone formed in a belt, approximately 80–250 km in extent, adjacent to the retreating Wisconsinan ice sheet. A number of wedges, reported from beyond the maximum limit of the Wisconsinan ice sheet in a broad zone extending through Montana (Schafer, 1949), the northern Great Plains (Clayton and Bailey, 1970), Wisconsin (Black, 1964) and Illinois (Wayne, 1967), are believed to be between 15 000 and 20 000 years old (Fig. 12.10). As the Wisconsinan ice sheet began to retreat, exposed till sheets became perennially frozen and ice and sand wedges were able to develop north of the Wisconsinan glacial border. Ice wedges developed in the damper environments of southern Ontario (Morgan, 1972), southern Quebec (Dionne, 1971), and the Maritime Provinces (Borns, 1965; Brooks, 1971), and sand wedges in the

Fig. 12.8 *Composite wedge developed in fossil soil in loess, Miechow Plateau, northeast of Krakow, southern Poland. The upper, inner, wedge is formed by primary infilling; the lower part shows evidence of secondary infilling.*

drier parts of the Canadian prairies (Berg, 1969). A number of ice wedges have also been reported from New England but recent investigations do not confirm this; in all probability, the climate of New England was too mild for ice-wedge formation as the ice sheet withdrew (Brown and Péwé, 1973, p. 91).

When compared to Europe, the permafrost zone of North America, as indicated by ice- and sand-wedge casts, was much narrower than that of Europe during the last glaciation. This probably reflected the more southerly limit of the ice sheet in North America, and the steeper climatic gradient to the south. In some areas, the treeline was very close to the ice margin. Another difference between Europe and North America was that the time duration of the late-glacial permafrost conditions in North America was short on account of the development of extensive proglacial water bodies. For example, Morgan (1972)

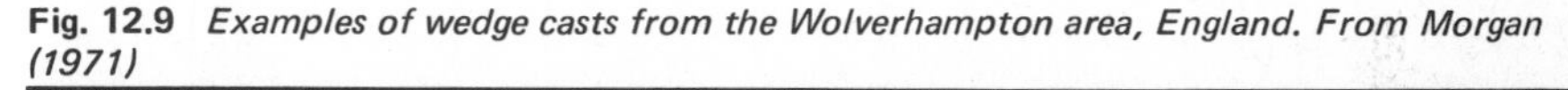

Fig. 12.9 *Examples of wedge casts from the Wolverhampton area, England. From Morgan (1971)*

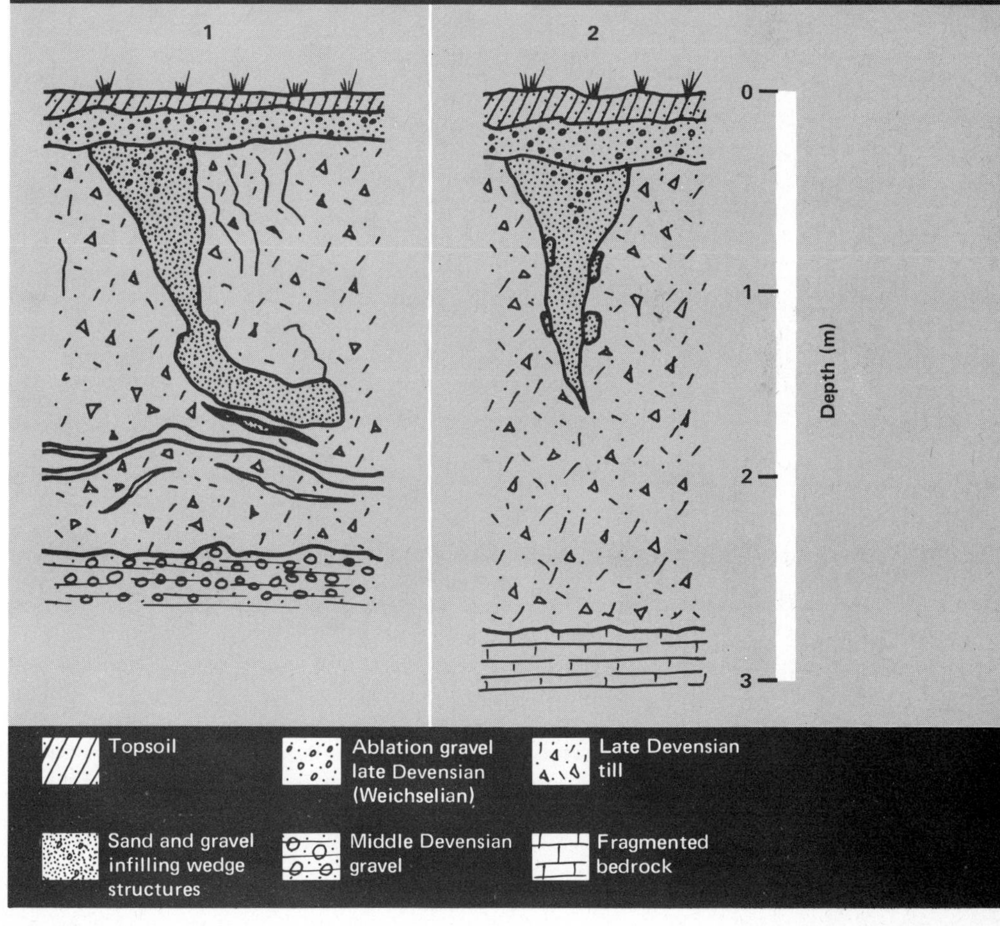

Fig. 12.10 *Distribution of wedge casts and other periglacial features in Wisconsin, U.S.A. and their relation to the glacial limit. From Black (1964)*

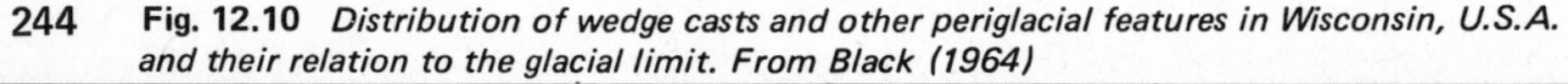

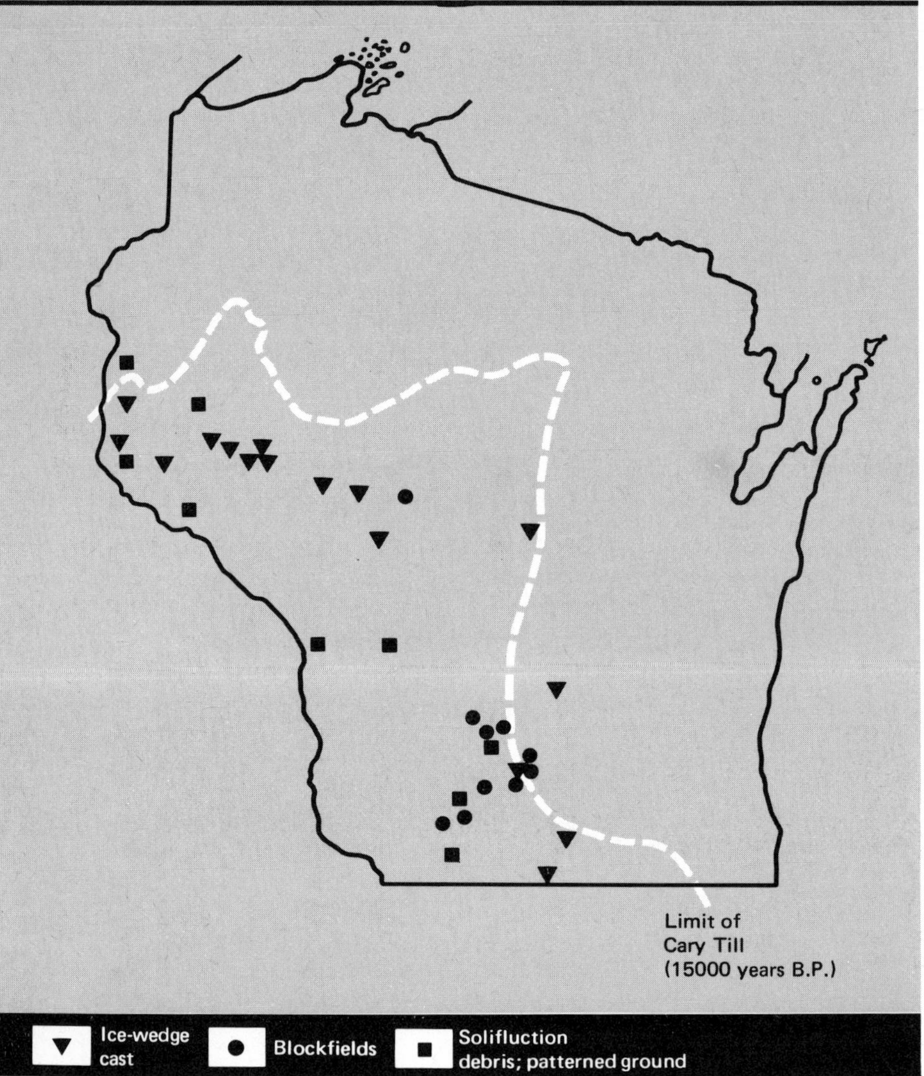

attributes the 'incompleteness' of the polygonal network in the Kitchener area of southern Ontario to the short-lived permafrost regime; in that area, suitable cold climate conditions existed only between approximately 14 000 and 13 000 years B.P., immediately after the retreat of the late Wisconsinan ice and immediately prior to the formation of Glacial Lake Whittlesey. This time period was not sufficient to allow either the development of a thick permafrost body or the wedges to propagate laterally to fully intersect and provide a complete polygon pattern. Similar proglacial water bodies existed in many parts of Canada following the retreat of the Wisconsinan ice sheet, especially in the St Lawrence

Lowlands and the central prairies. They undoubtedly restricted the land surface area exposed to periglacial conditions and ameliorated the climate of adjacent terrain.

Pingo remnants and related forms The remnants of pingos are certain evidence of the previous existence of permafrost. In terms of morphology, one might expect fossil pingos to possess a raised rim or rampart formed by the movement of material down the side of the pingo by solifluction and creep, and a central depression in which the ice cores had developed and from which the overburden had moved towards the rim. The rampart is the most significant feature since it enables one to distinguish pingo depressions from those of a simple thermokarst origin (see below). The height of the ramparts will vary considerably, depending upon the size of the initial pingo. If modern pingos and pingo remnants are a guide (see pp. 93–104), the ramparts can vary from as little as 0.5 m to over 5.0 m in height, and the diameter of the depression within the ramparts may be as great as 200–300 m. One group of hollows and ramparts thought to be remnants of a group of pingos in eastern England is illustrated in Fig. 12.11.

As our understanding of modern pingos and pingo-like forms in present permafrost regions increases, the recognition and interpretation of fossil pingos

Fig. 12.11 ***Oblique air photograph of part of Waltham Common, Norfolk, England, showing pingo hollows and ramparts. Photo by permission of the University of Cambridge.***

in mid-latitudes becomes relatively easier. The variety of forms and shapes of pingos in present periglacial environments implies that a similar diversity existed in Pleistocene environments, and many features initially regarded as not 'typical' of pingos can now be reconciled with an ice-cored origin. There still remains the necessity, however, of eliminating alternative, non-periglacial, explanations for mid-latitude features. On the Chalk of southern England and northern France for example, the presence of marl pits (e.g. Prince, 1961) and of Chalk solutional forms (e.g. Sparks ***et al.,*** pp. 333–4) may also produce enclosed depressions similar at first sight to pingo depressions. In glaciated terrain, a kettle origin has to be considered. The possibility that some depressions may be the remnants of palsas has also to be taken into account. However, palsas merely reflect the presence of permafrost bodies beneath localised peaty bodies. Their degeneration does not take place through a collapsing of the summit, as is the case with pingos, but through a caving and slipping process at the base of the palsa. Thus, no ramparts would develop and, in the absence of a large ice core, there is no central depression upon thawing.

Bearing these considerations in mind, fossil pingos and pingo-like features have been identified from both within and outside the glacial limits in western Europe. They include localities in Great Britain (Pissart, 1963; Watson, 1971; Sparks ***et al.,*** 1972), Ireland (Mitchell, 1973), Belgium (Pissart, 1965), Holland (Maarleveld and Van der Toorn, 1955), Germany (Weigand, 1965), and Norway (Svensson, 1969). Surprisingly, few fossil pingos have been recognised in Poland, and other parts of eastern Europe.

The form and distribution of many of the pingo remnants in western Europe suggest that the pingos were of the open system type (see pp. 95–8). For example, the majority of pingo remnants occur in clusters (Fig. 12.12). The ramparts are often irregular in plan and 'mutually interfering', thus giving a complex alignment and distribution. Only a minority of the ramparts are clear-cut circular or oval forms. Many are semicircular and open in an upslope direction while others are elongate in the direction of slope. In terms of morphology, these characteristics are typical of open system pingos in which new generations of pingos are repeatedly born at the site of older ones by the continual movement of water to the surface. In terms of their distribution, many of the pingo remnants occur in typical open system localities, such as on lower valley side slopes, as at Llangurig in Wales (Watson, 1971), or at spring line locations, as at Waltham Common, Norfolk (Fig. 12.11).

The age of various fossil pingos has been investigated by detailed stratigraphic and palynologic studies of the infilling material within the central depressions. These depressions were natural collecting grounds for pollen and organic sediments since they constituted small ponds during the decay stages of the pingo. In Belgium, the material infilling the pingos of the Hautes Fagnes is stratified clay, gyttja mud and peat (Mullenders and Gullentops, 1969). Pollen analyses suggest that the pingos date from the Late Dryas (10 500–11 000 years B.P.). In East Anglia, Sparks ***et al.*** (1972) have concluded that Zone III of the Late Glacial was the most likely phase of cold conditions for the Waltham Common features, while in Wales, the basal peat in one of the Llangurig pingo depressions has been assigned to a Zone III/IV transition (Watson, 1971, p. 391), also suggesting a Zone III age or earlier for the pingos. In general therefore, it would appear that the fossil pingos of western Europe formed in late glacial

Fig. 12.12 *Plan of pingo ramparts of the Brackvenn, Hantes Fagnes, Ardennes, Belgium, as identified by A. Pissart (1965)*

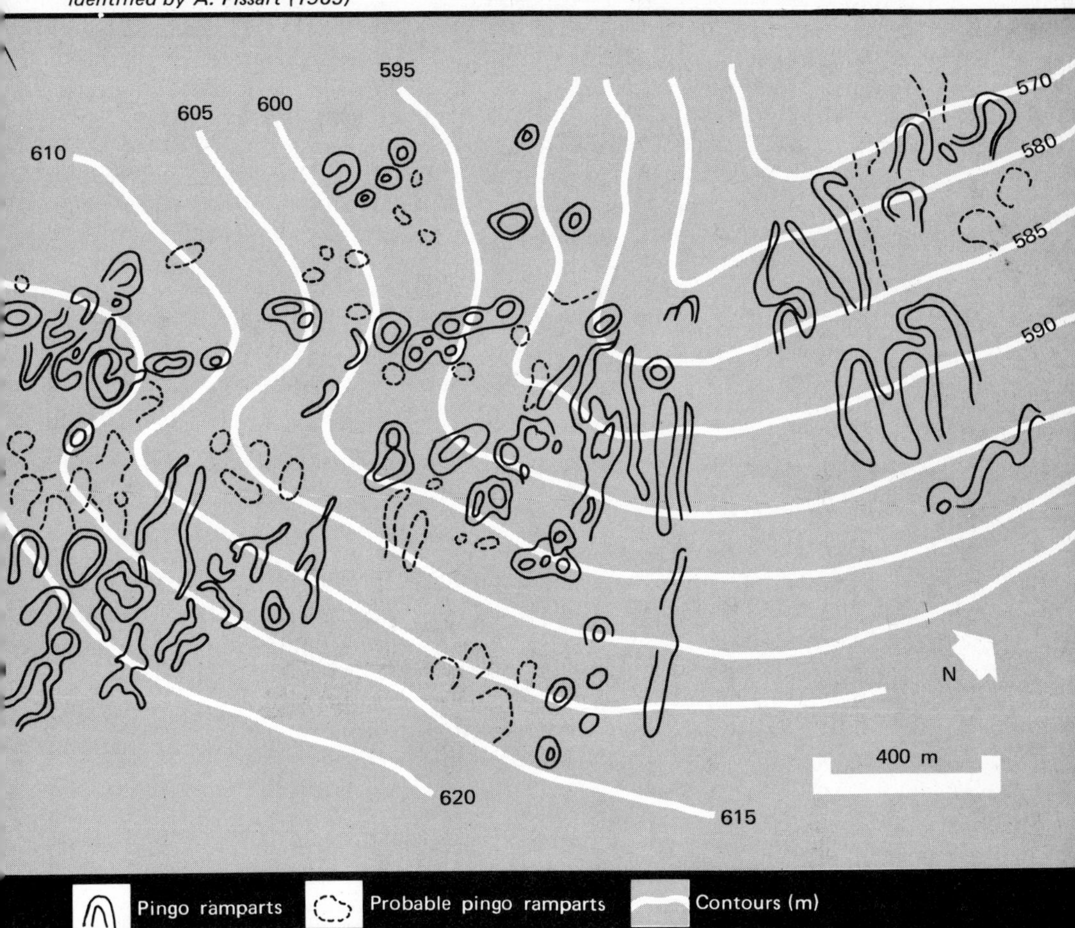

times and represent the most recent period of periglacial conditions. In all probability, these conditions were much less rigorous than those which existed at the maximum of the last glaciation.

The significance of an open system interpretation for the fossil pingos of western Europe lies in the fact that open system pingos require either discontinuous permafrost or continuous permafrost which is thinning out. In present-day periglacial environments, open system pingos are rarely found in areas of thick and continuous permafrost. During the late glacial period (i.e. Zone III) in western Europe, therefore, the evidence from fossil pingos indicates that permafrost was shallow and localised.

There appears to be little regularity to the overall distribution of fossil pingos in western Europe. For example, Watson (1971, p. 384) has commented how, at Llangurig, no pingos are to be found in an adjacent area of exactly similar elevation, aspect and lithology. The apparently random distribution of fossil pingos may be explained in terms of the rather demanding hydrologic conditions of open system pingo growth (see pp. 67–8). The water flowing beneath or within the permafrost must be small in amount and close to 0°C. If the volume is too large it will not freeze, while if the water temperature is considerably in excess of 0°C, it will form a spring, not a pingo. Also, if the water temperature drops below 0°C, the water will be sealed and the pingo will cease to grow. It follows therefore, that conditions suitable for open system pingo growth were limited, both spatially and temporally. They were favoured in late glacial times when the climate ameliorated and permafrost became less widespread and only in those localities where hydrologic conditions were appropriate. The apparent absence of any remnants of closed system pingos from western Europe is puzzling; one possibility is that they may have formed at earlier times under conditions of more extensive permafrost and subsequently been destroyed.

Collapsed pingo features, consisting of the typical circular ramparts and central depression filled with pond and vegetation debris, have not been reported from the temperate areas of North America. The reason for this is not clear. Equally puzzling is the presence of several hundreds of pingo-like mounds found in the non-permafrost regions of western Canada (Bik, 1969) and the DeKalb Mounds of north-central Illinois (Flemal ***et al.,*** 1973). The characteristic of all these mounds is the excess of material that appears to have moved into the mound from the surroundings. Thus, the mound shape has remained after the permafrost has disappeared, unlike a typical ice-cored pingo. In Illinois, where over 500 mounds occur, the majority are circular or elliptical. They rise 1–5 m above the ground surface and are flat topped with slightly depressed centres. The mounds are composed of lacustrine silts and clays surrounded by a sandy rim 30–100 m in diameter. It is believed that the sediments formed within the lakes of pingo craters and, as such, the mounds are the remnants of a large pingo field of mid-Wisconsinan age (22 000–12 500 years B.P.). On the other hand, the prairie mounds of southern Alberta possess few characteristics of fossil pingo remains and until further evidence is forthcoming, it seems best to regard them as a type of ablation till phenomenon (e.g. Stalker, 1960). Both the DeKalb Mounds and the prairie mounds of western Canada (if of a periglacial origin), are of interest since they have no modern analogue, as far as known, to present-day pingos. With respect to permafrost distribution in North America during the late Pleistocene, one must conclude that pingo remnants give little or no information at the moment.

Thermokarst forms Other features which permit an understanding of the former distribution of permafrost include the presence of various thermokarst phenomena. However, their recognition in temperate latitudes is particularly difficult. Thermokarst forms can assume a variety of shapes and sizes, all of which will suffer modification, particularly infilling from the sides, during their development and during the subsequent period of climatic amelioration. Furthermore, few distinct structures or casts are associated with thermokarst subsidence. Finally, as the permafrost surrounding the initial thermokarst form degrades, the feature may cease to exist or, at best, become less apparent in the landscape. For these reasons, fossil thermokarst phenomena are little known or studied, and their positive recognition in present temperate latitude landscapes rare. However, if identified and if proof of their origin is available, they provide certain evidence of the previous existence of permafrost.

Enclosed and shallow depressions which lack surrounding ramparts have frequently been interpreted as fossil thermokarst resulting from the melting of localised ground ice bodies. Present-day analogues are, presumably, the thaw lakes and depressions (see pp. 122–5). In the European literature, such depressions have been referred to as 'mares', 'mardelles' or 'solle' (Cailleux, 1956; Troll, 1962). They occur frequently in the lowlands of northern France, especially in the Beauce and Brie areas where their density may reach 35/km^2 (Cailleux, 1956). Pissart (1958; 1960) has interpreted these depressions as being thermokarstic in origin. A human origin for these depressions is also possible. For example, some of the pits, ponds and other shallow depressions in Norfolk and northern France are certainly marl pits dug by man (e.g. Prince, 1961) in which the underlying Chalk was brought to the surface to be mixed with the overlying Boulder Clay and till deposits. However, old marl pits are generally deeper and smaller in dimensions than many of the enclosed depressions, and the frequency of depressions, often exceeding one per field, argues against a human origin for all of the hollows. A thermokarstic origin has been given to essentially similar features occurring in Denmark (Cailleux, 1957), where a kettle moraine origin is also ruled out. The possibility that the Breckland Meres of eastern England may also be of a thaw lake origin has been raised, but not proven, by Sparks *et al.* (1972, p. 340). In North America, a number of enclosed depressions in New Jersey (Wolfe, 1953) may be of a periglacial nature, while in the Ottawa Valley of eastern Canada, Hamelin (1971) has hypothesised that certain areas of undulating lowland terrain may be fossil thermokarst.

Other forms of thermokarst, such as alas depressions, thermokarst valleys, ground ice slumps, and the various thermo-erosional forms have rarely been identified in the fossil form. In most cases, the evidence is equivocal and a thermokarst origin merely one of several alternatives. For example, the large (100 ha^2) enclosed depression of Leau, approximately 50 km east of Brussels in Belgium, is unexplained. A periglacial origin has recently been proposed by Mullenders and Gullentops (1969). Pollen studies of the sediments in the floor of the depression indicate it formed at the end of the last glaciation during the Older Dryas period. The possibility that the depression may have been an immense pingo or group of pingos is seriously considered, while a second periglacial interpretation is that it developed as an alas depression.

One must conclude that the nature of the thermokarst process makes the resulting topographic forms difficult to identify in the fossil state. This is unfortunate since they are certain indicators of the previous existence of permafrost. Generally speaking, the separation of thermokarst depressions as thermokarst phenomena from the multitude of enclosed depressions of all sizes and shapes of other origins which may exist in the landscape seems almost impossible.

Evidence for Pleistocene wind action

In addition to frost action and the presence of continuous or discontinuous permafrost in many areas, the Pleistocene periglacial environments are believed to have been characterised by particularly intense wind action. Some of the reasons for this have been outlined earlier in the previous chapter (p. 220).

In 1942, A. Cailleux published the results of a study of the wind modified form of over 3 000 sand samples collected from across Europe. Sand grains which have undergone transportation and saltation are rounded and possess a matt or 'frosted' appearance while those which have not are subangular and shiny. Cailleux found that frosted grains occurred widely in a broad belt extending eastwards from European Russia through the lowlands of central Europe into northern France. Moreover, he found that the percentage of frosted grains increased eastwards and reached nearly 100 per cent in eastern Europe in areas to the south of the glacial limits (Fig. 12.13). He advanced the idea, therefore, that periglacial conditions were most severe, and wind action most intense, in these areas. Further west, as the percentage of frosted grains decreased, so did the intensity of wind action.

Since then, the importance of wind action within the Pleistocene periglacial environment has become widely accepted. A number of other features and deposits are now thought to be the result of periglacial wind action. These include the whole range of wind-erosional forms, commonly termed 'ventifacts', windblown or eolian sands, often referred to as 'cover sands', and windblown silts, known as 'loess'. It must immediately be stated, however, that a periglacial

Fig. 12.13 *The distribution and frequency of wind modified sand grains (0.4-1.0 mm) in Europe. From Cailleux (1942)*

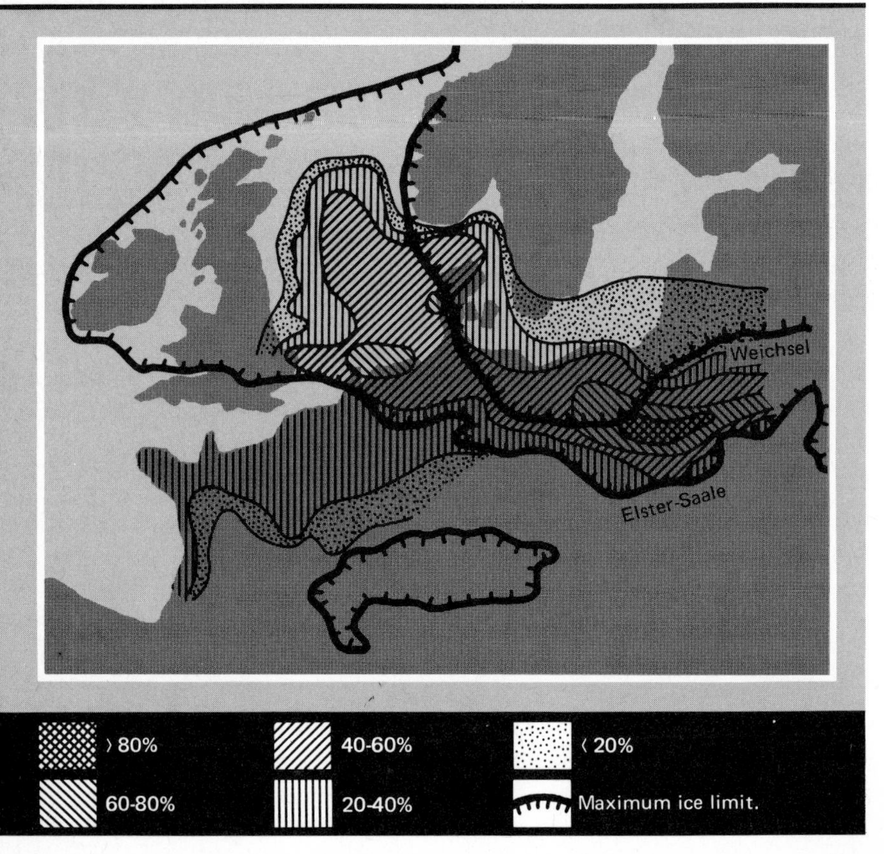

origin for all of these phenomena is not always the case, since these features and deposits can form in other climatic environments besides the periglacial. In all cases, the assignment of a periglacial origin must be made individually, in the light of all available local evidence.

Ventifacts form when a pebble or boulder projects from the ground surface and is exposed to strong wind carrying abrasive material. As a result, the pebble assumes a fluted, faceted or grooved appearance on its exposed side. An absence of vegetation to protect the pebble, an exposed location, a suitable supply of abrasive material, and dry conditions during part of the year are the essential requirements. Pleistocene ventifacts are particularly numerous in the mid-latitude regions of North America and Europe, particularly those areas which experienced continental and dry conditions such as eastern Europe (e.g. Dylik, 1956) and the north-central United States (e.g. Sharpe, 1949). In the more maritime locations, such as western Europe, ventifacts are not so common and reflect, probably, the influence of a more continuous tundra vegetation. In most cases, ventifacts are found in association with aeolian sands and silts. They are often found buried by younger windblown sediments and where ***in situ***, may form a layer of coarser pebbles from which the finer particles have been winnowed out. The present-day analogue is the 'desert pavement' of the polar deserts (see p. 204). Ventifacts provide evidence of Pleistocene wind directions only in special cases. Grooving and fluting of a constant orientation is the best criteria but is rarely found. Moreover, it has to be proven that the stone is *in situ* and has not moved. During the period of periglacial conditions, the pebble may have been moved by the wind itself or by frost heave on the ground surface. Only very large sand blasted boulders are reliable for wind direction inferences.

Aeolian sands are usually in the size range, 0.06–1.00 mm in diameter. They commonly form an irregular formless mantle deposit or 'cover sand' but in some places they also form dunes. Ventifacts may be found with aeolian sands and in certain instances, silts may also be incorporated. Occasionally, cover sands show a certain degree of stratification, involving alternating loamy and sandy layers. These are usually interpreted as being niveo-aeolian in origin, similar to those forming in present-day periglacial environments (see p. 205). In general, cover sands can be related to a nearby source of sediment. Their widespread development in the Netherlands for example (e.g. Maarleveld, 1960), may reflect the proximity of the exposed delta of the Rhine River during glacial times when sea level was lowered and the continental shelf of Europe was greatly increased in extent. Under favourable conditions, wind direction can be inferred from dune morphology and orientation, the dip of the forest beds, and the relation of dunes to known sediment source areas. For example, Maarleveld (1960) has concluded that wind directions in the Netherlands in late glacial times were predominantly north-westerly in the Early Dryas period but changed to south-west and west in the Late Dryas. In North America, Pleistocene dunes occur in the north-western United States, in the central interior, especially Nebraska, and around the shores of the Great Lakes (Smith, 1964). Although they may not all be of periglacial origin, they seem to indicate wind systems essentially similar to those of today.

Aeolian silt, generally referred to as 'loess', is the most widespread product of periglacial action. It is a buff coloured, unstratified and relatively well sorted deposit of dominant grain size between 0.01 and 0.05 mm in diameter. Not all loess is related to periglacial conditions, however, since it occurs widely in and around the margins of warm deserts. In northern China, loess thicknesses of several hundreds of metres are reported to mantle extensive areas. Pleistocene periglacial loess is thought to be derived from the frost comminuted glacially derived outwash deposits laid down at the margins of the retreating ice sheets (e.g. Troll, 1944). During the middle and late summer months, as these outwash plains dried out, strong winds winnowed out the finer silt particles forming dust clouds probably similar to those observed today on recent valley sandar in Arctic regions (see pp. 204–5). The sediments were deposited eventually some distance from the ice margin where tundra vegetation was able to trap and stabilise the silt particles.

The nature of loess provides useful, but general, information as to the climatic conditions at the time of its formation. First, it is generally agreed that loess formed in essentially dry environments. This is indicated by the fact that if wetted, the loess experience shrinkage and compaction, proving that the loess was not soaked by water or deposited by it. Also, unweathered loess is calcareous and calcium carbonate may be segregated into nodules, indicating no leaching of any significance. Second, loess often contains large quantities of faunal remains, especially land snails suited to open conditions of vegetation as are now found in cold steppe/tundra regions. It is generally accepted, therefore, that loess formed in glacial rather than interglacial times.

In both Europe and North America, loess is concentrated in a broad east–west belt to the south of the glacial limits. The European distribution is illustrated in Fig. 12.14. In terms of both extent and thickness, loess is greatest in central and eastern Europe where it mantles extensive areas of both upland

Fig. 12.14 *The generalised distribution of loess in Europe From West, (1968)*

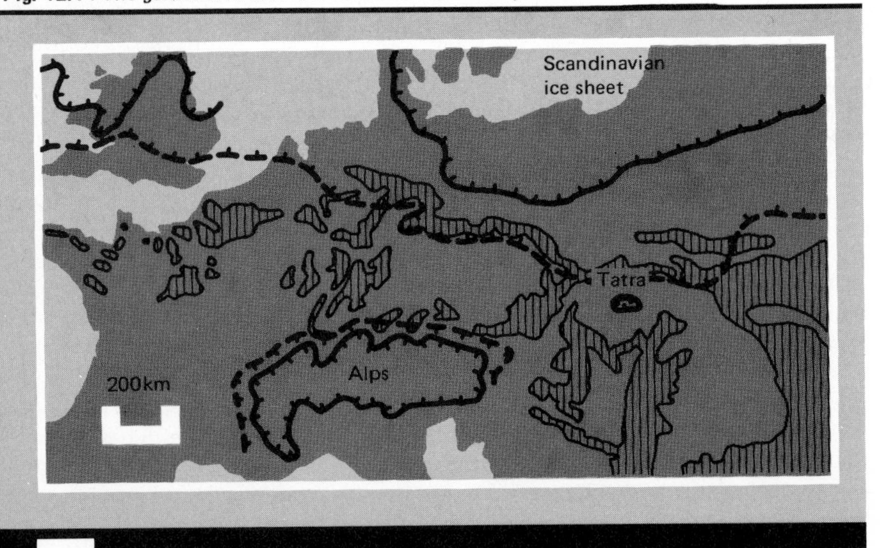

plateaus and lowland plains. In places, it attains a thickness in excess of 30 m. From a stratigraphic point of view, the loess of eastern Europe is of considerable importance. In southern Poland for example, at least five Late Pleistocene fossil soils and several different horizons of periglacial structures are preserved within the loess deposits, and provide evidence of environmental changes throughout much of the last glaciation. Further west, in Germany and northern France, loess deposits are not so thick or extensive. The source area for these deposits was probably the exposed floor of the North Sea and the English Channel during the glacial low water levels. In southern England, true loess is rare, but many of the older drift deposits which mantle the Chalk uplands, such as the Plateau Drift (e.g. Loveday, 1962), and some of the lowland valley infill deposits, such as the brickearths, clearly possess a loess fraction. In all probability, earlier episodes of loess deposition left a veneer of silty sediments on the surface. Subsequently, these were either incorporated into underlying materials through frost action and solifluction, or reworked by meltwaters and then redeposited on valley side slopes and in valley bottoms.

In North America, extensive loess deposits occur in the north-central United States, mainly in the Ohio–Missouri–Mississippi drainage basins (Smith, 1964). They attain thicknesses of over 35 m and extend over a zone several hundreds of kilometres south of the glacial border. There is convincing evidence that the loess was derived mainly from large proglacial streams which, in late summer, exposed broad areas of alluvial sediments on the higher bars between the major channels. In many areas, the loess is thickest near the main valleys and progressively decreases with distance from the valleys. According to Leonard and Frye (1954), the greater thicknesses of loess near to the valleys is the result of a forested belt which existed on the moister and more sheltered locations adjacent to the rivers, while on the uplands, a grassland vegetation with only occasional trees developed. Thus, the forested belt reduced wind velocities and trapped windblown sediments. In some areas, the source for the loess comprised till as well as stream outwash sediments. In Iowa for example, the mineral content of the loess resembles that of the corresponding size fraction of the till in the same region. Various episodes of loess deposition are recognised in the central United States. Loess sheets of Kansan, Illinoian and Wisconsinan age are now correlated to the advance and retreat of the continental ice sheets (e.g. Frye, Willman and Glass, 1968). In general terms, there is also evidence that the most recent loess deposition in central North America occurred under slightly moister conditions than in eastern Europe. For example, the Peoria loess of Iowa, of Wisconsinan age, contains faunal remains which indicate moister conditions than today. A forest–grassland vegetation cover is to be inferred for central North America, as opposed to a predominantly tundra vegetation cover for Europe during the last glacial stage.

Asymmetrical valleys

The presence of asymmetrical valleys in mid-latitudes has frequently attracted attention. In many cases, the asymmetry is thought significant in terms of either past frost action, permafrost, or wind action, or various combinations of these conditions. Before a periglacial interpretation can be accepted however, there is the necessity to consider a variety of non-periglacial causes of asymmetry. For example, asymmetry can result from structural controls, from

Table 12.1 *Some examples of slope asymmetry attributed to Pleistocene periglacial conditions in Europe*

Location	Reference	Orientation of steeper slope	*Processes involved*[1] (1) Differential insolation and freeze–thaw
United Kingdom			
Chiltern Hills	Ollier and Thomasson, 1957	W/SW	*X*
Hertfordshire	Thomasson, 1961	W	*X*
Chalk, southern England	French, 1972, 1973	W/SW	*X*
France			
Gascony	Taillefer, 1944	W	
Gascony	Faucher, 1931	W	
North France	Gloriad and Tricart, 1952	W	x
Netherlands			
Haspengouw	Geukens, 1947	W	*X*
Veluwe	Edelman and Maarleveld, 1949	W/SW	
Belgium			
Hesbaye	Grimberbieux, 1955	W	*X*
Ardennes	Alexandre, 1958	W	
Germany			
Erzegebirge	Losche, 1931	W	x
South Germany	Budel, 1944, 1953	W/SW	
Muschelkalk	Helbig, 1965	W	
Czechoslovakia			
Bohemia	Czudek, 1964		
Poland			
Lódź plateau	Klatkowa, 1965	W	*X*

[1] *X* – dominant process; x – secondary process.

differences in local slope erosional environments, or from different micro-climatic conditions existing under present temperate conditions (e.g. Hack and Goodlett, 1960).

In general terms, asymmetrical valleys pose a number of problems of slope and valley evolution. First, asymmetry can develop through a number of different mechanisms. For example, the decline of the gentler slope, the steepening of the steeper slope, or a combination of both mechanisms will all produce an asymmetrical valley form. One must also consider the fact that the asymmetry can develop in-phase with valley downcutting or subsequent to the formation of

Table 12.1 – *continued*

Processes involved[1]				*Mechanism involved*	
(2) Differential solifluction	**(3) Wind and snow**	**(4) Wind and loess**	**(5) Lateral stream erosion**	**Decline of N and E facing slope**	**Steepening of S and W facing slope**
x			x		x
x			x		x
x	x		x		x
x	X		x	x	x
		X	x		x
X			x		x
x	x		x	x	x
x	X		x	x	x
	x		x		x
X			x	x	x
			X		x
x	x	X	x	x	x
X			x		x
X				x	
x	x		x	x	x

the valley. Second, valley asymmetry involves the detailed consideration of the interaction between micro-climate and geomorphic processes, and between basal stream activity and slope evolution, not to mention structural and lithological considerations. Since the steeper slope can face, theoretically, in all directions, and since the slope may evolve in the different ways outlined, there are numerous possible interpretations of asymmetrical valleys.

Asymmetrical valleys attributed to a periglacial origin occur widely throughout Europe (Table 12.1) and in certain parts of North America (e.g. Smith, 1949b). In western Europe, the steeper slope commonly faces towards the west or southwest, and this is regarded as the 'normal' for western Europe. In North America, the steeper slopes frequently face towards the north. Since there are few detailed studies of periglacial slope asymmetry in mid-latitude North America, and since the asymmetry is sometimes interpreted in terms of present climatic conditions (e.g. Hack and Goodlett, 1960), the following discussion concentrates upon western Europe, where there is broad agreement.

The various interpretations of the 'normal' periglacial asymmetry of western Europe are summarised in Table 12.1. The most popular explanation involves differential insolation and freeze–thaw processes operating on exposed south- and west-facing slopes while the 'colder' north- and northeast-facing slopes remained frozen and geomorphologically inert (e.g. Ollier and Thomasson, 1957). A second group of explanations involve the differential deposition of either snow or loess on leeward slopes, resulting in greater solifluction on those slopes (e.g. Taillefer, 1944; Budel, 1944). This reduces the slope in angle while the greater amount of debris arriving at the basal channel from that slope causes the stream to migrate to the far slope, which it undercuts and steepens. A third interpretation involves a combination of differential insolation and freeze–thaw on exposed south- and west-facing slopes, and snow- or loess-induced solifluction on the leeward slopes together with a migrating stream in the valley bottom (e.g. Geukens, 1947; Edelman and Maarleveld, 1949).

The 'normal' asymmetry of western Europe can be compared to the asymmetry currently developing in present periglacial environments (see Table 8.3 and p. 179). The most striking difference is that in high latitudes the steeper slope, irrespective of orientation, is the colder slope (e.g. Currey, 1964; French, 1971a) while in western Europe, the steeper slope is the warmer (e.g. Geukens, 1947; Ollier and Thomasson, 1957). Presumably, this relates to the fact that in mid-latitudes freeze–thaw processes were dominant on the 'warmer' slopes, leading to their steepening. In high latitudes by contrast, the 'warmer' slopes usually give rise to thicker active layers and greater solifluction movement (see p. 178) rather than greater frequencies of freezing and thawing. This is because of the dominant seasonal, as opposed to diurnal, rhythm of climate.

Our understanding of present periglacial environments enables two further points to be made which are of relevance to the asymmetrical valleys of Europe. First, since greater solifluction occurs below the site of large and permanent snowbanks in high latitudes today (see pp. 135–40), the importance attached to wind and snow by Taillefer (1944) and Edelman and Maarleveld (1949) appears justifiable. Second, in view of the importance of fluvial activity in high latitudes (see pp. 167–76), the role given to lateral stream erosion by several investigators, notably Losche (1931), seems also justified.

In summary, it must be emphasised that the presence of asymmetrical valleys in temperate latitudes is not a reliable indicator of past periglacial conditions since there are a number of non-periglacial causes of asymmetry. If of a periglacial origin however, asymmetrical valleys provide circumstantial evidence of the importance of differential insolation and freeze–thaw activity, and/or differential snow distributions and solifluction activity, and fluvial action. Asymmetrical valleys provide little or no direct evidence for the previous existence of permafrost.

13 Periglacial landscape modification

Since periglacial conditions existed in mid-latitudes at several times during the Late Pleistocene, one must ask how much of the present landscape is in disequilibrium with the prevailing temperate conditions. Unfortunately, there is no easy answer to this question. Not only did periglacial conditions vary in intensity and duration in different areas, but different rock types would have reacted differently. Furthermore, the probability of preservation of relic features is highest in areas of gentle slope and lowland relief since, in areas of steeper slopes and greater relative relief, present-day processes would have been better able to modify the landscape and destroy the periglacial legacy.

In an attempt to give some perspective to this problem, this final chapter briefly describes the nature and extent of periglacial landscape modification in two areas of Europe, the Chalklands of southern England, and the lowlands of central Poland.

The Chalk landscapes of southern England

Chalk and Chalky Drift materials comprise a large area of southern and eastern England, ranging in topography from upland plateaus and escarpments rising to over 200 m in elevation through to undulating lowland plains. The extent of the Chalk outcrops and Chalky Drift materials and the limits of the last glaciation are indicated on Fig. 13.1.

For a number of reasons, the Chalk terrain is well suited for the preservation of relic periglacial phenomena. First, Chalk and Chalky Drift are some of the most frost-susceptible materials exposed in southern England. Second, the desiccation of the Chalk plateaus has meant that present-day fluvial modification has been limited, and the probability of the preservation of relic forms greatly increased over other lithologies. Third, nearly all of the Chalk and Chalky Drift outcrops lay to the south of the maximum limit of the Weichselian/Devensian ice sheet. They were exposed, therefore, to periglacial climatic conditions for the duration of the last cold stage.

Periglacial deposits

Much of the Chalk is covered with a veneer of periglacial deposits which, on faunal and stratigraphic grounds, can be assigned to the middle and late stages of the last, Weichselian, glaciation (e.g. Kerney, 1963). In most cases, these deposits appear to be derived, through frost shattering, from the Chalk. They have been laid down by subaerial or aeolian transport. At least three broad types of deposits can be recognised. First, solifluction debris or 'coombe rock' consists of a heterogeneous material composed of coarse angular nodules of Chalk held within a fine calcareous matrix. It is widespread on lower valley side slopes and in valley bottoms where it may attain thicknesses of several metres. A second type of deposit consists of sorted or stratified muds and solifluction (Taele) gravels. In that they are stratified they resemble fluvial deposits but in some, the presence of a terrestrial fauna of land mollusca indicates a subaerial origin. It is thought that the muds

Fig. 13.1 ***The extent of Chalk Drift materials in south and east England, and the distribution of wedge casts, involutions and patterned ground features south of the Weichselian glacial limit. Modified from Williams (1969)***

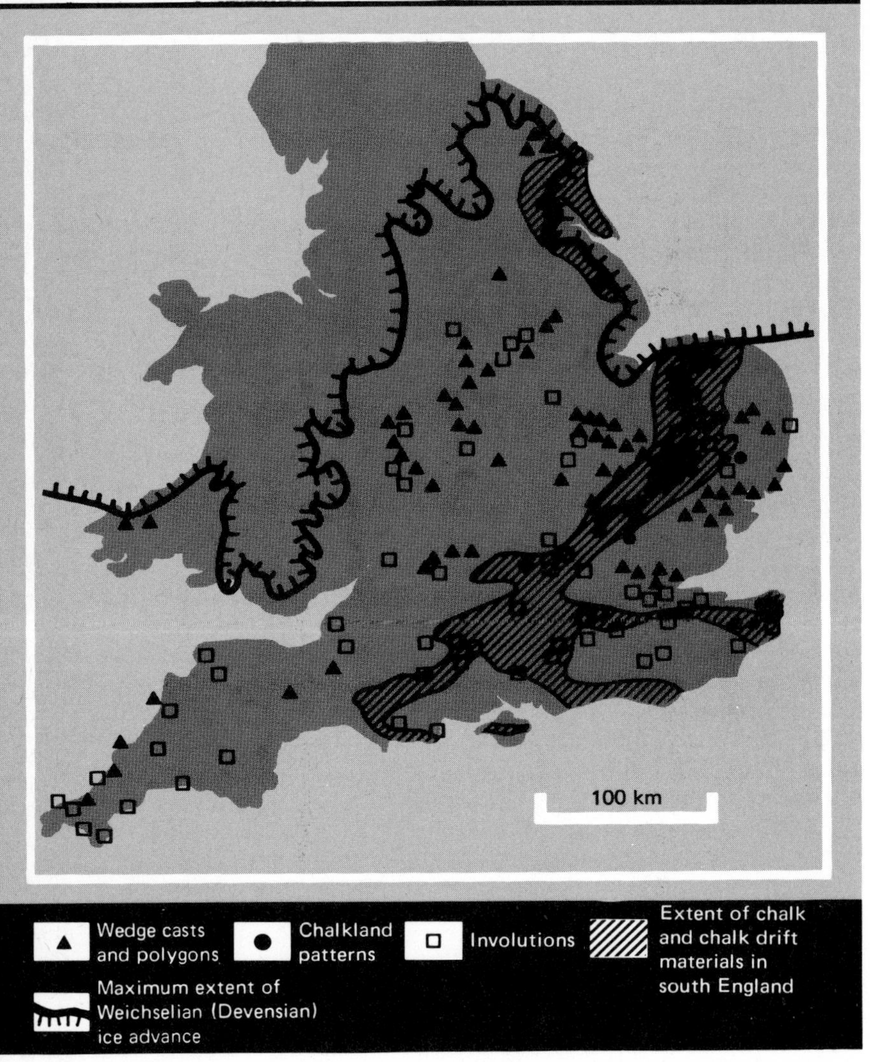

resulted from the frost shattering of the Chalk in association with the release of meltwater from beneath snowpatches and the melting of frozen ground each spring. The Taele gravels resulted from the solifluction of earlier till deposits, notably of the penultimate ice advance, during the beginning of the latest cold stage. The third category of periglacial deposits are those of a loessic nature which veneer exposed upland surfaces and testify to drier conditions than either of the previous two types of deposits. Where present in valleys, the loess has usually been reworked by either solifluction or meltwaters and is termed 'brick-earth'. For convenience flinty loams are included in this category. Although not derived from the Chalk, they often occur on valley side slopes. Their parent materials are various Tertiary sediments, notably clay-with-flints, which have undergone movement through solifluction while loessic materials have been incorporated through frost action.

Table 13.1 *Generalised sequence of periglacial conditions and associated deposits on chalklands of southern England during the Middle and Late Weichselian. Modified from Evans (1968)*

Stage	Zone	Period	Periglacial conditions and deposits	Date (^{14}C years)
Post Glacial	IV	Pre-Boreal	Amelioration of climate, soil formation. Birch and then hazel	10 300
Late Weichselian	III	Younger Dryas	Rapid physical weathering; cutting of coombes; snow meltwater deposits; involutions and loess. Park tundra (Dwarf birch)	10 800
	II	Allerod	Soil formation. Birch	12 000
	Ic	Older Dryas	Physical weathering; chalk detritus; snow meltwater deposits. Tundra	12 300
	Ib	Bolling	Weak soil formation	
	Ia	Earliest Dryas	Physical weathering; chalk detritus; snow meltwater deposits	14 000
Middle Weichselian			Dry cold climate with some solifluction; loessic brickearths; non-calcareous flinty loams; involutions	
			Cold and wet; intense physical weathering; coombe rock formation; fluvial activity and asymmetrical valley modification; permafrost in eastern England. Involutions and Taele gravels	*c.* 50 000
Last Interglacial				>100 000

The dating of these deposits, and their climatic implications, are summarised in Table 13.1. The solifluction and coombe deposits are thought to relate to cold and damp conditions during the Middle Weichselian or earlier. The formation of the Taele gravels of East Anglia probably date from the end of the Saale glaciation or the beginning of the Weichselian. The brickearths, flinty loams, and loessic deposits relate to a late Middle Weichselian period of relatively drier conditions but which were still sufficiently moist for their deposition in valley bottoms and on slopes by solifluction and meltwaters. The Chalk meltwater muds are primarily of a Late Weichselian age and reflect the higher temperatures that occurred at that time.

These deposits can be related to a number of landform modifications which took place during the last glacial stage.

Asymmetrical valley development Many of the dry valleys which dissect the Chalk plateaus and dipslopes of southern England are asymmetrical in cross profile (Fig. 13.2), with the steeper slope commonly facing west or southwest (e.g. Ollier and Thomasson, 1957). In addition to their slope asymmetry, these valleys also possess an asymmetrical pattern of soils, deposits and smaller tributaries. Similar valleys are known to occur on the Chalk plateaus of northern France (e.g. Gloriad and Tricart, 1952), Belgium (e.g. Grimberbieux, 1955), and northern Germany (e.g. Helbig, 1965).

The most likely explanation of the asymmetry of the Chalk valleys is that it developed during the various periglacial periods, when the Chalk was able to support a transitory surface drainage on account of a frozen subsoil. This is not

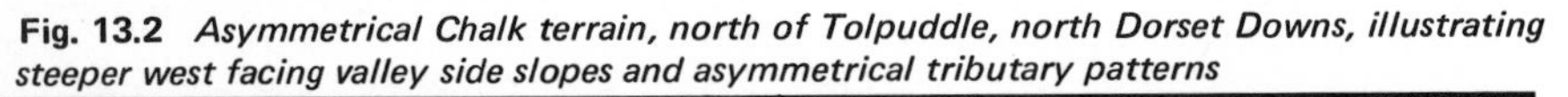

Fig. 13.2 *Asymmetrical Chalk terrain, north of Tolpuddle, north Dorset Downs, illustrating steeper west facing valley side slopes and asymmetrical tributary patterns*

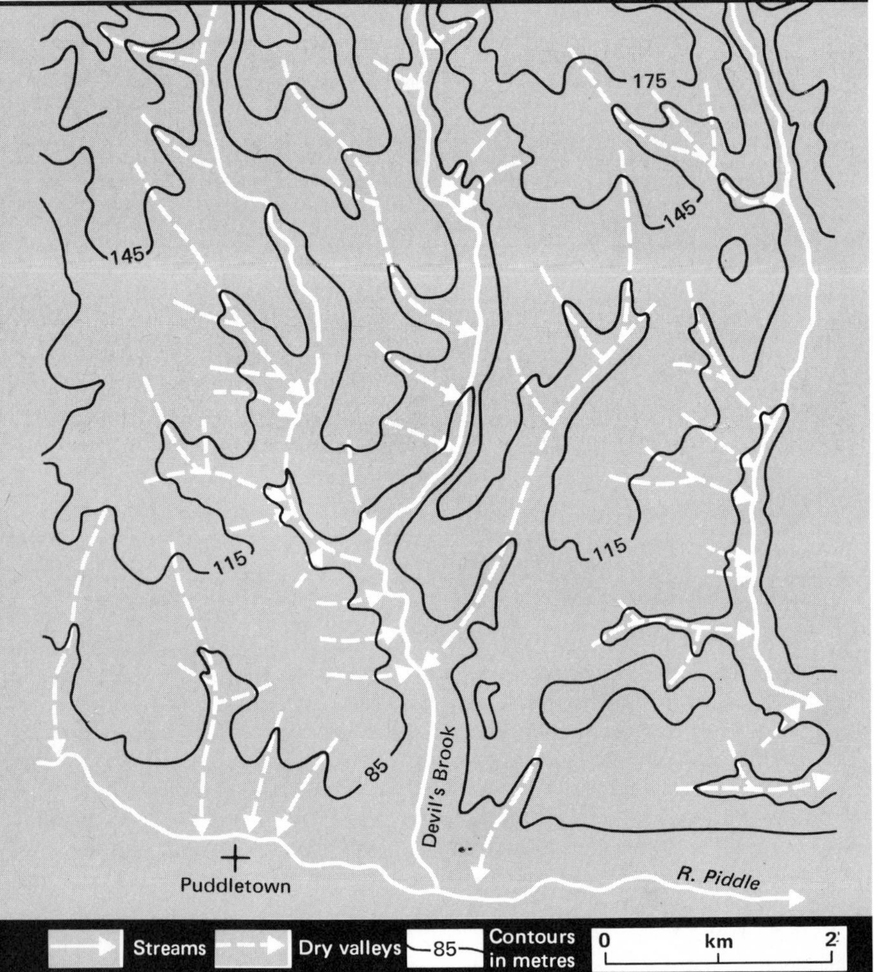

to admit that all Chalk valleys were formed in a periglacial fashion, by higher surface run-off and meltwater at a time when the Chalk was frozen and impermeable, as suggested by Reid (1887) and Bull (1940). Many are undoubtedly polycyclic since their long profiles show evidence of successive rejuvenations, and it is inconceivable that the Chalk possessed no valleys before the onset of periglacial conditions. In all probability, the major Chalk valleys developed through normal stream action at a time of higher water tables. Then, as the water table gradually fell throughout the Pleistocene, the smaller valleys became successively drier from their upper parts downwards. If this were so, the down-valley sections may have been active while the upvalley sections were dry and fossilised. During periglacial periods, therefore, this normal sequence of progressive desiccation was reversed, as the upper sections were reactivated by increased surface run-off over the impermeable (i.e. frozen) substratum.

If this general assessment of Chalk morphology is correct, the asymmetrical development of the valleys provides illustration of the nature of periglacial landscape modification. In essence, one must envisage three conditions; (1) perennially frozen ground on colder, east-facing slopes, (2) intense physical weathering and erosion through repeated freeze–thaw activity on the west-facing slopes as the result of greater insolation, and (3) running water in the valley bottoms for at least part of the year.

Fig. 13.3 ***Relationship between slopes and surficial materials in the Devil's Brook Valley, 2 km north of Dewlish, north Dorset Downs. From French (1973)***

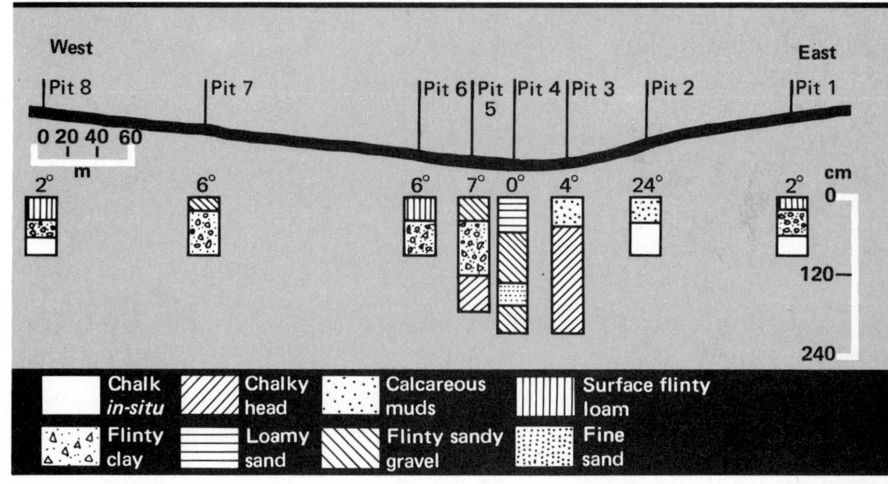

The relationship between slopes, soils and deposits in asymmetrical Chalk valleys has been investigated by Ollier and Thomasson (1957) in the Chiltern Hills and is illustrated here with reference to the Devil's Brook, in north Dorset (Fig. 13.3). On the upper part of the steeper slope, Chalk *in situ* is exposed. Lower down, the slope is mantled with a veneer of coarse chalky rubbles and coombe rock which grades into unweathered Chalk at depths of up to 3 m at the foot of the slope. In the flat valley bottom, at least 3 m of well sorted fluviatile sand, silt, and flinty gravels ('nailbourne deposits') are present, and testify to considerable discharge in the past. On the east-facing slope, the surficial deposits are more continuous than on the west-facing slope. Non-calcareous flinty loams and clays, derived from clay-with-flints and Plateau Drift deposits, mantle the lower valley side slope. Towards the bottom of the gentler slope, these materials overly chalky rubbles. The nature of the deposits and their stratigraphic positions within the valley make it clear that the calcareous materials have been derived from the erosion of the steeper, west-facing slope. Furthermore, their presence on the lower part of the gentler slope suggests a degree of deepening and lateral migration of the basal channel at the same time as the steeper slope was eroded. Finally, the flinty loams and clays on the gentle slope indicate a form of non-turbulent creep or solifluction, since there is no mixing with the underlying calcareous materials.

In detail, the asymmetry probably developed through two stages (Fig. 13.4). In the first stage, downcutting and lateral migration of the stream coincided with the steepening, through frost shattering, of the west-facing slope. The lateral movement of the stream reflected the increased erodibility of the west-facing slope. There is no evidence, in the form of extensive solifluction deposits at the foot of the gentler slope, that the stream was 'pushed' to the far slope; instead, the mere presence of different active layer thicknesses on the two slopes would have been sufficient to favour the lateral movement of the stream since one bank would have been more easily eroded and undercut than the other. Fluvio-thermal erosion may also have been effective. In the second stage, lateral migration of the stream became less important as the amount of frost shattered debris arriving at the foot of the slope would have been sufficient to prevent further undercutting. At this stage, the position of the stream would have become fully adjusted to the debris arriving from the two slopes. Probably, this stage coincided with the steepening of the west-facing slope to angles of 19–22° and the extension of the east-facing slope at angles of between 5 and 9° (French, 1972). Subsequently, the steeper slope evolved through parallel retreat, since gently concave cryopediments, or basal slopes of transportation (French, 1973), exist at the foot of many of the steeper west-facing slopes.

Thus, as in the case of the asymmetrical valleys of northwest Banks Island (see pp. 178–80), the asymmetry of the Chalk valleys reflected an equilibrium form in which the two slopes were continually adjusting to each other and to the stream channel, in overall response to different geomorphic and micro-climatic conditions on the two slopes.

No absolute age for the asymmetrical modification of the Chalk valleys can yet be given. However, the asymmetry appears intimately related to the chalky rubbles and coombe rock which mantles the foot of the steeper slope. A Middle Weichselian age is the most likely (Table 13.1), although one cannot rule out the possibility that the asymmetry and the cryopediments reflect more than one period of periglacial conditions.

The 'rock-streams' of Wiltshire and Dorset There are a number of more specific features of the Chalk landscape which indicate periglacial modification during Middle Weichselian or earlier times. The most striking are the periglacial 'rock-streams' or 'sarsen streams' which occur within certain of the Chalk valleys of Wiltshire and Dorset (e.g. Williams, 1968; Small ***et al.,*** 1970). The sarsen stones themselves are large blocks of silicified sand and flint conglomerate, usually regarded as being of early Tertiary age. Presumably, they originated on the

Fig. 13.4 *Probable mechanism of development of the asymmetry of the Chalk valleys. From French (1972)*

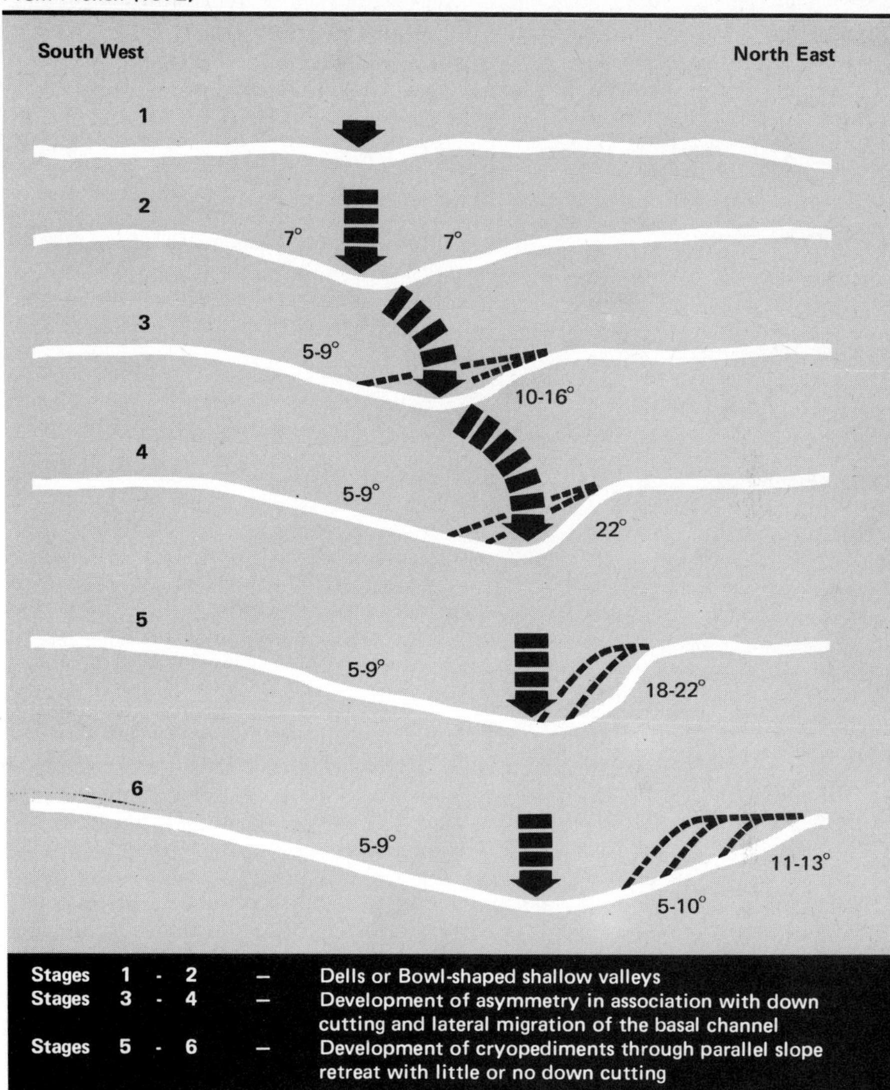

upland surfaces of the Chalk as a form of 'duricrust' or silcrete. Being extremely resistant, they have managed to survive the Pleistocene epoch.

Many of the sarsens have moved considerable distances from their initial positions on summits and interfluves, first downslope and then downvalley in the form of 'stone streams' (Fig. 13.5). In the Marlborough area for example, some may have moved as much as 4 000 m on an average gradient of between 1 and 3°. An explanation involving their movement by early man into the valley bottom is not now generally accepted. The more likely explanation is that the stones, often weighing several tons, were rafted downslope by solifluction as a

Fig. 13.5 *The sarsen 'rock-stream' in the Valley of Stones, near Blackdown, in southern Dorset.*

form of 'ploughing block' (see p. 139). Typically, the sarsens are concentrated near the surface in a non-calcareous flinty loam, probably derived from Eocene remaniés and clay-with-flints which caps the higher elevations. Underlying are deposits of coombe rock and chalky rubbles, occasionally containing small sarsens, which grade into Chalk ***in situ*** at depth. At Clatford Bottom, Wiltshire, Small ***et al.*** (1970) report solifluction debris in excess of 3 m in the valley bottom. The valleys in which the sarsens have accumulated are often shallow, reflecting this valley infill.

Frequently, the occurrence of sarsen accumulations in valley bottoms is associated with a striking asymmetry of the valley which is confined to the point where the sarsens entered the valley. Often, the asymmetry disappears before the downvalley termination of the stone stream. The asymmetry is not of the simple climatic asymmetry, as described above, since the steeper slope faces in various directions. It is, moreover, confined to the lower part of the slope as a form of undercutting. The most likely explanation is that the movement of the sarsens into the valley bottom pushed an ephemeral and small stream across the valley floor to undercut and steepen the opposite slope. A meltwater stream flowing over still frozen ground during the spring thaw would not have been deeply incised. Such a stream could easily be 'pushed' laterally, while solifluction later in the summer would obliterate its bed without destroying the asymmetry (Williams, 1968). Each year, the stream would locate itself at the point of lowest elevation at the foot of the undercut section. One might envisage, therefore, a

Fig. 13.6 *Possible development of Clatford Bottom, Wiltshire, according to Small, Clark and Lewin (1970)*

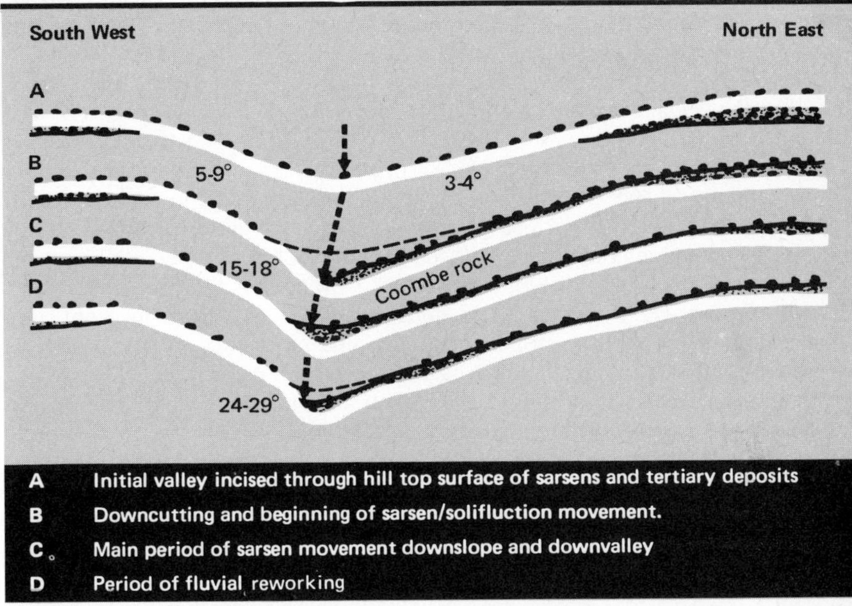

A	**Initial valley incised through hill top surface of sarsens and tertiary deposits**
B	**Downcutting and beginning of sarsen/solifluction movement.**
C	**Main period of sarsen movement downslope and downvalley**
D	**Period of fluvial reworking**

moving tongue of solifluction debris with rafted sarsen stones seasonally alternating with a small stream at the foot of the undercut section.

The evolution and age of the 'rock-streams' has been considered by both Williams (1968) and Small *et al.* (1970). At Clatford Bottom, where detailed excavations have been undertaken, no late glacial sediments were found and the flinty loam and coombe rock deposits are assigned to an early or main Weichselian period. Small *et al.* (1970) prefer to regard the flinty loam and underlying coombe rock deposits as representing one deposit, the upper part of which has been partially removed by fluvial action, and further reduced in volume by decalcification. One episode of solifluction and sarsen movement is envisaged (Fig. 13.6). An alternative interpretation is that the deposits represent two successive events, and that the sarsens, incorporated within the surface flinty loam, arrived relatively late in the valley bottom. In this case, an early stage of intense physical weathering resulted in the deposition of the coombe rock in the valley bottom. This was then followed by the influx of non-calcareous flinty loam laden with sarsens. In either case, the 'rock-streams' of Wiltshire and Dorset provide convincing proof of the operation of solifluction processes on the Chalk, probably during the last cold stage. In terms of landscape modifications, they indicate denudation from upland surfaces and the transportation and deposition of debris in valley bottoms.

Patterned ground features of the East Anglian Chalk Equally striking is the abundance of large polygons and stripes which occur widely on the very gentle terrain of the East Anglian Chalk Drift. They are sometimes referred to as 'Breckland' polygons and stripes (e.g. Williams, 1964). Their origin is not known with certainty; they are attributed a periglacial origin simply because of the lack of an adequate alternative explanation and because, in East Anglia, there is convincing evidence of other cold climate phenomena such as involutions, ice-wedge casts and pingo remains.

In terms of the polygons and stripes described from high latitudes (see p. 187), the East Anglian features are large, the average distance between polygon centres being approximately 10 m and that between the stripes being approximately 7.5 m (Williams, 1964). Typically, the polygons form on flat terrain while the stripes develop on very gentle slopes of 1–3° in angle. A significant fact is that when the stripes reach level terrain, the pattern either disappear or turns to polygons once again. This suggests that solifluction was unimportant in their formation since solifluction does not stop as the slope flattens out but continues at extremely low angles of slope (see Fig. 7.2). Also significant is the departure of some of the stripes from the line of maximum slope, and the meandering, bifurcation and joining by cross partition of others. This suggests a form of surface wash was important in their formation. In cross section, the polygons and stripes form broad troughs infilled with sandy materials overlying the Chalk or Chalky Drift. As the stripes disappear downslope, thin waste mantles of partly stratified but still largely unsorted sands and sandy gravels veneer the lower slopes. These deposits sometimes grade into extensive unsorted and stratified gravels which form low terraces 1–2 m above the level of rivers such as the Cam and the Little Ouse near Thetford (Williams, 1968).

The East Anglian patterns are particularly puzzling since no exact equivalents are known from present Arctic regions. In the lowlands of the western Arctic, the most similar features known to the writer occur on the undulating morainic terrain of eastern Banks Island. The stripes and patterns are similar but much smaller, usually less than 2–3 m in width (Fig. 9.5) and slope angles are a little higher, between 3 and 5°. The Banks Island features undoubtedly reflect rillwash or linear zones of concentrated surface and subsurface wash with lichen and moss identifying the lines of greater moisture. In general therefore, there is support for the interpretation of the East Anglian patterns as reflecting a stage of relative landscape stability, or 'maturity', in which sheetwash processes assumed dominance on slopes of extremely low angle (e.g. Williams, 1968). Probably, the patterns developed wherever the slope was reduced to such low angles that solifluction and erosion practically ceased. The very small amounts of chalky debris incorporated into the periglacial slope deposits of East Anglia suggests that surface wash, and not solifluction, was the main transporting process. The poorly stratified and weakly sorted sands which infill the patterns and which mantle the lower slopes of many areas also indicate wash, rather than solifluction, processes. It follows that the glacial legacy in East Anglia, which gave a subdued relief to much of the Chalk north of the Chiltern Hills, was very important with respect to subsequent periglacial modification. During the last cold stage, the low and undulating terrain was quickly reduced to slopes of extremely low angle. Despite the intensity of cold which eastern England experienced, drastic landscape modification did not take place. Instead, rillwash, sheetwash and probably thermo-erosional wash assumed dominance on the very low angled slopes and merely led to a further flattening and levelling of the landscape.

Periglacial valleys and dells In spite of the fact that the majority of dipslope valleys probably existed prior to their asymmetrical modification, there are valleys, both on the dip and scarp slopes of the Chalk outcrops, which are true periglacial valleys in the sense that they originated under periglacial conditions. Typically, these valleys are small, being less than 1–2 km in length, and are closely spaced with no apparent integration into the larger valley networks.

The most striking of these valleys are deeply incised escarpment valleys or 'coombes'. They possess steep, almost rectilinear, slopes, commonly 30–35° in angle, and abrupt headwall. While the angular course of some of these valleys, such as Rake Bottom, Hampshire, probably reflect spring sapping and the headward extension of the valleys along major joint lines in the Chalk (e.g. Small, 1965), in others, there is evidence that their excavation was associated with intense physical weathering in combination with meltwater movement over seasonally frozen ground. These valleys of a periglacial origin do not appear to be regularly distributed along the Chalk escarpments but occur in groups or as isolated features. Usually, they possess relatively straight courses, and are unrelated to the joint patterns of the Chalk. One group of these periglacial escarpment valleys occurs on the south-facing escarpment of the North Downs near Brook, Kent (e.g. Kerney, Brown and Chandler, 1964) and a second group occurs on the Chiltern escarpment near Ivinghoe, in Buckinghamshire (Brown, 1969). At Brook, the largest coombe, the Devil's Kneadingtrough, possesses a flat floor and an elongated triangular ground plan with a steep but smooth headwall. The valley infill, which forms the flat bottom, is composed primarily of calcareous chalk rubbles and muds which, on faunal and stratigraphic grounds, are assigned to a Late Weichselian (Zone III) age. These deposits extend, in the form of a large apron, over the lowlands at the foot of the escarpment. Since the muds are well sorted and often stratified, and since they have been transported over slopes of very low angle at the foot of the escarpment, they must have been associated with considerable meltwater activity. It is thought that the majority of the erosion of the coombes at Brook took place during a relatively short period of time, approximately 500 years, when the climatic deterioration of Zone III resulted in a specific combination of humidity and repeated freeze–thaw action. Probably, groundwater seepage, as indicated by springs today, was the factor which localised the frost shattering. The Chalk debris would have been moved into the widening and deepening coombe partly by solifluction and partly by water released from melting snowbanks on the escarpment summit and flowing over seasonally frozen ground. Transportation out of the coombe was accomplished by surface meltwaters and also by flow from the scarp foot springs.

It is highly unlikely that all the escarpment dry valleys of the southern England Chalk are the result of intense physical weathering during late Weichselian times. Aspect and local hydrological conditions were probably extremely important in determining whether such erosion took place or not. Some have certainly evolved through spring sapping in post glacial times. It seems reasonable to assume that in the marginally periglacial environments of the Late Weichselian, intense physical weathering was only favoured on south- and west-facing slopes or in localities of abundant moisture.

Periglacial valleys also exist on the dipslopes of the Chalk, but are of a slightly different nature. They occur on gentle valley side slopes, particularly the gentler slopes of asymmetrical valleys, as shallow, paddle-shaped or bowl-like depressions (French, 1972). They are analogous to the 'dells' (vallons en berceau or niecki denudayjne) reported in the European literature (e.g. Klatkowa, 1965). On contour maps, the dells give the crenulated appearance to many of the contours. In England, no detailed studies of such features on the Chalk have yet been undertaken. In all probability, however, the depressions developed through concentrated wash processes beneath and below snowpatches when the subsoil was either permanently or seasonally frozen. If wind directions during the colder periods were not substantially different to those of today, and westerly winds dominated, this would account for the preferential development of dells on the east- and northeast-facing slopes, since these would have been lee slopes. Typically, the Chalk dells are extremely shallow, with slopes of 1–3°, and broad, often 100–200 m in diameter. In places, their density is surprisingly high, often being less than 300–400 m apart. In size, frequency of occurrence and distribution, they appear unrelated to the larger dry valley network and a periglacial origin is the most likely explanation.

The periglacial legacy in southern England

There is no doubt that much of the Chalk landscapes of southern and eastern England reflect the previous existence of periglacial climatic conditions. This thesis, first developed in a systematic manner for southern England nearly 20 years ago by Te Punga (1957), is now gaining general acceptance for the Chalk landscapes. It is debatable, however, whether a similar interpretation is valid for adjacent non-Chalk terrain in southern England. The Chalk, by virtue of its desiccation, has been fossilised and protected from the post glacial fluvial modification which has occurred in other areas. On the other hand, it cannot be denied that if periglacial conditions affected the Chalk, they must also have affected adjacent areas.

The evidence for periglacial activity on non-Chalk lithologies is less clear. Here, only some of the more significant facts are briefly mentioned. For example, asymmetrical valleys essentially similar to those of the Chalk terrain occur in southeast Hertfordshire and north Middlesex on drift materials overlying London Clay (Thomasson, 1961). North and west of the Chalk outcrops, there is abundant evidence for permafrost conditions during the last cold stage, as indicated by the presence of ice- or sand-wedge casts in the lowlands of the English Midlands (e.g. Shotton, 1960; Morgan, 1971). If such conditions existed here, they must also have existed on the Chalk uplands. The absence of wedge casts reported from the Chalk is not surprising, however, in view of the permeable and fissured nature of the Chalk. Instead of regular thermal contraction cracking, the intense cold probably saw the opening up of joints and fissures in the Chalk (Te Punga, 1957). Equally predictable is the absence of thermokarst phenomena from the Chalk uplands since ice segregation by the movement of water to the freezing plane from below would have been unlikely in view of the permeability of the Chalk and the low level of the water table. The only examples of thermokarst features are the hollows developed in the almost drift free areas of lowlying Chalk between the boulder clay uplands of East Anglia and the Fenlands (e.g. Sparks ***et al.,*** 1972). Some, located at spring line sites at the foot of the Chalk escarpment, are probably pingo remnants formed by groundwater moving towards the surface and freezing. On other, more fine grained lithologies, conditions would have been more conducive for widespread ice segregation and one might expect to find more numerous thermokarst features. However, as stressed in the previous chapter (pp. 248–9), their recognition is extremely difficult.

In the west of England beyond the limits of the Chalk outcrops, there is also evidence of the previous operation of periglacial conditions. For example, the small sarsen stream at Corscombe, northwest of the high hill mass of Beaminster Down in northwest Dorset, indicates previous solifluction activity, head deposits occur widely throughout the southwest peninsula (e.g. Mottershead, 1970), and on Dartmoor, tors, blockstreams and altiplanation terraces have all been recognised (e.g. Waters, 1965). In the Otter drainage basin of southwest Devon, numerous small depressions or dells, similar to those of the Chalk, are found to occur on outcrops of sandstone and marl, and a periglacial origin advanced (Gregory, 1971).

Based upon the distribution of ice-wedge casts, involutions and other patterned ground features (see Fig. 13.1), Williams (1969) has concluded that permafrost occurred widely over central and eastern England during the last cold stage. In the milder areas of southwest England however, it was probably restricted to upland areas of Dartmoor, and the plateaus of east Devon and Dorset. Frost shattering and solifluction may have been particularly effective in these damper environments.

Williams (1968) has also speculated upon the regional variations in periglacial erosion that may have occurred in southern England. Based upon the movement of blockstreams and other solifluction debris, he concluded that periglacial erosion was less in the west than in the east. The greatest movement of solifluction debris is found in the Taele gravels of Cambridgeshire which have moved distances of up to 5–10 km from their source areas on the Chalk uplands on slopes averaging less than 1°. By contrast, the movement of granite and gabbro debris on Dartmoor indicates maximum movement of less than 1 km on slopes as high as 8–11° in angle. In the areas in-between, such as the north Wiltshire Chalk, the sarsen stones indicate intermediate amounts of movement. Whether one is justified in making such sweeping generalisations is debatable, however, since movement may have occurred at several different times throughout the Pleistocene.

In summary, there is no doubt that periglacial conditions have affected much, if not all, of southern England. The degree to which any landscape reflects periglacial conditions ultimately depends upon an assessment of the underlying rock and its susceptibility to frost action, and the degree of postglacial modification which has occurred. It must also be stressed that the initial form of the ground prior to periglaciation greatly influenced the degree of landform modification.

The landscape of central Poland

The classic example of relic periglacial terrain is to be found in central Poland where geomorphologists interpret whole landscapes as having been fashioned under periglacial conditions. The impact of periglaciation is so clear that Polish researchers have been traditionally among the leaders in the development of modern periglacial geomorphology.

Fig. 13.7 *Map showing the limits of the last glaciation and the periglacial zone of central Poland*

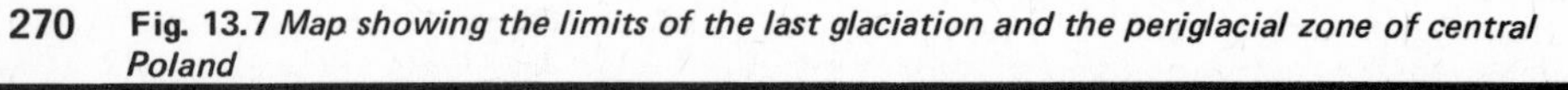

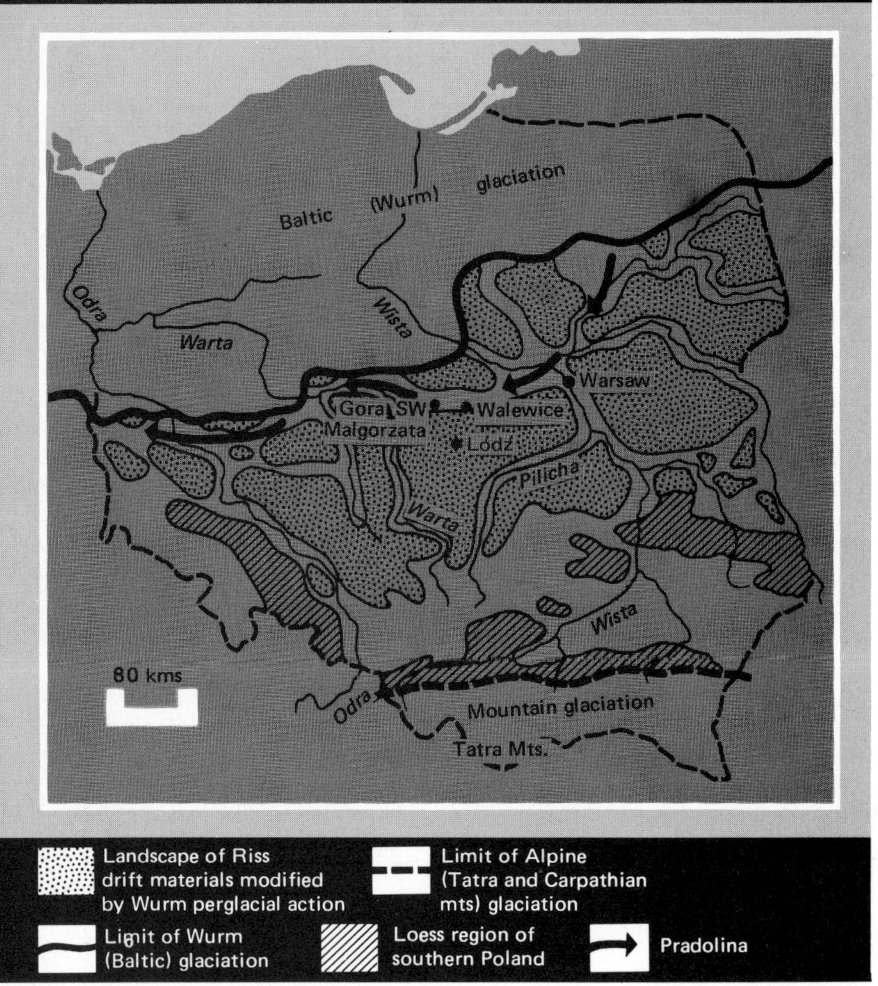

Typically, the relief of central Poland consists of extremely flat or gently undulating lowland terrain developed in drift materials of Middle Polish (Riss) age. During the last cold stage, central Poland lay beyond the limits of the Baltic (Wurm) ice sheet (Fig. 13.7). Thus, an extensive periglacial zone developed in the area between the ice sheet to the north and the mountain glaciation of the Carpathian Mountains to the south. In places, this periglacial zone was over 300 km wide.

To illustrate the variety of periglacial landscape modifications that occurred in central Poland, three examples are described from the Lódź Plateau and the adjacent lowland terrain of the Warsaw–Berlin pradolina.

The Lódź Plateau is an undulating low plateau, 200–300 m in elevation, lying to the north of the loess terrain of southern Poland and the south of the recent (Wurm) glacial terrain of northern Poland. It is dissected by a number of broad shallow valleys which drain to the Pilicha and Warta rivers. Slopes are long and gentle, usually less than 5–7° in angle. The plateau is underlain by a thick cover of Quaternary sediments which rest upon a shallow platform of Jurassic and Cretaceous bedrock. The surface sediments are sands, gravels and boulder clay of the Middle Polish (Riss) glaciation. The northern part of the Lódź Plateau descends to the main Warsaw–Berlin pradolina through a series of broad levels, each separated by a zone of steeper slopes, where angles of 10–15° may occur.

The Warsaw–Berlin pradolina is a large, east–west trending depression, between 15 and 20 km wide, which lies to the north of the Lódź plateau. During the last (Wurm) glaciation, the maximum limit of the ice lay 35–40 km to the north and the depression acted as a channel for the evacuation of meltwater from the ice front. A number of low terraces within the pradolina indicate that it was in operation on a number of occasions.

The dry valleys of the Lódź plateau The northern half of the Lódź plateau, and in particular, the steeper slopes along its northern edge are dissected by numerous shallow depressions and small valleys which are now dry (Fig. 13.8). They occur in groups or as isolated features. On morphological grounds, they can be divided into two groups. The shorter forms are less than 300 m in length, are straight in plan and occur most frequently on the steeper slopes, between

Fig. 13.8 *Periglacial dell, northern edge of the Lódź plateau near Smardzew, central Poland.*

6 and 10° in angle. They are termed 'niecki denudacyjne' (Klatkowa, 1965) and are clearly analogous to the paddle-shaped or bowl-like depressions ('vallons en berceau') which occur on the Chalk terrain of southern England (see pp. 267–8). The second type of valley is longer, between 300 and 1 500 m in length, and both deeper and broader in dimensions. These valleys are more irregular in plan, are best developed on slopes of between 2–4° in angle, and correspond, in a general sense, to the dry valleys of other areas, such as the Chalklands of southern England. Some are asymmetrical in cross profile, with the steeper slope facing west or northwest. It should be emphasised, however, that the dry valleys of the Lódź plateau do not constitute a complete or integrated drainage of the terrain, as is the case for many of the dry valley networks of the Chalk plateaus. Instead, they appear to represent a finer drainage network which has been superimposed upon the rather coarser network of the present-day drainage (Fig. 13.9).

Polish geomorphologists interpret these valleys as having formed under periglacial conditions. The smaller valleys formed through concentrated slopewash erosion during the early and middle stages of the last cold stage, probably beneath and below sites of snowbank accumulation, when the subsoil was frozen. The larger valleys are the result of more than one period of periglacial conditions. According to Klatkowa (1965), they were initiated probably during the late stages of the penultimate (Riss) glaciation, were deepened by normal fluvial activity during the following interglacial period, and were further modified during the last cold stage. Their asymmetrical modification occurred during the latter stage. Then, with the thawing of ground at the end of the last cold stage and the increase in infiltration which resulted, the drainage network contracted leaving the dells and valleys dry.

Gora Sw Malgorzata The isolated hill of Gora Sw Malgorzata is situated in the axial part of the Warsaw–Berlin pradolina, and rises 20 m above the flat bottom of the broad depression. The hill possesses smooth concave slopes, grading from 25–35° in angle in their upper parts to 10° in their lower parts. The hill is composed of sands, silts and gravels of a glaciofluvial nature which overlie till, and it is interpreted as a kame which originated during the deglaciation of the penultimate ice sheet (Riss) of the Warta stage (Dylik, 1963).

The foot of the hill is mantled by slope deposits which formed during the last (Wurm) cold stage. They enable a reconstruction of the various slope modifications which took place during that period of time. A section through the lower part of the slope is illustrated in Fig. 13.10. The lowermost deposits are fine grained fluvially bedded sands indicating the previous existence of a braided stream at the foot of the slope. Overlying and obliterating the relief of these channel deposits is a thickness of 5–10 m of solifluction debris and rhythmically stratified sediments, all clearly derived from the glaciofluvial sediments which comprise the hill. Lobe-like solifluction structures constitute the lower section of the deposits. The change to rhythmically bedded sediments indicates a change to sheet solifluction combined with slopewash on progressively lower-angled slopes. A second series of lobate deformations occur in an upper layer of stratified sands. Overlying all these sediments and extending to the foot of the present hillslope, is a layer of cross bedded sands and gravels with numerous stone and boulder beds.

Fig. 13.9 *Map of part of the northern section of the Lódź plateau, central Poland, illustrating distribution of periglacial valleys and dells and their relation to the present drainage pattern. From Klatkowa, (1965), Map 1.*

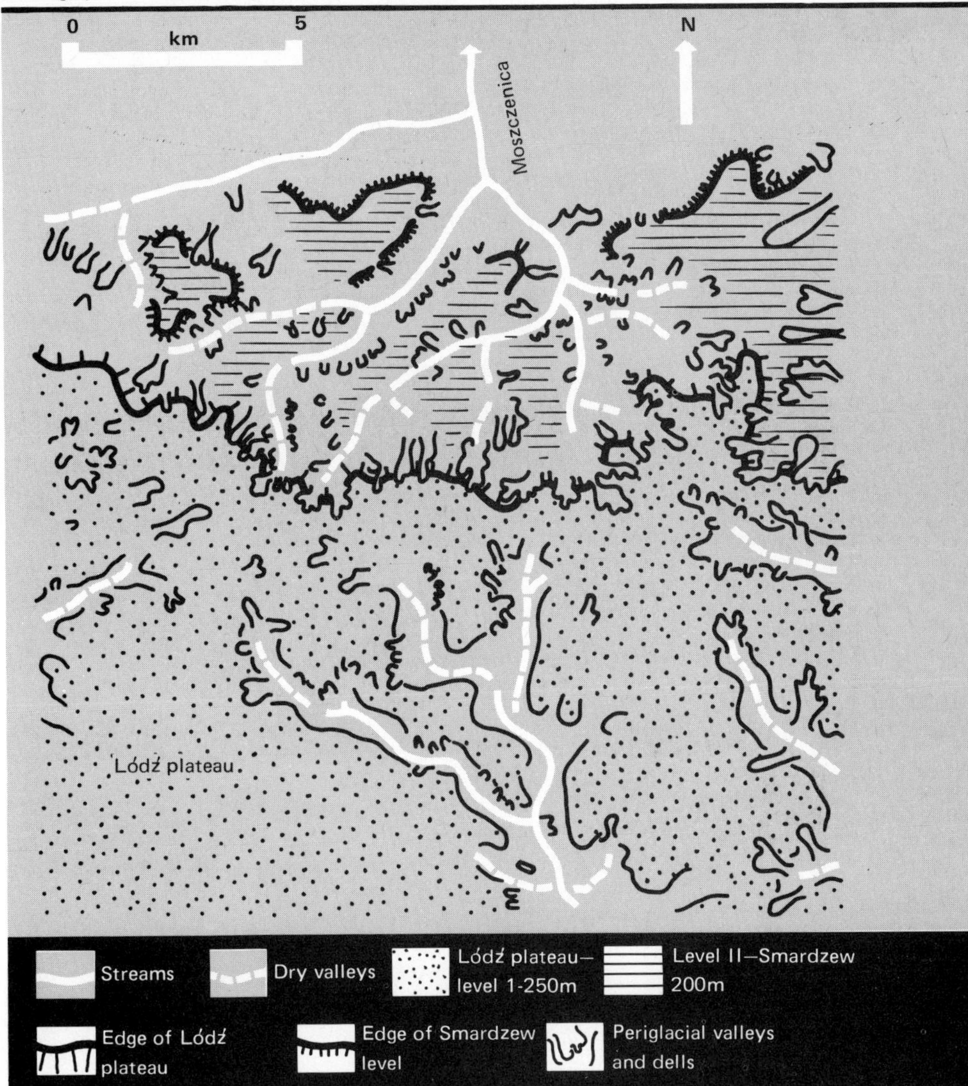

The various deposits indicate a progressive change in the slope form of Gora Sw Malgorzata (Fig. 13.10). In the earliest phase of periglacial conditions, probably at the beginning of the Wurm, solifluction and then slopewash obliterated the undulating surface produced by the old system of braided rivers. As a result, the slope became longer and gentler. To begin with, solifluction on the steep slope produced lobate movement which infilled the channel depressions and gave a step-like profile to the slope. Then, as the slope gradient decreased,

 Fig. 13.10 *Slope evolution at Gora SW Malgorzata, Central Poland.* **A** – *Slope deposits.* **B** – *Schematic illustration of slope evolution. From Dylik (1963; 1969a)*

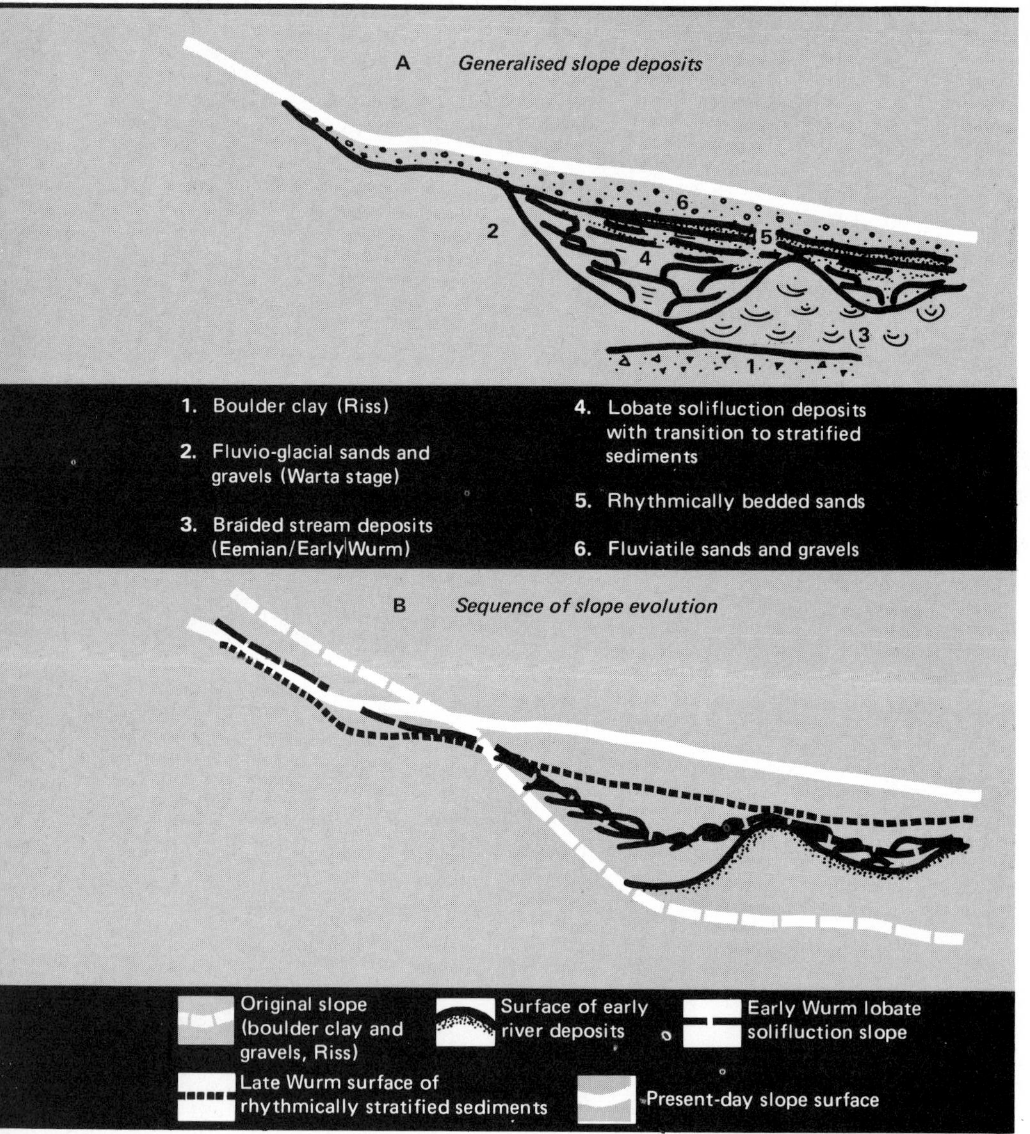

the importance of slopewash and stratified sheet-like deposition increased, and the slope assumed a smoother profile. Ultimately, the old channel deposits and relief were completely obliterated by a gentle depositional slope of 2–4° in angle on which slopewash processes dominated. During the climatic amelioration of the latter stages of the last cold period, the slope experienced a second phase of solifluction movement, probably resulting from the deep thawing and saturation of the sands. Further slope evolution was terminated however, by the deposition of the bedded gravels which covered all the preceding deposits and rose to the foot of the present-day hill. They suggest that the pradolina came into operation once again.

Thus, the slopes of Gora Sw Malgorzata are thought to have experienced active periglacial modification throughout the last cold stage. As climatic conditions deteriorated, solifluction and then sheetwash processes progressively reduced the lower slope in angle and extended it in length, while the upper part suffered erosion and retreat. Ultimately, a near-stable slope was attained. Then, in the climatic amelioration which followed, a second phase of solifluction was initiated before the lower slope was buried beneath the fluvial sediments of the pradolina drainage.

Walewice A more complex example of periglacial slope modification has been described by Dylik (1969a; 1969b) in which the various mass wasting processes operated in conjunction with thermal erosion and the presence of frost fissures. The immediate site was that of a long and extremely gentle slope extending away from the edge of the 10–20 m terrace of the Warsaw–Berlin pradolina near the village of Walewice. The top of the terrace is mantled with a layer of boulder clay derived from the Middle Polish (Riss) glaciation. As at Gora Sw Malgorzata, the last glaciation did not extend to this area. Therefore, the modifications of the terrace edge took place during the last cold phase.

A section nearly 200 m in length was excavated at right angles to the edge of the present terrace and extended to the valley floor (Fig. 13.11). The majority of the sediments which form the slope are rhythmically bedded silts and sands derived from upslope and interpreted as slopewash deposits. They appear to have buried and fossilised two older and steeper slopes. The older of the fossil slopes represents the original terrace edge, first covered with a veneer of boulder clay which had moved down the slope from the terrace edge and then buried by extensive stratified sediments. Further down the slope the stratified sediments are truncated by a younger fossil slope, at the foot of which is a block of rhythmically stratified sediments apparently upturned and buried by non-stratified earth-slide deposits. The block appears to have fallen in a frozen state and to have been quickly buried since if it had thawed, it would not have been preserved. The most likely explanation is that it formed when a stream undercut and temporarily steepened the slope. The development of a thermo-erosional niche caused the collapse of the block along the line of weakness presented by a frost fissure. Sloughing-off of the new bank then led to its rapid burial. In the excavations, at least two generations of frost fissures could be identified; an older set associated with the collapse of the block and a younger set developed in the material which fossilised the fallen block. They prove that permafrost was present throughout this period of slope modification. Subsequently, the younger fossil slope was covered by further rhythmically stratified slope deposits, and then braided river sediments were laid down. The older fossil slope was cut probably during the last (Eem) interglacial or during one of the older Wurm interstadials. The burial of the original terrace edge by solifluction and slope-wash deposits, and the steepening, undercutting and burial of the younger fossil slope all took place during the climax of the last cold stage.

Thus, the end result of periglacial modification at Walewice was the elimination of the terrace edge and the development of a smooth, low-angled slope of

Fig. 13.11 *Slope evolution at Walewice, central Poland.* **A** - *Initial slope;* **B** - *Fluvio-thermal modification;* **C** - *Present slope form and deposits. From Dylik (1969a; 1969b)*

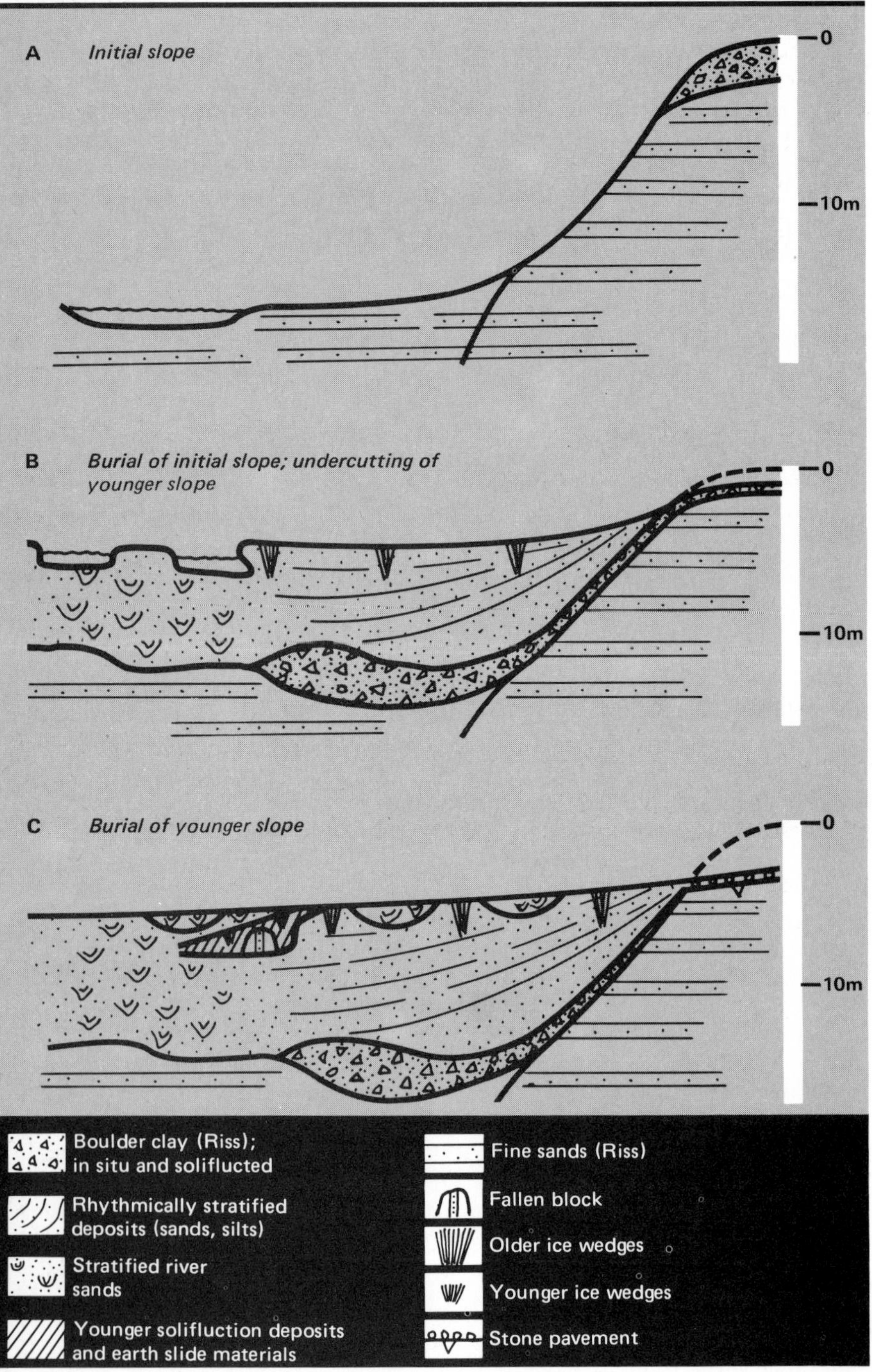

accumulation beneath. Slope angles declined from 20–25° along the original terrace edge to 3–4° on the depositional surface, and the landscape experienced a progressive reduction in relative relief through the movement of material downslope into the valley bottom. Although a temporary steepening of the slope did occur, the overall evolution of the slope was that of its extension in length and reduction in angle.

Summary

The similarities in periglacial landscape modifications which occurred in central Poland and southern England are clear and need little emphasis.

The 'rock-streams' and solifluction deposits of southern England and the evolution of slopes at Gora Sw Malgorzata and Walewice illustrate how mass wasting of upper slopes led to a flattening of landscapes and the production of smooth, gently concave slopes of low angle. Towards their lower ends, these erosional surfaces merged into depositional surfaces formed by the infilling of original valleys and depressions. In both East Anglia and the lowlands of central Poland the already undulating drift terrain and the unconsolidated nature of the drift deposits quickly led to the reduction of relief and the planation of the landscape. In the initial stages, when steeper slopes existed, solifluction was important but as the slopes attained gentler gradients the various slopewash processes assumed dominance.

In addition to frost action, there is abundant evidence to suggest that running water was an important agent in landscape modification. Streams were effective both as transporting and eroding agents. The streams were shallow and braided and reflected the high load and variable discharge characteristics of their environment. In upland areas, such as the Lódź plateau and the Chalk escarpments, periglacial conditions resulted in an accentuation of relief. Intense physical weathering on climatically favoured south- and west-facing slopes saw the asymmetrical modification and further deepening of existing valleys. Furthermore, on valley side slopes and on steeper terrain adjacent to upland plateaus numerous small depressions and valleys were initiated on the frozen subsoils. Typically, they developed below snowpatches through concentrated wash processes or at sites where physical weathering was accentuated by favourable hydrological conditions. Their combined effect on the landscape was a temporary increase in the density of the drainage network.

References

Aleshinskaya, Z. V., Bondarev, L. G. and Gorbonov, A. P. (1972). Periglacial phenomena and some palaeo-geographical problems of Central Tien-Shan, *Biuletyn Peryglacjalny*, No. 21, 5–14.

Alexandre, J. (1958). Le modèle quaternaire de l'Ardennes Centrale, *Annalles, Société Géologique de Belgique,* **81**, 213–331.

Andersson, J. G. (1906). Solifluction; a component of subaerial denudation, *Journal of Geology,* **14**, 91–112.

Andrews, J. T. (1961). The development of scree slopes in the English Lake District and Central Quebec–Labrador, *Cahiers de Géographie de Québec,* **10**, 219–30.

Anisimova, N. P., Nikitina, N. M., Piguzova, V. M. and Shepelyev, V. V. (1973). *Water sources of Central Yakutia*, Guidebook, Second International Permafrost Conference, Yakutsk, USSR, 47 pp.

Are, F. (1972). The reworking of shores in the permafrost zone, in *International Geography*, Vol. 1 (eds W. P. Adams and F. Helleiner), University of Toronto Press, pp. 78–9.

Arnborg, L., Walker, H. J. and Peippo, J. (1967). Suspended load in the Colville River, Alaska, *Geografiska Annaler,* **49A**, 131–44.

Avery, B. W. (1964). The soils and landuse of the district around Aylesbury and Hemel Hempstead, *Memoirs, Soil Survey of Great Britain*, HMSO.

Baranov, I. Ya. (1959). Geographical distribution of seasonally frozen ground and permafrost, in *General Geocryology* (Moscow, USSR; V. A. Obruchev Institute of Permafrost Studies, Academy of Science), Part I, Ch. 7, 193–219. National Research Council of Canada Technical Translation No. 1121 (1964).

Baranov, I. Ya. and Kudryatsev, V. A. (1966). Permafrost of Eurasia, in *Proceedings, 1st International Permafrost Conference*, National Academy of Science – National Research Council of Canada, Publication 1287, 98–102.

Benedict, J. B. (1970). Downslope soil movement in a Colorado alpine region; rates, processes and climatic significance, *Arctic and Alpine Research,* **2**, 165–226.

Berg, T. E. (1969). Fossil sand wedges at Edmonton, Alberta, Canada, *Biuletyn Peryglacjalny*, No. 19, 325–33.

Beschel, R. L. (1966). Hummocks and their vegetation in the high Arctic, in *Proceedings, 1st International Permafrost Conference*, National Academy of Science – National Research Council of Canada, Publication 1287, 13–20.

Bik, M. J. J. (1969). The origin and age of the prairie mounds of southern Alberta, *Biuletyn Peryglacjalny*, No. 19, 85–130.

Bird, J. B. (1967). *The Physiography of Arctic Canada*, The Johns Hopkins Press, Baltimore, 336 pp.

Black, R. F. (1952). Growth of ice wedge polygons in permafrost near Barrow, Alaska, *Bulletin, Geological Society of America,* **63**, 1235–6.

Black, R. F. (1960). *Ice Wedges in Northern Alaska*, Abstract of Paper, 19th International Geographical Congress, Stockholm, p. 26.

Black, R. F. (1963). Les coins de glace et le gel permanente dans le nord de L'Alaska, *Annales de Géographie,* **72**, 257–71.

Black, R. F. (1964). Periglacial phenomena of Wisconsin, north central United States, Vol. IV, *Report of VI INQUA Congress*, Warsaw, 1961. Lódź, 1964, 21–8.

Black, R. F. (1969a). Thaw depressions and thaw lakes; a review, *Biuletyn Peryglacjalny*, No. 19, 131–50.

Black, R. F. (1969b). Climatically significant fossil periglacial phenomena in north central United States, *Biuletyn Peryglacjalny*, No. 20, 225–38.

Black, R. F. (1973). Growth of patterned ground in Victoria Land, Antarctica, in *Permafrost; North American Contribution, Second International Permafrost Conference, Yakutsk, USSR*, National Academy of Science Publication 2115, 193–203.

Black, R. F. and Barksdale, W. L. (1949). Oriented lakes of northern Alaska, *Journal of Geology,* **57**, 105–18.

Bobov, N. G. (1969). The formation of beds of ground ice, *Soviet Geography: Review and Translation*, II, No. 6, 456–63.

Borns, H. W. (1965). Late glacial ice wedge casts in northern Nova Scotia, *Science,* **148**, 1223–5.

Brink, V. C., Mackay, J. R., Freyman, S. and Pearce, D. G. (1967). Needle ice and seedling establishment in southwestern British Columbia, *Canadian Journal of Plant Science,* **47**, 135–9.

Brooks, I. A. (1971). Fossil ice wedge casts in western Newfoundland, *Maritime Sediments,* **7**, 118–22.

Brown, E. H. (1969). Jointing, aspect and the orientation of scarp-face dry valleys, near Ivinghoe, Buckinghamshire, *Transactions, Institute of British Geographers,* **48**, 61–73.

Brown, J. (1966). Massive underground ice in northern regions, *Proceedings, Army Science Conference*, 14–17 June 1966; Washington, D.C.: Office, Chief of Research and Development, Department of the Army, **1**, 89–102.

Brown, J. and Johnson, P. L. (1965). *Pedo-ecological Investigations, Barrow, Alaska*, US Army CRREL, Technical Report 159, 32 pp.

Brown, J., Rickard, W. and Vietor, D. (1969). *The Effect of Disturbance on Permafrost Terrain*, US Army CRREL, Special Report 138, 13 pp.

Brown, R. J. E. (1960). The distribution of permafrost and its relation to air temperature in Canada and the USSR., *Arctic,* **13**, 163–77.

Brown, R. J. E. (1966). The relation between mean annual air and ground temperatures in the permafrost regions of Canada, in ***Proceedings; 1st International Permafrost Conference***; National Academy of Science – National Research Council of Canada, Publication 1287, 241–6.

Brown, R. J. E. (1967a). ***Permafrost in Canada***, Map 1246A, Geological Survey of Canada, National Research Council of Canada, Ottawa.

Brown, R. J. E. (1967b). Comparison of permafrost conditions in Canada and the USSR, ***Polar Record,* 13**, 741–51.

Brown, R. J. E. (1968). ***Permafrost Investigations in Northern Ontario and Northeastern Manitoba***, Technical Paper 291, Division of Building Research, National Research Council of Canada, Ottawa, 40 pp.

Brown, R. J. E. (1969). Factors influencing discontinuous permafrost in Canada, in ***The Periglacial Environment*** (ed. T. L. Péwé), McGill-Queen's University Press, Montreal, pp. 11–53.

Brown, R. J. E. (1970). ***Permafrost in Canada; its Influence on Northern Development***, University of Toronto Press, 234 pp.

Brown, R. J. E. (1972). Permafrost in the Canadian Arctic Archipelago, ***Zeitschrift für Geomorphologie***, Supplement No. 13, 102–30.

Brown, R. J. E. (1973a). Influence of climate and terrain factors on ground temperatures at three locations in the permafrost region of Canada, in ***Permafrost; North American Contribution, Second International Permafrost Conference, Yakutsk, USSR***, National Academy of Science Publication 2115, pp. 27–34.

Brown, R. J. E. (1973b). Permafrost distribution and relation to environmental factors in the Hudson Bay lowlands, in ***Proceedings; Symposium on the Physical Environment of the Hudson Bay Lowland***, University of Guelph publication, pp. 35–68. (Research Paper No. 576, Division of Building Research, National Research Council of Canada, Ottawa.)

Brown, R. J. E. (1974a). Ground ice as an initiator of landforms in permafrost regions, in ***Research in Polar and Alpine Geomorphology. Proceedings, 3rd. Guelph Symposium on Geomorphology***, 1973 (eds B. D. Fahey and R. D. Thompson), Guelph, Ontario, pp. 25–42.

Brown, R. J. E. (1974b). ***Some aspects of Airphoto Interpretation of Permafrost in Canada***, Technical Paper 409, Division of Building Research, National Research Council of Canada, Ottawa, 20 pp.

Brown, R. J. E. and Péwé, T. L. (1973). Distribution of permafrost in North America and its relationship to the environment; A review 1963–1973, in ***Permafrost; North American Contribution, Second International Permafrost Conference, Yakutsk, USSR***, National Academy of Science Publication 2115, pp. 71–100.

Brown, R. J. E. and Williams, G. P. (1972). ***The Freezing of Peatlands***, Technical Paper 381, Division of Building Research, National Research Council of Canada, Ottawa, 24 pp.

Brunnschweiller, D. (1962). The periglacial realm in North America during the Wisconsin glaciation, *Biuletyn Peryglacjalny*, No. 11, 15–27.

Bryan, K. (1948). Cryopedology – the study of frozen ground and intensive frost action with suggestions of nomenclature, *American Journal of Science,* **244**, 622–42.

Budel, J. (1944). Die morphologischen Wirkungen des Eiszeitklimas in Gletsherfrein Gebeit, *Geologische Rundschau,* **34**, 482–519.

Budel, J. (1951). Die klimazonen des Eiszeitalters, *Eiszeitalter und Gegenwart,* **I**, 16–26. (English translation; *International Geology Review*, **I(9)**, 72–9, 1959.)

Budel, J. (1960). *Die Frostschott-Zone südorst Spitzbergen*, Colloquium Geographica, Bonn, No. 6, 105 pp.

Bull, A. J. (1940). Cold conditions and landforms in the South Downs, *Proceedings, Geologist's Association,* **51**, 63–71.

Butrym, J., Cegla, J., Dzulynski, S. and Nakonieczny, S. (1964). New interpretation of 'periglacial' structures, *Folia Quaternaria*, No. 17, 34 pp., Polska Akademia Nauk, Krakow.

Butzer, K. W. (1973). Pleistocene 'periglacial' phenomena in southern Africa, *Boreas,* **2**, 1–12.

Cailleux, A. (1942). *Les Actions Eoliennes Périglaciaires en Europe*, Memoir 46, Société Géologique France, 176 pp.

Cailleux, A. (1956). Mares, mardelles et pingos, *Comptes Rendus, Académie des Science, Paris,* **242**, 1912–14.

Cailleux, A. (1957). Les mares du sud-est de Sjaelland (Danemark), *Comptes, Rendus, Académie des Sciences, Paris,* **245**, 1074–6.

Cailleux, A. (1961). Mares et lacs ronds et loupes de glace du sol, *Biuletyn Peryglacjalny*, No. 10, 35–41.

Cailleux, A. and Calkin, P. (1963). Orientation of hollows in cavernously weathered boulders in Antarctica, *Biuletyn Peryglacjalny*, No. 12, 147–50.

Caine, T. N. (1963). Movement of low angle scree slopes in the Lake District, northern England, *Revue de Géomorphologie Dynamique,* **14**, 171–7.

Caine, T. N. (1967). The tors of Ben Lomond, Tasmania, *Zeitschrift für Geomorphologie*, No. 11, 418–29.

Caine, T. N. (1969). A model for alpine talus slope development by slush avalanching, *Journal of Geology,* **77**, 92–100.

Carson, C. E. and Hussey, K. M. (1962). The oriented lakes of Arctic Alaska, *Journal of Geology,* **70**, 417–39.

Carson, C. E. and Hussey, K. M. (1963). The oriented lakes of Arctic Alaska; a reply, *Journal of Geology,* **71**, 532–3.

Carson, M. A. and Kirkby, M. J. (1972). *Hillslope Form and Process*, Cambridge University Press, England, 475 pp.

Chambers, M. J. G. (1966). Investigations of patterned ground at Signy Island, South Orkney Islands. II: Temperature regimes in the active layer, ***Bulletin, British Antarctic Survey,*** **10**, 71–83.

Chambers, M. J. G. (1967). Investigations of patterned ground at Signy Island, South Orkney Islands. III: Miniature patterns, frost heaving and general conclusions, ***Bulletin, British Antarctic Survey,*** **12**, 1–22.

Church, M. (1972). Baffin Island sandurs; a study of Arctic fluvial processes, ***Geological Survey Canada***, Bulletin 216, 208 pp.

Church, M. (1974). Hydrology and permafrost with reference to northern North America, in ***Permafrost Hydrology; Proceedings of Workshop Seminar, 1974***, Canadian National Committee, International Hydrological Decade, Environment Canada, Ottawa, pp. 7–20.

Clayton, L. and Bailey, P. K. (1970). Tundra polygons in the northern Great Plains, ***Abstracts, Geological Society of America,*** **2(6)**, p. 382.

Cook, F. A. (1956). Additional notes on mud circles at Resolute Bay, N.W.T., ***Canadian Geographer***, No. 8, 9–17.

Cook, F. A. (1967). Fluvial processes in the high Arctic, ***Geographical Bulletin,*** **9**, 262–8.

Cook, F. A. and Raiche, V. G. (1962a). Freeze–thaw cycles at Resolute, N.W.T., ***Geographical Bulletin***, No. 18, 64–78.

Cook, F. A. and Raiche, V. G. (1962b). Simple transverse nivation hollows at Resolute, N.W.T., ***Geographical Bulletin***, No. 18, 79–85.

Corbel, J. (1959). Erosion en terrain calcaire; vitesse d'erosion morphologie, ***Annales de Géographie,*** **68**, 97–120.

Corte, A. E. (1966). Particle sorting by repeated freezing and thawing, ***Biuletyn Peryglacjalny***, No. 15, 175–240.

Corte, A. E. (1971). Laboratory formation of extrusion features by multicyclic freeze–thaw in soils, in ***Etude des phénomènes périglaciaires en laboratoire***, Colloque International de Géomorphologie, Liège–Caen, 1971, Centre de Géomorphologie à Caen, Bulletin No. 13-14-15, pp. 157–82.

Corte, A. E. and Higashi, A. (1971). Growth and development of perturbations on the soil surface due to the repetition of freezing and thawing, in ***Etude des phénomènes périglaciaires en laboratoire***, Colloque International de Géomorphologie à Caen, Bulletin No. 13-14-15, pp. 117–31.

Cruickshank, J. and Colhoun, E. A. (1965). Observations on pingos and other landforms in Schuchertdal, Northeast Greenland, ***Geografiska Annaler,*** **47**, 224–36.

Currey, D. R. (1964). A preliminary study of valley asymmetry in the Ogotoruk Creek area, Northwest Alaska, ***Arctic,*** **17**, 85–98.

Czudek, T. (1964). Periglacial slope development in the area of the Bohemian Massif in Northern Moravia, ***Biuletyn Peryglacjalny***, No. 14, 169–94.

Czudek, T. and Demek, J. (1970a). Thermokarst in Siberia and its influence on the development of lowland relief, ***Quaternary Research,*** **1**, 103–20.

Czudek, T. and Demek, J. (1970b). Pleistocene cryopedimentation in Czechoslovakia, ***Acta Geographica Lodziensia***, No. 24, 101–8.

Czudek, T. and Demek, J. (1973). The valley cryopediments in Eastern Siberia, ***Biuletyn Peryglacjalny***, No. 22, 117–30.

Dahl, R. (1966). Blockfields and other weathering forms in the Narvik Mountains, ***Geografiska Annaler*, 48A**, 224–7.

Danilova, N. S. (1956). Soil wedges and their origin, in ***Data on the principles of the study of the frozen zones in the earth's crust***, issue 111. (Moscow, USSR; V. A. Obruchev Institute of Permafrost Studies, Academy Science.) National Research Council of Canada, Technical Translation No. 1088, Ottawa, 1964, pp. 90–9.

De La Beche, H. T. (1839). ***Report on the Geology of Cornwall, Devon, and West Somerset***, Memoirs, Geological Survey, United Kingdom.

Demek, J. (1964). Castle koppies and tors in the Bohemian Highland (Czechoslovakia), ***Biuletyn Peryglacjalny***, No. 14, 195–216.

Demek, J. (1968). Cryoplanation terraces in Yakutia, ***Biuletyn Peryglacjalny***, No. 17, 91–116.

Demek, J. (1969a). ***Cryoplanation Terraces, their Geographical Distribution, Genesis and Development***, Rozpravy Ceskoslovenske Akademie Ved, Rad Matematickych A Prirodnich Ved, Rocnik 79, sesit 4, 80 pp.

Demek, J. (1969b). Cryogene processes and the development of cryoplanation terraces, ***Biuletyn Peryglacjalny***, No. 18, 115–25.

Derbyshire, E. (1972). Tors, rock weathering, and climate in southern Victoria Land, Antarctica, in ***Polar Geomorphology***, Institute of British Geographers Special Publication No. 4, pp. 93–105.

Derbyshire, E. (1973). Periglacial phenomena in Tasmania, ***Biuletyn Peryglacjalny***, No. 22, 131–48.

Dines, H. G., Hollingworth, S. E., Edwards, W., Buchan, S. and Welch, F. B. A. (1940). The mapping of head deposits, ***Geological Magazine*, 77**, 198–226.

Dionne, J-C. (1971). Fente de cryoturbation tardiglaciaire dans la région de Québec, ***Revue de Géographie de Montréal***, No. 25, 245–64.

Dostovalov, B. N. (1960). ***The Developmental Laws of Tetragonal Systems of Ice and Soil Veins in Dispersed Strata***, Perigliatsial'nye iavlenija na territorii SSSR, Moscow State University Publishing House, 1960.

Dostovalov, B. N. and Kudryacev, W. A. (1967). ***Obszceje mierzlotowiedienije*** (***General Permafrost Science***), Moscow State University.

Dostovalov, B. N. and Popov, A. I. (1966). Polygonal systems of ice wedges and conditions of their development, in ***Proceedings, 1st International Permafrost Conference***, National Academy Science–National Research Council of Canada, Publication 1287, pp. 102–5.

Dunbar, M. and Greenaway, K. R. (1956). ***Arctic Canada from the Air***, Ottawa, Queen's Printer, 541 pp.

Dylik, J. (1956). Coup d'oeil sur la Pologne périglaciaire, *Biuletyn Peryglacjalny*, No. 4, 195–238.

Dylik, J. (1957). Tentative comparison of planation surfaces occurring under warm and under cold semi-arid climatic conditions, *Biuletyn Peryglacjalny*, No. 5, 175–86.

Dylik, J. (1960). Rhythmically stratified slope waste deposits, *Biuletyn Peryglacjalny*, No. 8, 31–41.

Dylik, J. (1963). Periglacial sediments of the Sw. Malgorzata hill in the Warsaw–Berlin pradolina, *Bulletin de la Société des Sciences et des Lettres de Lódź*, **xiv**, 1–16.

Dylik, J. (1964a). Eléments essentiels de la notion de 'périglaciaire', *Biuletyn Peryglacjalny*, No. 14, 111–32.

Dylik, J. (1964b). Le thermokarst, phénomène négligé dans les études du Pleistocene, *Annales de Géographie*, **73**, 513–23.

Dylik, J. (1966). Problems of ice wedge structure and frost fissure polygons, *Biuletyn Peryglacjalny*, No. 15, 241–91.

Dylik, J. (1968). Thermokarst, in *Encyclopedia of Geomorphology* (ed. R. W. Fairbridge), Reinhold Book Co., pp. 1149–51.

Dylik, J. (1969a). Slope development under periglacial conditions in the Lódź region, Biuletyn Peryglacjalny, No. 18, 381–410.

Dylik, J. (1969b). Slope development affected by frost fissures and thermal erosion, in *The Periglacial Environment* (ed. T. L. Péwé), McGill-Queen's University Press, Montreal, pp. 365–86.

Dylik, J. (1971). L'érosion thermique actuelle at ses traces figées dans le paysage de la Pologne Centrale, *Bulletin de l'Académie Polonaise des Science, Série des Sciences de la Terre*, **xix**, 55–61.

Dylik, J. (1972). Rôle du ruisellement dans le modèle périglaciaire, in *Sonderdrück aus Heft 60 der Hans-Poser-Festschrift* (eds J. Hoverman and G. Oberbeck), Gottingen, pp. 169–80.

Dylik, J. and Maarleveld, G. C. (1967). Frost cracks, frost fissures and related polygons, *Mededelingen van de Geol. Stichting, nieuwe serie*, No. 18, 7–21.

Dylikowa, A. (1961). Structures de pression congelistatique et structures de gonflement par le gel de Kartarzynow pres de Lódź, *Bulletin de la Société des Science et des Lettres de Lódź*, **xii**, 9, 1–22.

Eakin, H. M. (1916). The Yukon–Koyukuk region, Alaska, *United States Geological Survey Bulletin*, No. 631, pp. 67–88.

Edelman, C. H. and Maarleveld, G. C. (1949). De asymmetrische dalen van de Veluwe, *Tijdschrift Kon.ned aardrijksk. Genoot*, **66**, 143–6.

Eden, M. J. and Green, C. P. (1971). Some aspects of granite weathering and tor formation on Dartmoor, England, *Geografiska Annaler*, **53A**, 92–9.

Embleton, C. and King, C. A. M. (1968). *Glacial and Periglacial Geomorphology*, E. Arnold Ltd, London, 608 pp.

Evans, J. G. (1968). Periglacial deposits on the Chalk of Wiltshire, *Wiltshire Archaeological and Natural History Magazine*, **63**, 12–26.

Everett, K. R. (1966). Slope movement and related phenomena, in *The Environment of the Cape Thompson Region, Alaska* (ed. N. J. Wilimovsky), United States Atomic Energy Commission, PNE-481, pp. 175–220.

Everett, K. R. (1967). *Mass Wasting in the Taseiaq area, West Greenland. Meddelelser om Gronland*, **165**(5), 30 pp.

Fahey, B. D. (1973) An analysis of diurnal freeze–thaw and frost heave cycles in the Indian Peaks region of the Colorado Front Range, *Arctic and Alpine Research*, **5**, 269–81.

Fahey, B. D. (1974). Seasonal frost heave and frost penetration measurements in the Indian Peaks region of the Colorado Front Range, *Arctic and Alpine Research*, **6**, 63–70.

Fahnestock, R. K. (1963). Morphology and hydrology of a glacial stream – White River, Mt Rainier, Washington, *United States Geological Survey, Professional Paper*, 422-A, 70 pp.

Faucher, D. (1931). Note sur la dissymetrie des vallons de l'Armagnal, *Bulletin Soc. H.N. Toulouse*, **61**, 262–8.

Ferrians, O. (1965). Permafrost map of Alaska, *United States Geological Survey, Miscellaneous Map*, 1–445.

Ferrians, O., Kachadoorian, R. and Green, G. W. (1969). Permafrost and related engineering problems in Alaska, *United States Geological Survey, Professional Paper 678*, 37 pp.

Fisher, O. (1866). On the warp, its age, and probable connection with post geological events, *Quarterly Journal of the Geological Society*, **22**, 553–65.

Flemal, R. C., Hinkley, K. C. and Hesler, J. L. (1973). The DeKalb Mounds; A possible Pleistocene (Woodfordian) pingo field in north central Illinois, *Geological Society of America, Memoir 136*, 229–50.

Flint, R. F. (1971). *Glacial and Quaternary Geology*, John Wiley and Sons Inc., New York, 892 pp.

Fraser, J. K. (1959). Freeze–thaw frequencies and mechanical weathering in Canada, *Arctic*, **12**, 40–53.

Freidman, J. D., Johansson, C. E., Oskarsson, N., Svensson, H., Thorarinsson, S. and Williams, R. S. (1971). Observations on Icelandic polygon surfaces and palsa areas. Photo interpretation and field studies, *Geografiska Annaler*, **53A**, 115–45.

French, H. M. (1970). Soil temperatures in the active layer, Beaufort Plain, *Arctic*, **23**, 229–39.

French, H. M. (1971a). Slope asymmetry of the Beaufort Plain, northwest Banks Island, N.W.T., Canada, *Canadian Journal of Earth Sciences*, **8**, 717–31.

French, H. M. (1971b). Ice cored mounds and patterned ground, southern Banks Island, Western Canadian Arctic, ***Geografiska Annaler,*** **53A**, 32–8.

French, H. M. (1972). Asymmetrical slope development in the Chiltern Hills, ***Biuletyn Peryglacjalny***, No. 21, 51–73.

French, H. M. (1973). Cryopediments on the Chalk of Southern England, ***Biuletyn Peryglacjalny***, No. 22, 149–56.

French, H. M. (1974a). Mass wasting at Sachs Harbour, Banks Island, N.W.T., Canada, ***Arctic and Alpine Research,*** **6**, 71–8.

French, H. M. (1974b). Active thermokarst processes, eastern Banks Island, Western Canadian Arctic, ***Canadian Journal of Earth Sciences,*** **11**, 785–94.

French, H. M. (1975a). Pingo investigations and terrain disturbance studies, Banks Island, District of Franklin, ***Geological Survey Canada***, Paper 75-1A, 459–64.

French, H. M. (1975b). Man-induced thermokarst, Sachs Harbour airstrip, Banks Island, N.W.T., Canada, ***Canadian Journal of Earth Sciences,*** **12**, 132–44.

French, H. M. and Egginton, P. (1973). Thermokarst development, Banks Island, Western Canadian Arctic, in ***Permafrost; North American Contribution, Second International Permafrost Conference, Yakutsk, USSR***, National Academy of Science, Publication 2115, pp. 203–12.

Fristrop, B. (1952). Wind erosion within the arctic deserts, ***Geogr. Tidsskr.,*** No. 52, 51–65.

Frye, J. C., Willman, H. B. and Glass, H. D. (1968). Correlation of midwestern loesses with the glacial succession, ***Proceedings, VII INQUA Congress,*** **12** (Loess and related eolian deposits of the world), 3–21.

Fyles, J. G. (1963). Surficial geology of Victoria and Stefansson Islands, District of Franklin, ***Geological Survey Canada***, Bulletin 101, 38 pp.

Gerasimov, I. P. and Markov, K. K. (1968). ***Permafrost and Ancient Glaciation***, Defence Research Board, Ottawa, Translation T499R, pp. 11–19.

Geukens, F. (1947). De asymmetrie der droge dalen van Haspengouw, ***Natuurw. Tijdschr.***, No. 29, 13–18.

Gloriad, Ā. and Tricart, J. (1952). Etude statistique des vallées asymmétriques de la feuille St. Pol, au 1:50,000, ***Revue de Géomorphologie Dynamique***, No. 3, 88–98.

Gold, L. W., Johnston, G. H., Slusarchuk, W. A. and Goodrich, L. E. (1972). Thermal effects in permafrost, in ***Proceedings, Canadian Northern Pipeline Conference, Ottawa, 2–4 February, 1972***, Associate Committee on Geotechnical Research, National Research Council of Canada, Technical Memorandum 104, pp. 25–45.

Gozdzik, J. (1973). Origin and stratigraphic position of periglacial structures in Middle Poland, ***Acta Geographica Lodziensia***, No. 31, 104–17.

Granberg, H. B. (1973). Indirect mapping of the snowcover for permafrost prediction at Schefferville, Quebec, in ***Permafrost; North American Contribution, Second International Permafrost Conference, Yakutsk, USSR***, National Academy of Science, Publication 2115, pp. 113–20.

Gravis, G. F. (1969). Fossil slope deposits in the northern Arctic asymmetrical valleys, ***Biuletyn Peryglacjalny***, No. 20, 239–57.

Gray, J. T. (1971). ***Processes and Rates of Development of Talus Slopes and Protalus Rock Glaciers in the Ogilvie and Wernecke Mountains, Yukon Territory, Canada***, unpublished Ph.D. Thesis, McGill University, Montreal.

Gregory, K. J. (1971). Drainage density changes in southwest England, in ***Exeter Essays in Geography*** (eds K. J. Gregory and W. Ravenhill), University of Exeter, England, pp. 33–53.

Grigoryev, N. F. (1966). Mnogoletnemerzlyye porody Primorskoy zony Jakutii, ***Nauka, Moscow***, USSR.

Grimberbieux, J. (1955). Origine et asymmétrie des vallees sèches de Hesbaye, ***Annales, Société Géologique de Belgique*, 78**, 267–86.

Guillien, Y. (1951). Les grèzes litées de Charente, ***Revue Géographique de Pyrénées et de Sud-Ouest*, 22**, 154–62.

Hack, J. T. and Goodlett, J. C. (1960). Geomorphology and forest ecology of a mountain region in the Central Appalachians, ***United States Geological Survey Professional Paper*** 347, 65 pp.

Harris, C. (1972). Processes of soil movement in turf-banked solifluction lobes, Okstindan, northern Norway, in ***Polar Geomorphology***, Institute of British Geographers Special Publication, No. 4, pp. 155–74.

Hamelin, L-E. (1971). Dans la plaine Laurentienne, la glace du sol aurait-elle contribue au façonnement des glissements et autre forms de relief en creux?, ***Cahiers de Géographie de Québec*, 36**, 439–65.

Haugen, R. K. and Brown, J. (1970). Natural and man-induced disturbances of permafrost terrain, in ***Environmental Geomorphology*** (ed. D. R. Coates), State University of New York, Binghampton, New York, pp. 139–49.

Heginbottom, J. A. (1973). ***Effects of Surface Disturbance upon Permafrost***, Report 73-16, Environmental–Social Committee Northern Pipelines, Task Force on Northern Oil Development, Information Canada, 29 pp.

Helbig, K. (1965). Asymmetrische Eiszeittaler in Süddeutschland und Ostereich, ***Wurzburger Geogr. Arbeiten*, 14**, 103 pp.

Holmes, G. W., Hopkins, D. M. and Foster, H. J. (1968). Pingos in central Alaska, ***United States Geological Survey Bulletin*** 1241-H, 40 pp.

Hopkins, D. M. (1949). Thaw lakes and thaw sinks in the Imuruk Lake area, Seward Peninsula, Alaska, ***Journal of Geology*, 57**, 119–31.

Hopkins, D. M. and Bond Taber (1962). Asymmetrical valleys in central Alaska, ***Geological Society of America, Special Papers***, No. 68, p. 116.

Hopkins, D. M. and Sigafoos, R. S. (1954). Role of frost thrusting in the formation of tussocks, *American Journal of Science,* **252**, 55–9.

Hopkins, D. M., Karlstrom, T. N. and others (1955). Permafrost and ground water in Alaska, *United States Geological Survey Professional Paper* 264-F, pp. 113–46.

Howarth, P. J. and Bones, J. G. (1972). Relationship between process and geometric form on high Arctic debris slopes, southwest Devon Island, Canada, in *Polar Geomorphology*, Institute of British Geographers, Special Publication No. 4, pp. 139–53.

Hughes, O. (1969). Distribution of open system pingos in central Yukon Territory with respect to glacial limits, *Geological Survey of Canada*, Paper 69-34, 8 pp.

Hughes, O. (1972). Surficial geology and land classification, in *Proceedings, Canadian Northern Pipeline Conference, Ottawa, 2–4 February, 1972*, Associate Committee on Geotechnical Research, National Research Council of Canada, Technical Memorandum 104, pp. 17–24.

Hume, J. D. and Schalk, M. (1964). The effects of ice push on Arctic beaches, *American Journal of Science,* **262**, 267–73.

Hume, J. D. and Schalk, M. (1967). Shoreline processes near Barrow, Alaska; a comparison of the normal and the catastrophic, *Arctic,* **20**, 86–103.

Hume, J. D., Schalk, M. and Hume, P. W. (1972). Short-term climatic changes and coast erosion, Barrow, Alaska, *Arctic,* **25**, 272–9.

Hussey, K. M. and Michelson, R. W. (1966). Tundra relief features near Point Barrow, Alaska, *Arctic,* **19**, 162–84.

Ives, J. D. (1966). Blockfields, associated weathering forms on mountain tops, and the nunatak hypothesis, *Geografiska Annaler,* **48A**, 220–3.

Ives, J. D. and Fahey, B. D. (1971). Permafrost occurrence in the Front Range, Colorado Rocky Mountains, USA, *Journal of Glaciology,* **10**, 105–10.

Jahn, A. (1960). Some remarks on evolution of slopes on Spitsbergen, *Zeitschrift für Geomorphologie*, Supplement I, 49–58.

Jahn, A. (1961). Quantitative analysis of some periglacial processes in Spitsbergen, *Nauka o Ziemi II*, Seria B, Nr 5, Warsaw, 3–34.

Jahn, A. (1962). The origin of granite tors, *Czasopismo Geograficzne,* **33**, 41–4.

Jahn, A. (1970). *Zagadnienia Strefy Peryglacjalnei*, Warsawa–Panstwowe Wydawnictwo Naukowe, 202 pp.

Johnsson, G. (1959). True and false ice wedges in southern Sweden, *Geografiska Annaler,* **41**, 15–33.

Johnston, G. H. and Brown, R. J. E. (1964). Some observations on permafrost distribution at a lake in the Mackenzie Delta, *Arctic,* **17**, 162–75.

Kachurin, S. P. (1962). Thermokarst within the territory of the USSR, ***Biuletyn Peryglacjalny***, No. 11, 49–55.

Kallio, A. and Reiger, S. (1969). Recession of permafrost in a cultivated soil of interior Alaska, ***Proceedings, Soil Science Society of America,*** **33**, 430–2.

Karrasch, H. (1970). Das phanomen der klimabedingten reliefasymmetrie in Mitteleuropa, ***Gottingen Geographische Abhandlungen,*** **56**, 299 pp.

Katasonov, E. M. (1973). Present day ground and ice veins in the region of the Middle Lena, ***Biuletyn Peryglacjalny***, No. 23, 81–9.

Katasonov, E. M. and Ivanov, M. S. (1973). ***Cryolithology of Central Yakutia***, Guidebook, Second International Permafrost Conference, Yakutsk, USSR, 38 pp.

Kennedy, B. A. and Melton, M. A. (1972). Valley asymmetry and slope forms in a permafrost area in the Northwest Territories, Canada, in ***Polar Geomorphology***, Institute of British Geographers Special Publication No. 4, pp. 107–21.

Kerfoot, D. E. (1972). Thermal contraction cracks in an Arctic tundra environment, ***Arctic,*** **25**, 142–50.

Kerfoot, D. E. (1974). Thermokarst features produced by man made disturbances to the tundra terrain, in ***Research in Polar and Alpine Geomorphology***, Proceedings, 3rd Guelph Symposium on Geomorphology, 1973 (eds B. D. Fahey and R. D. Thompson), pp. 60–72.

Kerney, M. P. (1963). Late glacial deposits on the Chalk of south-east England, ***Philosophical Transactions, Royal Society of London***, Series B, **246**, 203–54.

Kerney, M. P., Brown, E. H. and Chandler, T. J. (1964). The late glacial and post glacial history of the Chalk escarpment near Brook, Kent, ***Philosophical Transactions, Royal Society of London***, Series B, **248**, 135–204.

Klatkowa, H. (1965). Vallons en berceau et vallées sèches aux environs de Lódź, ***Acta Geographica Lodziensia***, No. 19, 124–42.

Kostyaev, A. G. (1969). Wedge and fold like diagenetic disturbances in Quaternary sediments and their palaeogeographic significance, ***Biuletyn Peryglacjalny***, No. 19, 231–70.

Krigstrom, A. (1962). Geomorphological studies of sandur plains and their braided rivers in Iceland, ***Geografiska Annaler,*** **44**, 328–46.

Kurfurst, P. J. (1973). Norman Wells, 96E/7, map 22; ***Terrain Disturbance Susceptibility Maps***, Environmental–Social Program of the Task Force on Northern Oil Development, Dept. Energy, Mines and Resources, Ottawa.

Lachenbruch, A. (1957). Thermal effects of the ocean on permafrost, ***Bulletin, Geological Society of America,*** **68**, 1515–29.

Lachenbruch, A. (1962). Mechanics of thermal contraction cracks and ice wedge polygons in permafrost, ***Geological Society of America***, Special Paper, **70**, 69 pp.

Lachenbruch, A. (1966). Contraction theory of ice wedge polygons; a qualitative discussion, in *Proceedings, 1st International Permafrost Conference, National Academy of Science*, National Research Council of Canada, Publication 1287, pp. 63–71.

Lachenbruch, A. (1968). Permafrost, in *Encyclopedia of Geomorphology* (ed. R. W. Fairbridge, Reinhold Book Co., New York, pp. 833–8.

Lachenbruch, A., Greene, G. W. and Marshall, B. V. (1966). Permafrost and the geothermal regimes, in *The Environment of the Cape Thompson Region, Alaska* (ed. N. J. Wilimovsky), United States Atomic Energy Commission, PNE-481, pp. 149–63.

Lamothe, C. and St-Onge, D. A. (1961). A note on a periglacial process in the Isachsen area, N.W.T., *Geographical Bulletin*, No. 16, 104–13.

Lautridou, J-P. (1971). Bilan des recherches de gélifraction expérimentale effectuées au Centre de Géomorphologie, in *Etude des phénomènes périglaciaires en laboratoire, Colloque International de Géomorphologie*, Liège–Caen, 1971, Centre de Géomorphologie à Caen, Bulletin No. 13-14-15, pp. 63–75.

Leffingwell, E de K. (1919). The Canning River region, northern Alaska, *United States Geological Survey Professional Paper* 109, 251 pp.

Leonard, A. B. and Frye, J. C. (1954). Ecological conditions accompanying loess deposition in the Great Plains region, *Journal of Geology,* **62**, 399–404.

Leopold, L. B., Emmett, W. W. and Myrick, R. M. (1966). Channel and hillslope processes in a semi-arid area, New Mexico, *United States Professional Paper* 352-G, pp. 193–253.

Leopold, L. B., Wolman, M. G. and Miller, J. P. (1964). *Fluvial Processes in Geomorphology*, Freemans, San Francisco, 522 pp.

Linton, D. L. (1955). The problem of tors, *Geographical Journal,* **121**, 470–87.

Linton, D. L. (1964). The origin of the Pennine tors; an essay in analysis, *Zeitschrift für Geomorphologie,* **8**, 5–24.

Losche, H. (1930). Lassen sich die diluvialen Breitenkreise aus klimabedingten, diluvialen Vorzeitformen rekonstruieren?, *Arch. Deutch. Seewarte* (Hamburg), **48**, 7, 39 pp.

Loveday, J. (1962). Plateau deposits of the southern Chiltern Hills, *Proceedings, Geologist's Association*, London, **73**, 83–102.

Lundquist, J. (1969). Earth and ice mounds; a terminological discussion, in *The Periglacial Environment* (ed. T. L. Péwé), McGill-Queen's University Press, Montreal, pp. 203–15.

Maarleveld, G. C. (1960). Wind directions and cover sands in the Netherlands, *Biuletyn Peryglacjalny*, No. 8, 49–58.

Maarleveld, G. C. (1965). Frost mounds; a summary of the literature of the past decade, *Nededelingen van de Geologische Stichting*, nieuwe serie, No. 17, 3–17.

Maarleveld, G. C. and van den Toorn, J. C. (1955). Pseudo-solle in Noord-Nederland, *Tijdschrift Kon. ned. aardrijksk. Genoot,* **72**, 334–60.

Mackay, J. R. (1953). Fissures and mud circles on Cornwallis Island, N.W.T., *Canadian Geographer*, No. 3, 31–8.

Mackay, J. R. (1958). A subsurface organic layer associated with permafrost in the Western Arctic, *Geographical Branch Paper*, No. 18, 1–12.

Mackay, J. R. (1962). Pingos of the Pleistocene Mackenzie River Delta area, *Geographical Bulletin*, No. 18, 21–63.

Mackay, J. R. (1963a). The Mackenzie Delta area, *Geographical Branch Memoir*, No. 8, 202 pp.

Mackay, J. R. (1963b). Notes on the shoreline recession along the coast of the Yukon Territory, *Arctic,* **16**, 195–7.

Mackay, J. R. (1965). Gas-domed mounds in permafrost, Kendall Island, N.W.T., *Geographical Bulletin,* **7**, 105–15.

Mackay, J. R. (1966). Segregated epigenetic ice and slumps in permafrost, Mackenzie Delta area, N.W.T., *Geographical Bulletin,* **8**, 59–80.

Mackay, J. R. (1970). Disturbances to the tundra and forest tundra environment of the Western Arctic, *Canadian Geotechnical Journal,* **7**, 420–32.

Mackay, J. R. (1971). The origin of massive icy beds in permafrost, Western Arctic coast, Canada, *Canadian Journal of Earth Sciences,* **8**, 397–422.

Mackay, J. R. (1972a). The world of underground ice, *Annals, Association of American Geographers,* **62**, 1–22.

Mackay, J. R. (1972b). Offshore permafrost and ground ice, southern Beaufort Sea, Canada, *Canadian Journal of Earth Sciences,* **9**, 1550–61.

Mackay, J. R. (1973). The growth of pingos, Western Arctic coast, Canada, *Canadian Journal of Earth Sciences,* **10**, 979–1004.

Mackay, J. R. (1974). Ice wedge cracks, Garry Island, N.W.T., *Canadian Journal of Earth Sciences,* **11**, 1366–83.

Mackay, J. R. and Black, R. F. (1973). Origin, composition, and structure of perennially frozen ground and ground ice; a review, in *Permafrost; North American Contribution, Second International Permafrost Conference, Yakutsk, USSR,* National Academy of Science Publication 2115, pp. 185–92.

Mackay, J. R., Rampton, V. N. and Fyles, J. G. (1972). Relic Pleistocene permafrost, Western Arctic, Canada, *Science,* **176**, No. 4041, 1321–3.

Malaurie, J. N. (1952). Sur l'asymmétrie des versants dans l'ile de Disko, Groenland, *Comptes Rendus, Académie des Sciences, Paris,* **234**, 1461–2.

Malaurie, J. N. and Guillien, Y. (1953). Le modèle cryo-nival des versants meubles de Skansen (Disko, Groenland). Interprétation géneral des grèzes litées, *Bulletin, Société Géologique de France,* **3**, 703–21.

Manley, G. (1959). The late glacial climate of northwest England, ***Liverpool and Manchester Geological Journal,*** **2**, 188–215.

McCann, S. B. (1972). Magnitude and frequency of processes operating on Arctic beaches, Queen Elizabeth Islands, N.W.T., Canada, in ***International Geography*** (eds P. W. Adams and F. Helleiner), University of Toronto Press, **1**, 41–3.

McCann, S. B. and Carlisle, R. J. (1972). The nature of the ice foot on the beaches of Radstock Bay, southwest Devon Island, N.W.T., Canada, in the summer of 1970, in ***Polar Geomorphology***, Institute of British Geographers Special Publication No. 4, pp. 175–86.

McCann, S. B. and Owens, E. H. (1969). The size and shape of sediments in three Arctic beaches, southwest Devon Island, N.W.T., Canada, ***Arctic and Alpine Research,*** **1**, 267–78.

McCann, S. B., Howarth, P. J. and Cogley, J. G. (1972). Fluvial processes in a periglacial environment; Queen Elizabeth Islands, N.W.T., Canada, ***Transactions, Institute of British Geographers,*** **55**, 69–82.

McDonald, B. C. and Lewis, C. P. (1973). ***Geomorphic and Sedimentologic Processes of Rivers and Coasts, Yukon Coastal Plain***. Report 73–39, Environmental–Social Committee Northern Pipelines, Task Force on Northern Oil Development, Information Canada, Ottawa, 245 pp.

Mitchell, G. F. (1973). Fossil pingos in Camaross Townland, Co. Wexford, ***Proceedings, Royal Irish Academy,*** **73B**(16), 269–81.

Morgan, A. V. (1971). Polygonal patterned ground of late Weichselian age in the area north and west of Wolverhampton, England, ***Geografiska Annaler,*** **54A**, 146–56.

Morgan, A. V. (1972). Late Wisconsinan ice wedge polygons near Kitchener, Ontario, Canada, ***Canadian Journal of Earth Sciences,*** **9**, 607–17.

Mullenders, W. and Gullentops, F. (1969). The age of the pingos of Belgium, in ***The Periglacial Environment*** (ed. T. L. Péwé), McGill-Queen's University Press, pp. 321–36.

Muller, F. (1959). ***Beobachtungen über pingos. Meddelelser om Gronland,*** **153**, 3, 126 pp. National Research Council of Canada Technical Translation 1073, 1963, 177 pp.

Muller, F. (1968). Pingos, modern, in ***The Encyclopedia of Geomorphology*** (ed. R. W. Fairbridge), Reinhold Book Co., pp. 845–7.

Muller, S. W. (1945). Permafrost or perennially frozen ground and related engineering problems, ***United States Geological Survey Special Report, Strategic Engineering Study*** 62, 2nd edn, 231 pp.

National Academy of Science–National Research Council of Canada (1966). ***Proceedings, 1st International Permafrost Conference, Purdue University, 1963***. National Academy of Science–National Research Council publication 1287, 563 pp.

National Academy of Science (1973). *Permafrost; North American Contribution, Second International Permafrost Conference, Yakutsk, USSR,* National Academy of Science Publication 2115, Washington D.C., 783 pp.

Nekrasov, I. A. and Gordeyev, P. P. (1973). *The Northeast of Yakutia.* Guidebook, Second International Permafrost Conference, Yakutsk, 46 pp.

Nichols, R. L. (1966). Geomorphology of Antarctica, in *Antarctic Soils and Soil Forming Processes* (ed. J. C. F. Tedrow), American Geophysical Union, Antarctic Research Series, **8**, 1–59.

Nicholson, F. H. and Granberg, H. B. (1973). Permafrost and snowcover relationships near Schefferville, in *Permafrost; North American Contribution, Second International Permafrost Conference, Yakutsk, USSR*, National Academy of Science Publication 2115, pp. 151–8.

Nicholson, F. H. and Thom, B. G. (1973). Studies at the Timmins 4 permafrost experimental site, in *Permafrost; North American Contribution, Second International Permafrost Conference, Yakutsk, USSR*, National Academy of Science Publication 2115, pp. 159–66.

O'Brien, R. (1971). Observations on pingos and permafrost hydrology in Schuchert Dal, northeast Greenland, *Meddelelser om Gronland,* **195 (1)**, 1–19.

Ollier, C. D. and Thomasson, A. J. (1957). Asymmetrical valleys of the Chiltern Hills, *Geographical Journal,* **123**, 71–80.

Outcalt, S. I. (1969). Weather and diurnal frozen soil structure at Charlotteville, Virginia, *Water Resources Research,* **5**, 1377–81.

Owens, E. H. and McCann, S. B. (1970). The role of ice in the Arctic beach environment with special reference to Cape Ricketts, Southwest Devon Island, N.W.T., Canada, *American Journal of Science,* **268**, 397–414.

Palmer, J. and Nielson, R. A. (1962). The origin of granite tors on Dartmoor, Devonshire, *Proceedings, Yorkshire Geological Society,* **33**, 315–40.

Palmer, J. and Radley, J. (1961). Gritstone tors of the English Pennines, *Zeitschrift für Geomorphologie,* **5**, 37–52.

Peltier, L. C. (1950). The geographic cycle in periglacial regions as it is related to climatic geomorphology, *Annals, Association of American Geographers,* **40**, 214–36.

Péwé, T. L. (1948). Origin of the Mima Mounds, *Scientific Monthly,* **46**, 293–6.

Péwé, T. L. (1954). Effect of permafrost upon cultivated fields, *United States Geological Survey Bulletin,* **989**, 315–51.

Péwé, T. L. (1955). Origin of the upland silt near Fairbanks, Alaska, *Geological Society of America, Bulletin,* **66**, 699–724.

Péwé, T. L. (1959). Sand wedge polygons (tesselations) in the McMurdo Sound region, Antarctica, *American Journal of Science,* **257**, 545–52.

Péwé, T. L. (1966a). Permafrost and its effect on life in the north, in ***Arctic Biology***, 2nd edn (ed. H. P. Hansen), Oregon State University Press, Corvallis, pp. 27–66.

Péwé, T. L. (1966b). Ice wedges in Alaska – classification, distribution and climatic significance, in ***Proceedings, 1st International Permafrost Conference***, National Academy of Science–National Research Council of Canada, Publication 1287, pp. 76–81.

Péwé, T. L. (1969). The periglacial environment, in ***The Periglacial Environment*** (ed. T. L. Péwé), McGill-Queen's University Press, Montreal, pp. 1–11.

Péwé, T. L., Church, R. E. and Andresen, N. J. (1969). Origin and palaeoclimatic significance of large scale polygons in the Donnely Dome area, Alaska, ***Geological Society of America Special Paper*** 109, 87 pp.

Péwé, T. L. and Paige, R. A. (1963). Frost heaving of piles with an example from the Fairbanks area, Alaska, ***United States Geological Survey Bulletin,*** **1111–1**, 333–407.

Pihlainen, J. A. and Johnston, G. H. (1963). ***Guide to a Field Description of Permafrost***, National Research Council of Canada, Associate Committee on Snow and Soil mechanics, Technical Memorandum 79 (National Research Council Publication 7 576).

Pissart, A. (1953). Les coulées pierreuses du Plateau des Hautes Fagnes, ***Annales, Société Géologique de Belgique,*** **76**, 203–19.

Pissart, A. (1958). Les dépressions fermées de la région parisienne, ***Revue de Géomorphologie Dynamique,*** **9**, 73–83.

Pissart, A. (1960). Les dépressions fermées de la région parisienne. Les difficultés d'admettre une origine humaine, ***Revue de Géomorphologie Dynamique,*** **11**, 12.

Pissart, A. (1963). Les traces de 'pingos' du Pays de Galles (Grande Bretagne) et du Plateau des Hautes Fagnes (Belgique), ***Zeitschrift für Geomorphologie,*** **7**, 147–65.

Pissart, A. (1964a). Vitesse des mouvements du sol au Chambeyron (Basse Alpes), ***Biuletyn Peryglacjalny***, No. 14, 303–9.

Pissart, A. (1964b). Contribution expérimentale a la connaissance de la genèse des sols polygonaux, ***Annales, Société Géologique de Belgique,*** **87**, 213–23.

Pissart, A. (1965). Les pingos des Hautes Fagnes. Le problème de leur genèse, ***Annales, Société Géologique de Belgique,*** **88**, 277–89.

Pissart, A. (1966a). Le rôle géomorphologique du vent dans la région de Mould Bay (Ile Prince Patrick, T.N–O, Canada), ***Zeitschrift für Geomorphologie,*** **10**, 226–36.

Pissart, A. (1966b). Etude de quelques pentes de l'ile Prince Patrick, ***Annales, Société Géologique de Belgique,*** **89**, 377–402.

Pissart, A. (1967a). Les modalités de l'écoulement de l'eau sur l'Ile Prince Patrick, ***Biuletyn Peryglacjalny***, No. 16, 217–24.

Pissart, A. (1967b). Les pingos de l'Ile Prince Patrick (76°N–120°W), *Geographical Bulletin,* **9**, 189–217.

Pissart, A. (1968). Les polygons de fente de gel de l'Ile Prince Patrick (Arctique Canadien, 76° Lat. N.), *Biuletyn peryglacjalny*, No. 17, 171–80.

Pissart, A. (1970). Les phénomènes physiques éssentiels liés au gel; les structures périglaciaires qui en résultent et leur signification climatique, *Annales, Société Géologique de Belgique,* **93**, 7–49.

Pissart, A. (1973). Resultats d'éxperiences sur l'action du gel dans le sol, *Biuletyn Peryglacjalny*, No. 23, 101–13.

Pissart, A. (1975). Banks Island; pingos, wind action, periglacial structures, *Geological Survey Canada*, Paper 75–1A, pp. 479–81.

Popov, A. I. (1956). Le thermokarst, *Biuletyn Peryglacjalny*, No. 4, 319–30.

Popov, A. I. (1962). *The Origin and Development of Massive Fossil Ice*, Issue 11, Academy of Sciences of the USSR, V. A. Obruchev Institute of Permafrost Studies, Moscow. National Research Council of Canada, Technical Translation 1006, 1962, pp. 5–24.

Popov, A. I. (1969). Underground ice in the Quaternary deposits of the Yana–Indigirka lowland as a genetic and stratigraphic indicator, in *The Periglacial Environment* (ed. T. L. Péwé), McGill-Queen's University Press, Montreal, pp. 55–64.

Popov, A. I. (1970). Underground ice of northern Eurasia, *Acta Geographica Lodziensia*, No. 24, 365–70.

Popov, A. I., Kachurin, S. P. and Grave, N. A. (1966). Features of the development of frozen geomorphology in northern Eurasia, in *Proceedings, 1st International Permafrost Conference*, National Academy of Science–National Research Council of Canada, Publication 1287, pp. 181–5.

Porsild, A. E. (1938). Earth mounds in unglaciated Arctic northwest America, *Geographical Review,* **28**, 46–58.

Porsild, A. E. (1955). The vascular plants of the western Canadian Arctic archipelago, *National Museum of Canada, Bulletin,* **135**, 226 pp.

Poser, H. (1948). Boden und klimaverhalnisse in Mittel und Westeuropa der Wurmeiszeit, *Erdkunde,* **2**, 53–68.

Potts, A. S. (1970). Frost action in rocks; some experimental data, *Transactions, Institute of British Geographers,* **49**, 109–24.

Presniakow, J. A. (1955). Les vallées asymmétriques en Siberie, *Questions de Géologie de l'Asie,* **11**, Moscow, 391–6.

Prest, V. K., Grant, D. R. and Rampton, V. N. (1968). *Glacial Map of Canada*, Geological Survey of Canada, Map 1253A.

Prestwich, J. (1892). The raised beaches and 'head' or rubble drift of the south of England, *Quarterly Journal of the Geological Society,* **48**, 263–343.

Price, L. W. (1972). *The Periglacial Environment, Permafrost, and Man*,

Resource Paper No. 14, Association of American Geographers, Commission on College Geography, Washington, 88 pp.

Price, L. W. (1973). Rates of mass wasting in the Ruby Range, Yukon Territory, in *Permafrost; North American Contribution, Second International Permafrost Conference, Yakutsk, USSR*, National Academy of Science Publication 2115, pp. 235–45.

Price, R. J. (1969). Moraines, sandar, kames, and eskers near Breidamerkurjokull, Iceland, *Transactions, Institute of British Geographers,* **46**, 17–43.

Price, W. A. (1963). The oriented lakes of Arctic Alaska; a discussion, *Journal of Geology,* **71**, 530–1.

Price, W. A. (1968). Oriented lakes, in *Encyclopedia of Geomorphology* (ed. R. W. Fairbridge), Reinhold Book Co., New York, pp. 784–96.

Prince, H. (1961). Some reflexions on the origin of hollows in Norfolk compared with those in the Paris region, *Revue de Géomorphologie Dynamique,* **12**, 110–17.

Rampton, V. N. (1974). The influence of ground ice and thermokarst upon the geomorphology of the Mackenzie–Beaufort region, in *Research in Polar and Alpine Geomorphology*. Proceedings, 3rd Guelph Symposium on Geomorphology, 1973 (eds B. D. Fahey and R. D. Thompson), pp. 43–59.

Rampton, V. N. and Mackay, J. R. (1971). Massive ice and icy sediments throughout the Tuktoyaktuk Peninsula, Richards Island, and nearby areas, District of Mackenzie, *Geological Survey of Canada*, Paper 71–21, 16 pp.

Rampton, V. N. and Walcott, R. I. (1974). Gravity profiles across ice cored topography, *Canadian Journal of Earth Sciences,* **11**, 110–22.

Rapp, A. (1960a). Recent development of mountain slopes in Karkevagge and surroundings, northern Sweden, *Geografiska Annaler,* **42**, 71–200.

Rapp, A. (1960b). Talus slopes and mountain walls at Tempelfjorden, Spitsbergen, *Norsk Polarinstitutt Skrifter,* **119**, 96 pp.

Rapp, A. (1967). Pleistocene activity and Holocene stability of hillslopes, with examples from Scandinavia and Pennsylvania, in *L'Evolution des Versants,* **40**, Les Congrès et Colloques de L'Université de Liège, Belgium, pp. 229–44.

Raup, H. (1966). Turf hummocks in the Mesters Vig District, northeast Greenland, in *Proceedings, 1st International Permafrost Conference*, National Academy of Science–National Research Council Publication 1287, pp. 43–50.

Reid, C. (1887). On the origin of dry valleys and of coombe rock, *Quarterly Journal of the Geological Society,* **43**, 364–73.

Richie, A. M. (1953). The erosional origin of the Mima Mounds of southwestern Washington, *Journal of Geology,* **61**, 45–50.

Robitaille, B. (1960). Géomorphologie du sud-est de l'Ile Cornwallis, T.N–O., *Cahiers de Géographie de Québec,* **8**, 359–65.

Rockie, W. A. (1942). Pitting on Alaskan farms; a new erosion problem, *Geographical Review,* **32**, 128–34.

Rudberg, S. (1962). A report on some field observations concerning periglacial geomorphology and mass movement on slopes in Sweden, ***Biuletyn Peryglacjalny***, No. 11, 311–23.

Rudberg, S. (1963). Morphological processes and slope development on Axel Heiberg Island, N.W.T., Canada, ***Nach. Akad. Wis., Gottingen***, KL.11, **14**, 218–28.

Rudberg, S. (1964). Slow mass movement processes and slope development in the Norra Storfjall area, southern Swedish Lapland, ***Zeitschrift für Geomorphologie***, supplement 5, 192–203.

St-Onge, D. A. (1959). Note sur l'érosion du gypse en climat périglaciaire, ***Revue Canadienne de géographie,*** **13**, 155–62.

St-Onge, D. A. (1965). La géomorphologie de l'Ile Ellef Ringnes, Territoires du Nord-Ouest, Canada, ***Etude Géographique, Direction de la Géographie***, No. 38, 46 pp.

St-Onge, D. A. (1969). Nivation landforms, ***Geological Survey Canada***, Paper 69–30, 12 pp.

Saunders, G. E. (1973). Vistulian periglacial environments in the Lleyn Peninsula, ***Biuletyn Peryglacjalny***, No. 22, 257–70.

Schafer, J. P. (1949). Some periglacial features in central Montana, ***Journal of Geology,*** **57**, 154–74.

Schumm, S. A. (1964). Seasonal variations of erosion rates and processes on hill-slopes in western Colorado, ***Zeitschrift für Geomorphologie***, Supplement 5, 215–38.

Sekyra, J. (1969). Periglacial phenomena in the oases and the mountains of the Enderby Land and the Dronning Maud Land (East Antarctica), ***Biuletyn Peryglacjalny***, No. 19, 277–89.

Selby, M. J. (1971). Salt weathering of landforms, and an Antarctic example, ***Proceedings, Sixth Geography Conference, Christchurch, New Zealand Geographical Society***, pp. 30–5.

Sharp, R. (1942). Periglacial involutions in north-eastern Illinois, ***Journal of Geology,*** **50**, 113–33.

Sharp, R. (1949). Pleistocene ventifacts east of the Bighorn Mountains, Wyoming, ***Journal of Geology,*** **57**, 175–95.

Shearer, J. M., McNab, R. F., Pelletier, B. R. and Smith, T. B. (1971). Submarine pingos in the Beaufort Sea, ***Science,*** **174**, 816–18.

Shostakovitch, W. B. (1927). Der ewig gefrorene boden Siberiens, ***Gessel. Erdkunde Berlin Zeitschrift***, pp. 394–427.

Shotton, F. W. (1960). Large scale patterned ground in the valley of the Worcestershire Avon, ***Geological Magazine,*** **97**, 404–8.

Shumskiy, P. A. (1964). *Ground (Subsurface) Ice*, National Research Council of Canada Technical Translation 1130, 118 pp.

Shumskiy, P. A. and Vtyurin, B. I. (1966). Underground ice, in *Proceedings, 1st International Permafrost Conference*, National Academy of Science–National Research Council of Canada, Publication 1287, pp. 108–13.

Sigafoos, R. S. (1951). Soil instability in tundra vegetation, *Ohio Journal of Science,* **51**, 281–98.

Small, R. J. (1965). The role of spring sapping in the formation of Chalk escarpment valleys, *Southampton Research Series in Geography*, No. 1, pp. 1–29.

Small, R. J. (1970). *The Study of Landforms*, Cambridge University Press, 486 pp.

Small, R. J., Clark, M. J. and Lewin, J. (1970). The periglacial rock-stream at Clatford Bottom, Marlborough Downs, Wiltshire, *Proceedings, Geologist's Association,* **81**, 87–98.

Smith, D. I. (1972). The solution of limestone in an arctic environment, in *Polar Geomorphology*, Institute of British Geographers, Special Publication No. 4, pp. 187–200.

Smith, H. T. U. (1949a). Physical effects of Pleistocene climatic changes in nonglaciated areas; eolian phenomena, frost action and stream terracing, *Bulletin, Geological Society of America,* **60**, 1485–516.

Smith, H. T. U. (1949b). Periglacial features in the driftless area of southern Wisconsin, *Journal of Geology,* **57**, 196–215.

Smith, H. T. U. (1962). Periglacial frost features and related phenomena in the United States, *Biuletyn Peryglacjalny*, No. 11, 325–42.

Smith, H. T. U. (1964). *Periglacial Eolian Phenomena in the United States*, Volume IV, VI INQUA Congress, Warsaw, 1961. Lódź, 1964, pp. 177–86.

Soloviev, P. A. (1962). Alasnye relyef Centralnoj Jakutii i ego proiskhozhdenie. (Alas relief and its origins in Central Yakutia), in *Mnoholetne-merzlyye Porody i Soputstvuyuhchiye yim Yavleniya na territorii JASSR*, Izdatelstvo AN SSSR, Moscow, pp. 38–53.

Soloviev, P. A. (1973a). *Alass Thermokarst Relief of Central Yakutia*, Guidebook, Second International Permafrost Conference, Yakutsk, USSR, 48 pp.

Soloviev, P. A. (1973b). Thermokarst phenomena and landforms due to frost heaving in Central Yakutia, *Biuletyn Peryglacjalny*, No. 23, 135–55.

Soons, J. M. and Rayner, J. M. (1960). Micro-climate and erosional processes in the Southern Alps, New Zealand, *Geografiska Annaler,* **50**, 1–15.

Souchez, R. (1966). Réflexions sur l'évolution des versants sous climat froid, *Revue de Géographie Physique et de Géologie Dynamique,* **viii**, 317–34.

Souchez, P. (1967). Gélivation et évolution des versants en bordure de l'Islandsis d'Antartides orientale, in *L'Evolution des Versants,* **40**, Les Congrès et Colloques de L'Université de Liège, pp. 291–8.

Sparks, B. W. and West, R. G. (1968). *The Ice Age in Britain*, Methuen and Co., Ltd, 302 pp.

Sparks, B. W., Williams, R. G. B. and Bell, F. G. (1972). Presumed ground ice depressions in East Anglia, ***Proceedings, Royal Society London***, Series A, **327**, 329–43.

Stalker, A. Mac. S. (1960). Ice-pressed drift forms and associated deposits in Alberta, ***Geological Survey of Canada Bulletin,*** **57**, 38 pp.

Svensson, H. (1964). Aerial photographs for tracing and investigating fossil tundra ground in Scandinavia, ***Biuletyn Peryglacjalny***, No. 14, 321–5.

Svensson, H. (1969). A type of circular lake in northernmost Norway, *Geografiska Annaler,* **51A**, 1–12.

Sverdrup, H. V. (1938). Notes on erosion by drifting snow and transport of solid material by sea ice, ***American Journal of Science,*** **35**, 370–3.

Taber, S. (1929). Frost heaving, ***Journal of Geology,*** **37**, 428–61.

Taber, S. (1930). The mechanics of frost heaving, ***Journal of Geology,*** **38**, 303–17.

Taillefer, F. (1944). La dissymétrie des vallées Gasgonnes, ***La Revue de Géographie de Pyrénnees et du Sud-Ouest,*** **xv**, 153–81.

Tedrow, J. C. F. (1966a). Arctic soils, in ***Proceedings, 1st International Permafrost Conference***, National Academy of Science–National Research Council Publication 1287, pp. 50–4.

Tedrow, J. C. F. (1966b). Polar desert soils, ***Proceedings, Soil Science Society of America,*** **30**, 381–7.

Tedrow, J. C. F. (1969). Thaw lakes, thaw sinks and soils in northern Alaska, ***Biuletyn Peryglacjalny***, No. 20, 337–45.

Tedrow, J. C. F. (1973). Soils of the polar regions of North America, ***Biuletyn Peryglacjalny***, No. 23, 157–65.

Tedrow, J. C. F. (1974). Soils of the high Arctic landscapes, in ***Polar Deserts and Modern Man*** (eds T. L. Smiley and J. H. Zumberge), University of Arizona Press, pp. 63–9.

Tedrow, J. C. F. and Cantlon, J. E. (1958). Concepts of soil formation and classification in Arctic regions, ***Arctic,*** **11**, 166–79.

Te Punga, M. T. (1956). Altiplanation terraces in southern England, ***Biuletyn Peryglacjalny***, No. 4, 331–8.

Te Punga, M. T. (1957). Periglaciation in southern England, ***Tijdschrift Kon. ned. aardrijksk. Genoot.,*** **74**, 400–12.

Thie, J. (1974). Distribution and thawing of permafrost in the southern part of the discontinuous zone in Manitoba, ***Arctic,*** **27**, 189–200.

Thomasson, A. J. (1961). Some aspects of the drift deposits and geomorphology of southeast Hertfordshire, ***Proceedings, Geologist's Association,*** **72**, 287–302.

Thompson, H. A. (1966). Air temperatures in northern Canada with emphasis upon freezing and thawing indices, in ***Proceedings, 1st International Permafrost Conference***, National Academy of Science—National Research Council of Canada Publication, 1287, pp. 18–36.

Thomson, S. (1966). Icings on the Alaska Highway, in ***Proceedings, 1st International Permafrost Conference***, National Academy of Science—National Research Council of Canada, Publication 1287, pp. 526–9.

Tricart, J. (1956). Etude expérimentale du problème de la gelivation, ***Biuletyn Peryglacjalny***, No. 4, 285–318.

Tricart, J. (1968). Periglacial landscapes, in ***Encyclopedia of Geomorphology*** (ed. R. W. Fairbridge), Reinhold Book Co., pp. 829–33.

Tricart, J. (1970). ***Geomorphology of Cold Environments*** (translated by E. Watson), Macmillan, St Martin's Press, New York, N.Y., 320 pp.

Tricart, J. and Cailleux, A. (1967). *Le modèle des régions périglaciaires,* **11**, *Traité de Géomorphologie*, SEDES, Paris, 512 pp.

Troll, C. (1944). Struckturboden, solifluktion und frostklimate de erde, ***Geologische Rundschau,*** **34**, 545–694.

Troll, C. (1958). ***Structure Soils, Solifluction and Frost Climates of the World***, United States Army Snow, Ice and Permafrost Research Establishment, Corps of Engineers, Wilmette, Ill., Translation 43, 121 pp.

Troll, C. (1962). 'Solle' and 'Mardelle', ***Erdkunde,*** **16**, 31–4.

Tyrtikov, A. P. (1964). ***The Effect of Vegetation on Perennially Frozen Soil***, National Research Council of Canada Technical Translation 1088, Ottawa, pp. 69–90.

Ugolini, F. C., Bockheim, J. G. and Anderson, D. M. (1973). Soil development and patterned ground evolution in Beacon Valley, Antarctica, in ***Permafrost; North American Contribution, Second International Permafrost Conference, Yakutsk, USSR***, National Academy of Science Publication 2115, pp. 246–54.

Veireck, L. A. (1965). Relationship of white spruce to lenses of perennially frozen ground, Mount McKinley National Park, ***Arctic,*** **18**, 262–67.

Veireck, L. A. (1973). Ecological effects of river flooding and forest fires on permafrost in the taiga of Alaska, in ***Permafrost; North American Contribution, Second International Permafrost Conference, Yakutsk, USSR***, National Academy of Science Publication 2115, pp. 60–7.

Wahrhaftig, C. (1965). Physiographic divisions of Alaska, ***United States Geological Survey Professional Paper*** 482, 52 pp.

Walker, H. J. and Arnborg, L. (1966). Permafrost ice wedge effect on riverbank erosion, in ***Proceedings, 1st International Permafrost Conference***, National Academy of Science—National Research Council Publication 1287, pp. 164–71.

Wallace, R. E. (1948). Cave-in lakes in the Nebesna, Chisana and Tanana river valleys, eastern Alaska, ***Journal of Geology,*** **56**, 171–81.

Washburn, A. L. (1956). Classification of patterned ground and review of suggested origins, ***Bulletin, Geological Society of America,*** **67**, 823–65.

Washburn, A. L. (1967). Instrumental observations on mass wasting in the Mesters Vig District, Northeast Greenland, ***Meddelelser om Gronland,*** **166**, 318 pp.

Washburn, A. L. (1969). Weathering, frost action and patterned ground in the Mesters Vig District, Northeast Greenland, ***Meddelelser om Gronland,*** **176**, 303 pp.

Washburn, A. L. (1970). An approach to a genetic classification of patterned ground, ***Acta Geographica Lodziensia***, No. 24, 437–46.

Washburn, A. L. (1973). ***Periglacial Processes and Environments***, E. Arnold, London, 320 pp.

Washburn, A. L., Smith, D. D. and Goddard, R. H. (1963). Frost cracking in a middle latitude climate, ***Biuletyn Peryglacjalny***, No. 12, 175–89.

Waters, R. S. (1962). Altiplanation terraces and slope development in West-Spitsbergen and southwest England, ***Biuletyn Peryglacjalny***, No. 11, 89–101.

Waters, R. S. (1964). The Pleistocene legacy to the geomorphology of Dartmoor, in ***Dartmoor Essays*** (ed. I. G. Simmons), The Devonshire Association, pp. 73–96.

Waters, R. S. (1965). The geomorphological significance of Pleistocene frost action in south-west England, in ***Essays in Geography for Austin Miller*** (eds J. B. Whittow and P. D. Wood), University of Reading, pp. 39–57.

Watson, E. (1971). Remains of pingos in Wales and the Isle of Man, ***Geological Journal,*** **7**, 381–92.

Wayne, W. J. (1967). Periglacial features and climatic gradient in Illinois, Indiana and western Ohio, east central United States, in ***Quaternary Palaeoecology*** (eds E. J. Cushing and H. E. Wright), **7**, VII INQUA Congress, Yale University Press, New Haven, pp. 393–414.

West, R. G. (1968). ***Pleistocene Geology and Biology***, Longmans Group Limited, London, 377 pp.

Weigand, G. (1965). Fossile pingos in Mitteleuropa, ***Wurzburger Geogr. Arbeiten***, No. 16, 152 pp.

Williams, J. E. (1949). Chemical weathering at low temperatures, ***Geographical Review,*** **4**, 432–42.

Williams, J. R. and van Everdingen, R. O. (1973). Groundwater investigations in permafrost regions of North America: a review, in ***Permafrost; North American Contribution, Second International Permafrost Conference, Yakutsk, USSR***, National Academy of Science Publication 2115, pp. 435–46.

Williams, P. J. (1961). Climatic factors controlling the distribution of certain frozen ground phenomena, ***Geografiska Annaler*, 43**, 339–47.

Williams, P. J. (1962). Quantitative investigations of soil movement in frozen ground phenomena, ***Biuletyn Peryglacjalny***, No. 11, 353–60.

Williams, P. J. (1967). ***Properties and Behaviour of Freezing Soils***, Publication No. 72, Norwegian Geotechnical Institute, Oslo.

Williams, R. G. B. (1964). Fossil patterned ground in eastern England, ***Biuletyn Peryglacjalny***, No. 14, 337–49.

Williams, R. G. B. (1968). Some estimates of periglacial erosion in southern and eastern England, ***Biuletyn Peryglacjalny***, No. 17, 311–35.

Williams, R. G. B. (1969). Permafrost and temperature conditions in England during the last glacial period, in ***The Periglacial Environment*** (ed. T. L. Péwé), McGill-Queen's University Press, Montreal, pp. 399–410.

Wiman, S. (1963). A preliminary study of experimental frost weathering, *Geografiska Annaler*, **45**, 113–21.

Wolfe, P. E. (1953). Periglacial frost–thaw basins in New Jersey, ***Journal of Geology*, 61**, 113–41.

Wright, H. E. (1961). Late Pleistocene climate of Europe; a review, ***Bulletin, Geological Society of America*, 72**, 933–84.

Young, A. (1972). ***Slopes***, Oliver and Boyd, Edinburgh, 288 pp.

Zoltai, S. C. (1971). Southern Limit of Permafrost Features in Peat Landforms, Manitoba and Saskatchewan, ***Geological Association of Canada***, Special Paper No. 9, pp. 305–10.

Zoltai, S. C. (1973). Vegetation, surficial deposits and permafrost relationships in the Hudson Bay lowlands, in ***Proceedings, Symposium on the Physical Environment of the Hudson Bay Lowlands***, University of Guelph, 1973, pp. 17–34.

Zoltai, S. C. and Pettapiece, W. W. (1973). ***Terrain, Vegetation and Permafrost Relationships in the Northern Part of the Mackenzie Valley and Northern Yukon***, Report 73–4, Environmental-Social Committee Northern Pipelines, Task Force on Northern Oil Development, Information Canada, Ottawa, 105 pp.

Zoltai, S. C. and Pettapiece, W. W. (1974). Tree distribution on perennially frozen earth hummocks, ***Arctic and Alpine Research*, 6**, 403–11.

Zoltai, S. C. and Tarnocai, C. (1971). Properties of a wooded palsa in northern Manitoba, ***Arctic and Alpine Research*, 3**, 115–19.

Zoltai, S. C. and Tarnocai, C. (1974). ***Soils and Vegetation of Hummocky Terrain***, Report 74–5, Environmental-Social Committee Northern Pipelines, Task Force on Northern Oil Development, Information Canada, Ottawa, 86 pp.

Index

Subjects indexed below are restricted to major references. Locations are not included unless they are either elaborated in the text or of broad significance. References to authors are not included.

Active layer, **45, 59–62, 106–9, 128, 131**
  failures, **118, 122, 145–7**
Aeolian sands, **202, 250–1**
Alas thermokarst relief, **111–16**
Alaska, **9, 23, 47, 51–3, 69, 73–4, 89, 91, 104, 116, 123–5, 128–9, 144, 208**
Alaska Highway, **47, 69**
Alpine periglacial climates, **9–10**
Altiplanation terraces, ***see*** Cryoplanation terraces
Antarctica, **8, 14–15, 17, 23, 29, 31, 85, 87, 148, 154–5, 199, 201, 203–4, 207, 212**
Arctic brown soils, **197–8**
Artesian pressure, and open-system pingos, **67–8, 96–7**
Aspect, effect of, **17, 53, 60, 138–9, 178–83, 202, 219, 254–6, 267**
Asymmetrical valleys, **108, 122, 178–83, 202, 253–6, 260–2, 264, 268, 272**
Atmospheric circulation, Pleistocene, **220–1**
Atterberg Limits, **76, 119, 133, 135**
Aufeis, ***see*** Icings

Backwearing thermokarst, **110–11, 119–25**
Badland thermokarst terrain, **118–19**
Banks Island, N.W.T., Canada, **17, 43, 60, 101, 102, 104, 108–9, 118–19, 121–2, 130–2, 138, 152–3, 176, 177, 178–80, 193, 235, 266**
Baydjarakhs, **112, 116, 118–19, 122, 130**
Beaches, **207–11**
Beaded drainage, **117**
Beaufort Plain, Banks Island, N.W.T., Canada, **17, 60, 178–80**
Block fields, **37, 229–31**
Block streams, ***see*** Rock streams
Blow-outs, **204**
Bog soils, **197–200**
Boulder fields, ***see*** Block fields
Boreal forest, North American, **61, 157**
Box-shaped valleys, ***see*** Valley forms
Braided streams, **174–6, 181–2, 272–4**
Breckland polygons and stripes, **265–6**
Brickearths, **253, 258–9**
Brodelböden, **227**
Brown wooded soils, **197–8**
Bugors, **101–2**
Bulgannyakhs, ***see*** Pingos

Cape Kellett, Banks Island, N.W.T., Canada, **210–11**
Cape Thompson, Alaska, **28, 59**
Caribou Hills, N.W.T., Canada, **180–2**
'Cemetary mounds', ***see*** Thermokarst mounds
Chalk
  deposits, **257–9**
  dry valleys, **260–8**
  terrain, southern England, **257–68**
Chambeyron, French Alps, **195**
Channel forms, **174–6**
Climates of low annual temperature range, **6, 10**
Closed system pingos, **68, 98–100**
Circles, sorted and non-sorted, **184–95, 199**
Clatford Bottom, Wiltshire, England, **264–5**
Climatic fluctuations, Late Pleistocene, **214–18**

Closed cavity ice, **79**
Coastal
  deposition, **211–12**
  retreat, **208, 211–12**
Coastlines, **206–7, 210–12**
Cold climate indicators, Pleistocene, **221, 225–6**
Collapsed pingos, **93, 96–100**
Colville River, Alaska, **126–7, 173**
Congelifraction, ***see*** Frost wedging
Congeliturbation, ***see*** Solifluction
Continental periglacial climates, **6, 9**
Coombe rock, **257, 259, 261, 264–5**
Coombes, southern England, ***see*** Chalk dry valleys
Coulées pierreuses, ***see*** Rock streams
Cover sands, **205–6, 250–1**
Cryopediments, **155–7, 262**
Cryoplanation, **145, 162–4**
  terraces, **143, 157–61, 232–4**
Cryostatic pressures, **30, 40–3, 192**
Cryoturbation, **34, 157, 197, 199, 227–9**

Dartmoor, southwest England, **39, 232–3, 269**
Dawson City, Yukon, Canada, **7, 57**
'Debris islands', ***see*** Circles, sorted and non-sorted
Debris mantled slopes, **152–5**
Deformations, density controlled, **43–4**
Degradation polygons, ***see*** 'Graveyard mounds'
Degree days, **13**
DeKalb Mounds, Illinois, USA, **248**
Dells, **156, 266–9, 271–2**
Desert pavements, **200, 202, 204, 251**
Desiccation cracking, **192, 194, 195**
Detachment failures, Mackenzie Valley, **147**
Devil's Brook Valley, Dorset, England, **261–2**
Devil's Cheesewing, Devon, England, **232**
Devil's Kneadingtrough, Kent, England, **267**
Downwash, ***see*** Slopewash
Downwearing thermokarst, ***see*** Thermokarst subsidence
'Drunken forests', **34, 156–7, 197**
Dry valleys, Lódź Plateau, Poland, **271–3**
Dujodas, Yakutia, **112**
Dunes, Pleistocene, **251**

Earth hummocks, **34, 186–8, 197, 200**
Éboulis ordonnés, ***see*** Grèzes litées
Ekalugad River, Baffin Island, N.W.T., Canada, **169, 171–3**
Engineering problems of frost action and permafrost, **69–74**
Epigenetic ice, **77**
Ellef Ringnes Island, N.W.T., Canada, **39, 161, 171, 177**
Experimental studies, frost action, **30–1, 35–7, 39–41, 43, 195–6**

Fairbanks, Alaska, **67, 128, 129, 206**
Felsenmeer, ***see*** Block fields
Fire, in boreal and taiga forests, **9, 62, 64, 109, 116, 147, 157**
Fissure polygons, ***see*** Frost fissure polygons
Fluvial
  processes, **167–76, 220–1**
  regimes, **168–71**
Fluvio-thermal erosion, **22, 105, 118–19, 125–8, 129, 173–4, 199, 275–7**
Free face and debris slopes, **150–2**
Freeze–thaw cycles, **13, 18–20, 40, 218–19**
  'Icelandic', **39**
  'Siberian', **40**
Freeze–thaw extrusion structures, **195**
Freeze-up, **16**
Freezing-induced pressures, ***see*** Cryostatic pressures
Freezing process, **12–13, 37**
Front Range, Colorado Rockies, USA, **17, 19, 28, 136**
Frost
  action
    Pleistocene, **218–21, 225–6, 227–36, 261–2, 267–9**
    processes, **19–44, 227–36**

climates, *see* Periglacial climates
cracks
filled with ice, *see* Ice wedges
filled with sand, *see* Sand wedges
seasonal, *see* Soil wedges
creep, **33, 135–7, 140, 219, 235–6**
fissure polygons, **21, 84, 237–45**
fissures, *see* Thermal contraction cracks
heaving, **27–37, 73–4, 100, 123, 192, 195, 197–8**
mounds, **33–7, 42, 101–4**
sorting, **35–7**
-rubble tundra zone, *see* 'Frostschuttzone'
susceptibility, of materials, **28, 70**
weathering, *see* Weathering processes, mechanical
wedging, **37–40, 147–8, 159–61, 162–4, 165–6, 219, 229–34, 269**
pressures, **37–9**
'Frostschuttzone', **224–5**
Frozen ground
perennial, *see* Permafrost
seasonal, *see* Active layer

Garry Island, N.W.T., Canada, **77, 87, 128**
Gelifluction, *see* Solifluction
Geothermal gradient, **47–9, 56**
Glei soils, **197**
Gleization, **198–9**
Golets terraces, *see* Cryoplanation terraces
Gora Sw Malgorzata, Poland, **272–5**
'Graveyard mounds', **112, 116, 118**
Gravity profiling, **82–3**
Great Whale River, P.Q., Canada, **63**
Grèzes litées, **143, 234–5**
Ground ice, **75–104**
quantitative description, **75–7**
slumps, **108, 119–22**
types, **76–80**
Ground veins, *see* Soil wedges
Groundwater
flow, permafrost, **65–9**
springs, **66–7**
supply, **65–7**

Hautes Fagnes, Ardennes, Belgium, **246–7**
'Head' deposits, **235–6**
High Arctic periglacial climates, **6–9**
Hummocks, *see* Tundra hummocks and Earth hummocks
Hydrolaccoliths, **101–4**
Hydrology
groundwater, *see* Permafrost, hydrology
surface run-off, *see* Fluvial regimes

Ice
coefficient of linear expansion, **21**
-cored mounds, *see* Hydrolaccoliths
-cored topography, **82–4**
-foot, **207, 208–9**
mounds, beach, **209**
pushing, **122, 209–11**
segregation processes, **27–37**
volume expansion, **37**
wedge
casts, **22–3, 236–45**
cracking, **21, 85–7**
polygons, **21, 87–93**
thermokarst relief, **116–19**
wedges, **21, 84–93, 199**
Icings, **68–9**
Icy sediments, *see* Massive ice
Injection ice, **95, 97–8, 100**
Insolation
differential, **17, 60, 178–80, 182, 218–19, 254–6**
diurnal variations, **17–18, 178–9, 182–3, 218–19**
seasonal variations, **14–17**
Intra-permafrost water, **66–9**
Intrusive ice, **80**
Inuvik, N.W.T., Canada, **16–17, 65, 72, 109**
Involuted hills, Mackenzie Delta, **83**
Involutions, *see* Periglacial involutions

Jokulhlaupes, **171**

Karkevagge Valley, Lapland, Sweden, **148–9, 151**
Karst, *see* Limestone solutional features

Kitchener, southern Ontario, Canada, **243–4**

Lacustrine action, **122, 209**
Landscape modification, Pleistocene, **260–77**
Lateral
  permafrost degradation, *see* Backwearing thermokarst
  stream migration, **108, 122, 174, 178, 182, 262**
Leau, Belgium, **249**
Lewis River, Baffin Island, N.W.T., Canada, **170–3**
Limestone solutional features, **143–4**
Llangurig, Wales, **246, 247**
Load-cast structures, **43–4**
Localised permafrost zone, **64**
Lódź Plateau, central Poland, **239, 270–3**
Loess, **202, 205–6, 250–3**
Lozinzki, Walery von, **2–3**

Mackenzie Delta, N.W.T., Canada, **77, 80, 83–4, 87, 90, 93, 98–100, 121, 128–9**
Mackenzie Valley, Canada, **34, 132–3, 147**
'Macro-beaded drainage', *see* Thermokarst valleys
Marine action, **206–12**
Mass wasting processes, **134–49**
Mass displacements, **40–4, 192, 194–5**
Massive ice, **77, 80–4**
Massive icy bodies, *see* Massive ice
Meltwater muds, **234, 259, 267**
Mesters Vig, East Greenland, **19–20, 28, 137**
Midlands, England, **241, 268**
Mima Mounds, Washington State, USA, **129**
Morphoclimatic reconstructions, Pleistocene, **223–5**
Morphogenetic region, periglacial, **5–6**
Mud
  boils, **185, 192–3**
  circles, *see* Mud boils
Muskeg, *see* Organic terrain

Naledi, *see* Icings
Needle ice, **33**
Nets, sorted and non-sorted, **184–5**
Nival flood, **168–70, 173**
Nivation, **144–5, 158–9, 202**
  terraces, *see* Cryoturbation terraces
Niveo-eolian deposits, **205, 251**
Non-sorted circles, *see* Circles
Non-sorted polygons, *see* Polygons
Non-sorted stripes, *see* Stripes

Offshore permafrost, **57–8**
Open cavity ice, **79**
Open system pingos, **67–9, 95–8, 246–8**
Organic terrain, **60–1, 63–4, 197, 199**
Oriented lakes, **124–5, 202**
Outwash plains, **206, 220–1, 224, 252–3**

Palsas, **63–4, 246**
Patterned ground
  forms, **184–9**
  origin, **189–96**
Peat
  plateaus, **63, 116**
  thermal effects, **60**
Pediments, *see* Cryopediments
Pedotranslocation, **198, 201**
Perennially frozen ground, *see* Permafrost
Pergelisol, *see* Permafrost
Periglacial
  climates, **5–10, 218–26**
  domaine, **2–3, 5–6, 222–6**
  involutions, **42–4, 227–9**
  sandar, **176–8**
  slopewash, *see* Thermo-erosional wash
  valleys, *see* Dells
  zone, **2–3, 5–6, 218, 223–6, 241–5, 270**
Permafrost, **45–74, 219–21, 226, 236–49, 268–9**
  climatic controls, **52–3**
  degradation, *see* Thermokarst
  distribution
    and glacial history, **56–9**
    Pleistocene, **219–26, 268–9**

present-day, **51–7**
hydrology, **65–9**
surface features, **62–5, 82–101, 111–25**
terrain conditions, **59–62**
thicknesses, **47–50, 53–9**
zones, continuous and discontinuous, **51–6**
Pingo-like features, **100–1, 248**
Pingos, **67–8, 93–101, 112–14, 245–8**
Pipkrakes, *see* Needle ice
Plateau Drift deposits, **253, 262**
Pleistocene chronology and time scale, **214–18**
Ploughing blocks, **139, 264**
Podzolisation, **197, 198, 199**
Point Barrow, Alaska, **87, 124, 208, 211**
Poland, periglacial terrain, **269–77**
Polar desert soils, **199–201**
Pollen stratigraphy, **217–18, 224, 246**
Polygons, sorted and non-sorted, **184–96, 199, 266**
Pore ice, **28, 77, 142, 235**
Pradolina, Warsaw–Berlin, **270–1**
Prairie mounds, Canada, **248**
Prince Patrick Island, N.W.T., Canada, **100–1, 102, 165, 177**
Proglacial lakes, Late Wisconsinan, **56, 224–5, 243–5**
Protalus ramparts, **229**

Radstock Bay, Devon Island, N.W.T., Canada, **152**
Relic
frost action phenomena, **227–36**
permafrost, **58–9**
phenomena, Pleistocene, **236–49**
pingos, **95, 101, 245–8**
thermokarst, **248–9**
Resolute Bay, N.W.T., Canada, **20, 207–8**
Retrograde movement, **135–6**
Retrogressive thaw flowslides, *see* Ground ice slumps
Rillwash, **141–2, 188, 266**
River
channels, *see* Channel forms
networks, **167**
Rock streams, **158, 229–30, 262–5, 269**
Rockfalls, **147–8, 151–2**
Roundness values, beach pebbles, **208–9**
Ruby Range, Yukon, Canada, **138–9**
'Ruissellement', *see* Slopewash
Runoff regimes, *see* Fluvial regimes

Sachs Harbour, Banks Island, N.W.T., Canada, **6–9, 72, 130–2, 138**
Salt
crusts, **200–1**
weathering, **203**
Sand
wedge casts, **25, 236–45**
wedges, **23–5, 201**
Sandar, *see* Periglacial sandar
Sarsens, **262–5, 269**
Schefferville, P.Q., Canada, **61–2**
Scree slopes, **151–2, 231, 234**
Sea ice, **207–9**
Seasonal frost cracks, *see* Soil wedges
Seasonally frozen ground, *see* Active layer
Sediment movement,
coasts, **208, 211–12**
rivers, **171–3**
slopes, **137–8, 141–3, 149**
Segregated ice, **27–8, 36, 77–9, 95, 98–100**
Seismic lines, Mackenzie Valley, **132**
'Self developing thermokarst', **106, 117**
Sheetwash, *see* Slopewash
Shergin's Well, Yakutsk, USSR, **45**
Siberia, USSR, **6–7, 9, 25–6, 45–7, 53–7, 59, 80–2, 92–3, 102, 108–9, 122, 156, 158, 212**
Signy Island, South Orkney Islands, **14–20, 29–31, 193**
Sill ice, **80**
Slope
evolution, **161–6, 260–2, 272–7**
forms, **149–61**
processes, **134–49, 263–6, 272–7**
replacement, **164–5**

Slopewash, **141–3, 151, 159, 188, 200, 234–5, 266, 272–7**
Slumping, ***see*** Ground ice slumps
Snow
  and debris avalanches, **149**
  relation to nivation and solifluction, **140, 144–5, 159, 161, 179–81, 202, 255–6**
Snowbanks, **144–5, 182, 199–200**
Snowcover and permafrost, **61–2, 129**
Snowfall effects, **8–10, 18, 53, 61–2, 129**
Snowline, Pleistocene, **222**
Snowmelt, **167–8**
Soil
  churning, ***see*** Cryoturbation
  wedges, **25–7**
Soils and frost action, **196–9, 227–8**
Solifluction, **135–41, 144–5, 159, 162–3, 263–5, 272–7**
  lobes and terraces, **139–41**
  sheets, **139**
  stripes, **158**
Solution, **143–5, 149**
Sorted circles, ***see*** Circles
Sorted polygons, ***see*** Polygons
Sorted stripes, ***see*** Stripes
Sorting, ***see*** Frost sorting
Spitsbergen, **6–8, 137, 143–4, 148–51**
Steps, sorted and non-sorted, **185**
Stone banked terraces and lobes, **141**
Stone pavement, ***see*** Desert pavement
Stone tilting, **32–3**
Stratified slope deposits, **143, 234–5, 272–7**
Stream flow, **4, 125–7, 167–71, 173–4, 182, 220–1, 260–2, 264–5, 267, 271–5**
Stripes, sorted and non-sorted, **184, 188–9, 193–4, 199, 266**
Sub-permafrost water, **66–8**
Suffosion, **142, 156**
Supra-permafrost water, **66, 68–9**
Syngenetic ice wedges, **85–8, 90–3, 122**

Taele gravels, **257–9, 269**
Taffoni, **203**
Taiga, Siberian, **61, 109, 156–7**
Taimyr polygons, ***see*** Ice wedge polygons
Taliks, **45, 65–8, 101, 176**
Talus
  shift, **152**
  slopes, ***see*** Scree slopes
Tasmania, Australia, **225–6**
Temperature
  curve, Middle and Late Pleistocene, **214–15**
  cycle
    annual, **14–17**
    short term, **17–19**
  depressions, Pleistocene, **219, 222, 226**
Terraces, ***see*** Cryoplanation terraces; Solifluction lobes and terraces
Terrain
  disturbance, permafrost, **62, 64, 70–4, 106–9, 128–33, 147**
  maps, **133**
'Tesselons', ***see*** Sand wedges
Thaw
  lakes and depressions, **112, 117, 122–5, 249**
  sinks, **112, 116, 124**
Thermal
  abrasion, ***see*** Fluvio-thermal erosion
  contraction cracks, **21–7, 236–45**
  erosion, ***see*** Fluvio-thermal erosion
Thermocirques, Siberia, **120, 122**
Thermo-erosional
  niches, **108, 126–7, 182, 211, 275–7**
  wash, **105, 142, 266**
Thermokarst
  causes of, **106–10**
  depressions, Pleistocene, **248–9, 268–9**
  lakes, ***see*** Thaw lakes and depressions
  man-induced, **70–2, 110**
  mounds, **116, 129–30**
  subsidence, **106, 110–19**
  terrain, **83–4, 105–31, 248–9**
  valleys, **114, 116**
Thermoplanation, **105, 142, 164**
Thurfurs, ***see*** Tundra hummocks
Tors, **37, 154–5, 157–61, 163–4, 231–4**

Treeline, **1–3, 5, 9, 56, 65, 110, 197, 223–34, 243**
Tuktoyaktuk, N.W.T., Canada, **79, 81, 83, 125, 211**
Tundra
  hummocks, **33–4, 42, 187, 199**
  polygon relief, ***see*** Ice wedge thermokarst relief
  polygons, ***see*** Ice wedge polygons
  ponds, ***see*** Thaw lakes and depressions
  soils, **197–200**
Turf banked lobes and terraces, **141**

Upfreezing, **30–2, 192**
Utilidors, **73**

Valley
  asymmetry, ***see*** Asymmetrical valleys
  forms, **175–83, 260–8, 271–2**
Valley of Stones, Dorset, England, **264**
Valleys, periglacial, ***see*** Dells
Vegetation
  changes, Late Pleistocene, **217–18**
  effects, **42, 60–2, 110, 128–32, 138–9, 141, 143, 202, 253**
Vehicle movement, permafrost terrain, **132–3**
Vein ice, **79–80**
Ventifacts, **202, 203, 250–1**

Walewice, central Poland, **234, 275–7**
Waltham Common, Norfolk, England, **245–6**
Water
  freezing induced pressures, **37–8**
  volume change upon freezing, **37**
Wave
  action, **207–8**
  transport, **208, 211**
Weathering processes
  mechanical, **3–4, 15, 37–40, 147–9, 165, 203, 219**
  chemical, **3–4, 15, 155, 203, 234**
Wind
  action, **183, 202–6, 220, 249–53**
  deflation, **204–6**
  erosion, **155, 203–4, 250–1**

Yakutia, USSR, **25, 53, 56, 66–7, 69, 80, 94, 96–8, 100, 105, 111–15, 142**
Yakutsk, central Yakutia, Siberia, **6–7, 9, 45–7, 72–3, 111**

Zero curtain, **14–17**